Wireless Flexible Personalised Communications

Wireless Flexible Personalised Communications

COST 259: European Co-operation in Mobile Radio Research

Luis M. Correia
Instituto Superior Técnico, Portugal

John Wiley & Sons
Chichester • New York • Weinheim • Brisbane • Singapore • Toronto

Other Wiley Editorial Offices

John Wiley & Sons, Inc., 605 Third Avenue,
New York, NY 10158-0012, USA

WILEY-VCH Verlag GmbH, Pappelallee 3,
D-69469 Weinheim, Germany

John Wiley & Sons (Australia) Ltd, 33 Park Road, Milton,
Queensland 4064, Australia

John Wiley & Sons (Asia) Pte Ltd, 2 Clementi Loop #02-01,
Jin Xing Distripark, Singapore 129809

John Wiley & Sons (Canada) Ltd, 22 Worcester Road,
Rexdale, Ontario M9W 1L1, Canada

Library of Congress Cataloging-in-Publication Data

Correia, L. M.
  Wireless Flexible Personalised Communications: COST 259, European Co-operation in Mobile Radio Research / L.M. Correia.
    p. cm.
  Includes bibliographical references.
  ISBN 0-471-49836- (alk. paper)
  1. Personal communication service systems—Standards—Europe. 2. Mobile communication systems—Standards—Europe. I. Title.
  TK5103.485.C67 2001
  621.3845'6'094—dc21

British Library Cataloguing in Publication Data

A catalogue record for this book is available from the British Library

ISBN   0 471 49836X

Image on cover of Nokia 3G terminal concept I was supplied by Nokia Corporate Communications, Finland.
Produced from PostScript files supplied by the editor.
Printed and bound in Great Britain by Antony Rowe Ltd.
This book is printed on acid-free paper responsibly manufactured from sustainable forestry,
in which at least two trees are planted for each one used for paper production.

People who live in the real world are people who think technology has suddenly been frozen today, and visionaries are talking about things that will not happen, or are so far away that there is no point in caring. If you want European mobile telecommunications industry to die listen to the people who live in the real world.

Panel on *Advanced Wireless and Mobility Concepts* Visionary Research for the Communications Society, February 1997

Editor's Preface

The area of mobile/wireless communications is in great expansion these days, not only due to the new systems and services that are being introduced and brought to the user, but also to the increased competition among telecommunications operators. As a consequence, many books are being published in this area, in an attempt to correspond to the demand for knowledge in its many topics. These books either bring a new perspective into well know subjects, or present results of recent developments coming from the author's work. This book contains the Final Report of *COST 259*, *'Wireless Flexible Personalised Communications'*. As such, it reports on the work performed and on the results achieved within the project by its participants, which makes it different from many other books. In fact, the material presented here corresponds to the results obtained in the last three years from collaborative work by more than 200 European researchers from more than 90 institutions (universities, operators, manufacturers, regulators, independent laboratories, and others) belonging to 19 countries in the area of mobile radio. In these circumstances, it was felt that publishing the Final Report as a book could contribute to advancement of knowledge in the area of mobile/wireless communications, as well as making it available to a wider audience.

The book is structured in a way similar to the project itself; that is to say, each chapter corresponds to a Working Group, hence guaranteeing the coherence of the results being presented. Chapters, sections and sub-sections were edited by some of the projects participants, and are based not only on their own contributions to the project, but on many others as well. These contributions were given in the form of Temporary Documents (TDs), to which many references appear in among the text. Since they were internal documents, their distribution cannot be carried out without the authors' permission, which could pose problems for the reader in getting access to them. To overcome this situation, references to TDs were substituted as much as possible by the corresponding ones in the open literature (e.g. papers in journals and in conferences); nevertheless, anyone interested can have access to the list of TDs and their authors via the project's web site, http://www.lx.it.pt/cost259. Much more information about *COST 259* can be found in this site, e.g. the description of its activities or the Technical Annex to the Memorandum of Understanding.

As already mentioned, this book is the result of the work of a lot of people. An acknowledgment is due to all *COST 259* participants, and to its Management Committee, for their effort and contributions. Without their continuing interest and participation in the project it would not have been possible to reach the goal of publishing this Final Report. Thanks also to the sections and subsections editors, as well as to all the other direct contributors, for volunteering to perform the task of putting together all the information in a uniform and coherent style. A word of recognition is due as well to the former Chairman, Mr. Eraldo Damosso, who started this project initially, to the Vice-Chairman, Dr. Jean-Frédéric Wagen, to the Working Group Chairmen and chapter editors, Prof. Stephen K. Barton, Prof. Ernst Bonek, Dr. Thomas Kürner and Dr. Alister Burr, and to the Sub-Working Group Chairmen, Mr. Martin Steinbauer, Mr. Mattias Schneider and Mr. Andreas Eisenblätter, for their co-ordination efforts and for help in running the project during its lifetime, and in editing this book.

Luis M. Correia

(Chairman, COST 259)

Lisbon, March 2001

Table of Contents

List of Acronyms

2G	Second Generation
3G	Third Generation
ACF	Auto-Correlation Function
ACI	Adjacent-Channel Interference
ACTS	Advanced Communications Technologies and Services in Europe
ADPS	Azimuth-Delay Power Spectra
AM	Amplitude Modulation
AMPS	Advanced Mobile Phone System
ANSI	American National Standards Institute
AoA	Angle of Arrival
APDP	Average Power Delay Profile
APS	Azimuth Power Spectra
APSK	Amplitude PSK
ARFCN	Absolute Radio Frequency Channel Number
ARQ	Automatic Retransmission Request
AS	Azimuth Spread
ASIC	Application Specific Integrated Circuit
ATM	Asynchronous Transfer Mode
AWGN	Additive White Gaussian Noise
BCCH	Broadcast Control Channel
BCH	Bose, Chaudhuri and Hocquenghen
BER	Bit Error Rate
BIC	Blind Interference Cancellation
B-ISDN	Broadband ISDN
BLAST	Bell Layered Space-Time
BPSK	Binary PSK
BS	Base Station
BTS	Base Transceiver Station
BWLAN	Broadband WLAN
CAC	Call Admission Control

CAD	Computer Aided Design
CAP	Cell Assignment Probabilities
CATV	Cable TV
CBR	Constant Bit Rate
CC	Convolutional Codes
CCI	Co-channel Interference
CDF	Cumulative Distribution Function
CDMA	Code Division Multiple Access
CEPT	Conference of European Postal and Telecommunications Administrations
CFIE	Combined Field Integral Equation
ChIR	Channel Impulse Response
CHT	Channel Holding Time
CIR	Carrier-to-Interference Ratio
CM	Compatibility Matrix
CNR	Carrier-to-Noise Ratio
COST	European Co-operation in the Field of Scientific and Technical Research
CPM	Continuous Phase Modulation
CR	Code Re-use
CRT	Cell Residence Time
CSM	Channel Sounder Measurement Format
CTS	Clear to Send
CW	Continuous Wave
DAB	Digital Audio Broadcast
D-AMPS	Digital-AMPS
DAPSK	Differential APSK (also known as Star-QAM)
DBS	Digital Broadcast System
DCA	Dynamic Channel Allocation
DCIR	Directional Channel Impulse Response
DCM	Directional Channel Model
DECT	Digital Enhanced Cordless Telecommunications
DER	Delay Error Rate
DFE	Decision Feedback Equaliser
DFT	Discrete Fourier Transform
DLC	Data Link Control

DoA	Direction of Arrival
DOS	Disk Operating System
DPCCH	Dedicated Physical Control Channel
DPSK	Differential PSK
DQ	Distributed Queuing
DQPSK	Differential QPSK
DS	Delay Spread
DS-CDMA	Direct Sequence - CDMA
DSP	Digital Signal Processing
DTX	Discontinuous Transmission
DVB-T	Digital Video Broadcast – Terrestrial
DXF	Data Exchange Format
EB	Exponential Back-off
EDGE	Enhanced Data Rates for GSM Evolution
EFIE	Electric Field Integral Equation
EGC	Equal Gain Combining
EM	Electro-Magnetic
EPL	Excess Path-loss
ERC	European Radiocommunications Committee
ESPRIT	Estimation of Signal Parameters via Rotational Invariance Techniques
ETSI	European Telecommunications Standards Institute
ETSI/BRAN	ETSI / Broadband Radio Access Network
ETSI/RES	ETSI / Radio Equipment & Systems
ETSI/SMG	ETSI / Special Mobile Group
EU	European Union
FBP	Forward Broadcast Path
FCA	Fixed Channel Allocation
FDD	Frequency Division Duplex
FDDSF	Field Delay-Direction Spread Function
FDE	Frequency Domain Equaliser
FDMA	Frequency Division Multiple Access
FEC	Forward Error Correction
FER	Frame Erasure Rate
FFT	Fast Fourier Transform
FH	Frequency Hopping

FMM	Fast Multipole Method
FR	Frequency Reuse
FSK	Frequency Shift Keying
FSPL	Free Space Path Loss
FWA	Fixed Wireless Access
GBN	Go Back N
GIS	Geographic Information System
GMSK	Gaussian MSK
GO	Geometrical Optics
GoS	Grade of Service
GP	Global Parameter
GPRS	General Packet Radio Service
GPS	Global Positioning System
GSCM	Geometry-based Stochastic Channel Model
GSM	Global System for Mobile Communications
GTD	Geometric Theory of Diffraction
HC	Half Compensation
HD-PIC	Hard-Decision Parallel Interference Cancellation receiver
HIPERLAN	High Performance Radio LAN
HO	Handover
HPA	High Power Amplifier
HSCSD	High Speed Circuit Switch Data
HSR	High Sensitivity Receiver
HTS	High Temperature Superconducting
IC	Interference Cancellation
ICI	Inter-Channel Interference (in OFDM)
IDFT	Inverse DFT
IFFT	Inverse FFT
IMT-2000	International Mobile Telecommunications – 2000
IR	Impulse Response
IRC	Interference Rejection Combining
ISDN	Integrated Services Digital Network
ISI	Inter-Symbol Interference
ISM	Industrial, Scientific and Medical (frequency bands)
ITS	Intelligent Transportation System
ITU	International Telecommunications Union

ITU-R	ITU – Radiocommunications sector
IUO	Intelligent Underlay Overlay
JD	Joint Detection
LA	Location Area
LAC	Location Area Code
LAN	Local Area Network
LDD	Linear Decorrelating Detector
LDI	Limiter-Discriminator-Integrator
LLC	Link Layer Control
LMDS	Local Multipoint Distribution System
LoS	Line-of-Sight
LP	Local Parameter
LS	Least Square
MAC	Medium Access Control
MAF	Measured Assignment Frequency
MAI	Multiple Access Interference
MAP	Maximum a Posteriori
MBS	Mobile Broadband Systems
MC	Monte Carlo
MC-CDMA	Multi-Carrier CDMA
MD	Mean Delay
MEG	Mean Effective Gain
MFIE	Magnetic Field Integral Equation
MFS	Maximum Frequency Separation
MIMO	Multiple-In, Multiple-Out
ML	Maximum Likelihood
MLSE	Maximum Likelihood Sequence Estimator
MMAC	Multimedia Mobile Access Communication systems
MMDS	Microwave Multichannel Distribution System
MMSE	Minimum Mean Square Error
mmw	millimetre waves
MPC	Multipath Component
MRC	Maximal Ratio Combining
MS	Mobile Station
MSK	Minimum Shift Keying
MUD	Multi User Detection

MUSIC	Multiple Signal Classification
MVDS	Multipoint Video Distribution System
MVU	Minimum Variance Unbiased
NACK	Negative Acknowledgment
NB	Narrow Band
NF	Noise Figure
NLoS	Non-LoS
NRP	Normalised Received Power
OFDM	Orthogonal Frequency Division Multiplexing
OLoS	Obstructed LoS
OTDF	Originated TDF
PA	Paging Area
PAP	Power Angular Profile
PC	Power Control
PCB	Printed Circuit Board
PCC	Parallel Concatenated Convolutional Codes
PDC	Personal Digital Cellular System
PDC-P	Personal Digital Cellular – Packet System
PDDP	Power Delay-Direction Profile
PDF	Probability Density Function
PDP	Power Delay Profile
PHS	Personal Handyphone System
PLL	Phase Lock Loop
PM	Phase Modulation
PO	Physical Optics
POTS	Plain Old Telephone System
PRMA	Packet Reservation Multiple Access
PSK	Phase Shift Keying
QAM	Quadrature Amplitude Modulation
QLoS	Quasi LoS
QoS	Quality of Service
QPSK	Quaternary PSK
RA	Rural Area
RACE	Research and Technology Development in Advanced Communication Technologies in Europe
RDN	Ray Density Normalisation

RE	Radio Environment
RF	Radio Frequency
RIP	Return Interaction Path
RMS	Root Mean Square
RNN	Recurrent Neural Network
RS	Reed Solomon
RSSI	Receiver Signal Strength Indicator
RT	Ray Tracing
RTS	Request to Send
RX	Receiver
SAGE	Space-Alternating Generalised Expectation
SAR	Specific Absorption Rate
SC	Selection Combining
SDL	Specification and Description Language
SDMA	Space Division Multiple Access
SDW	Sliding Delay Window
SFH	Slow Frequency Hopping
SFIR	Spatial Filtering for Interference Reduction
SHO	Soft HO
SINR	Signal to Interference plus Noise Ratio
SNR	Signal to Noise Ratio
SR	Selective Repeat
STD	Standard Deviation
SWG	Sub-Working Group
TCDMA	Time-CDMA
TCH	Traffic Channel
TD	Temporary Document
TDD	Time Division Duplex
TDF	Traffic Density Function
TDMA	Time Division Multiple Access
TETRA	Terrestrial Trunked Radio
THP	Tomlinson-Harashima Precoding
TIM	Tabulated Interaction Method
TLS	Total Least Squares
TRX	Transceiver
TTDF	Total TDF

TU	Typical Urban
TVTF	Time-Variant Transfer Function
TWT	Travelling Wave Tube
TX	Transmitter
UHF	Ultra High Frequency
UID	User Identification
ULA	Uniform Linear Array
ULBF	Uplink Beamformer
UMTS	Universal Mobile Telecommunications System
UPC	Universal Personal Communications
US	Uncorrelated Scattering
UTD	Uniform Theory of Diffraction
UTRA	UMTS Terrestrial Radio Access
VAF	Voice Activity Factor
VCDA	Virtual Cell Deployment Areas
VMS	Virtual Memory System
WAP	Wireless Application Protocol
WBCS	Wireless Broadband Communication Systems
WCDMA	Wideband CDMA
WG	Working Group
WLAN	Wireless LAN
WLL	Wireless Local Loop
WPN	Wireless Packet Networks
WSS	Wide-Sense-Stationary
WSSUS	Wide-Sense-Stationary Uncorrelated Scattering
XPD	Cross Polarisation Discrimination
ZF-BLE	Zero Forcing - Block Linear Equaliser

1

Introduction

Luis M. Correia

This chapter contains a discussion on the evolution of mobile communications, as well as a brief description of the activities of COST 259. For the former, a perspective on the evolution of mobile communications is given, focusing on the use of data and multimedia, and consequently on broadband communications. The latter intends to present to all those that are not familiar with the COST framework in general, and with COST 259 in particular, information on the way it worked and on the technical areas that were addressed.

1.1. Evolution of Wireless/Mobile Communications

The importance of wireless/mobile communications in today's telecommunications industry in general is indisputable. The last few years has witnessed an explosion of new operators and customers of cellular mobile communications in many countries of the world. This is partially due to the internal competition introduced in those countries, and partially because of the success of GSM [GSM00] at the worldwide level (allowing international roaming and many other features that, in some countries, were introduced before the fixed network ones, among other reasons). This success has created the necessary momentum in the whole telecommunications industry to speed up the development of third generation mobile communication systems, IMT-2000 [IMT00], expected to be operational in two years from now, for which standardisation is presently being finished in international bodies.

No justification is needed today as to why Research & Development should be carried out on UMTS (the European approach to IMT-2000) [UMTS00], but many people still argue about the need for studying/developing Wireless Broadband Communication Systems (WBCS) (no distinction is drawn here between 'wireless' and 'mobile' systems), those providing user data rates higher than 2 Mb/s and up to 155 Mb/s (for the time being), which can be seen as belonging to a fourth generation. Although, at the present time, there is no clear view of what WBCS will be, they will probably emerge as a combination of two different existing concepts: WLANs and cellular mobile systems. Many things need to be defined, especially which applications will be supported; many people argue that there is no need for such systems, due not only to the evolution of compression techniques, but also to the non-existence of applications really requiring such high data rates. Nevertheless, some things are already generally accepted, one of them being that WBCS will work at the millimetre waveband (or at the high microwave one, at least), since such high data rates cannot be accommodated at lower frequencies.

The concept of a WLAN is well known today: basically, it provides a wireless connection to LAN users, usually considered to work indoors. WLANs by themselves are not part of WBCS, since one can have WLANs at very low rates, but once they achieve data rates higher than 2 Mb/s, one can talk about BWLANs (Broadband WLANs), thus making them part of the group. The idea of Mobile Broadband Systems (MBS) is not as well known, having been launched in the European R&D framework of RACE [RACE00]: it is a cellular mobile system, thus covering both indoor and outdoor scenarios, enabling users to have a mobile extension of B-ISDN (Broadband-ISDN). The differences between the two concepts, assuming the common definitions that are used today for both basic systems, are also well known, and will not be listed here. Bearing in mind some of the services and application scenarios that are foreseen for WBCS, which will be addressed later on, it seems that these systems really need to present themselves as a merger of the two concepts, so throughout this chapter no distinction will be made between them. An example of this merger is the recent initiative from telecommunications and computer manufacturers (which operators have also joined), Bluetooth [Blue00], with the goal to promote the convergence between the computer and mobile communications worlds.

The development of mobile communications is intrinsically associated with the development of data transmission on mobile networks, and this is in fact what makes the difference between second and third generation systems: the former were designed for voice, while the latter are being designed for data. An exercise can be carried out on the evolution of data rates in the mobile communications industry. Considering the existing second generation cellular systems, and the incoming ones (Table 1.1.1), one can see that there is a clear trend to increase the user data rate, thus showing that wireless data communications are in fact still in their early days. A similar trend exists for WLANs and other wireless communication systems. One can also plot the maximum data rate versus the starting date of commercial operation (actual or foreseen) for cellular systems, Figure 1.1.1: a linear trend (when data rates are plotted on a log scale) is observed, even taking into consideration the forecasts for the next few years. If this trend is followed, services at 155 Mb/s will be operational around year 2010; whether this will happen or not remains to be seen.

Table 1.1.1: Evolution of data rates in cellular mobile communication systems.

System	Start date	Max. user data rate [kb/s]	Frequency band [GHz]	Reference
PDC	1993	2.4	0.8/1.5	[PDC00]
PDC	1995	9.6	0.8/1.5	[PDC00]
GSM	1995	9.6	0.9/1.8/1.9	[GSM00]
PDC-P	1997	28.8	1.5	[PDC00]
PHS	1997	32	1.9	[PHS00]
cdmaOne (IS-95A)	1998	14.4	0.8/1.9	[CDMA00]
D-AMPS (IS-136)	1999	19.2	0.8/1.9	[CTIA00]
PHS	1999	64	1.9	[PHS00]
GSM-HSCSD	1999	64	0.9/1.8/1.9	[GSM00]
cdmaOne (IS-95B)	2000	64	0.8/1.9	[CDMA00]
PHS	2000	128	1.9	[PHS00]
GSM-GPRS	2000	164	0.9/1.8/1.9	[GSM00]
cdmaOne (1XRTT)	2001	144	0.8/1.9	[CDMA00]
GSM-EDGE	2001	384	0.9/1.8/1.9	[GSM00]
UMTS	2002	2 000	2.0	[UMTS00]
MBS	?	155 000	40/60	[SAMB00]

The need for high data rates has been addressed in [CoPr97]. The reasons then given for that need became stronger, and today they make more sense than ever. The example of the computer industry was shown, demonstrating that memory and clock have evolved in a way that would not have been foreseen some years ago; that evolution has continued, with the same trend, showing that one should not be too conservative with regards to the needs and perspectives of future systems. As far as high data rates for mobile systems are concerned, one of the basic issues is to find services that will require the transmission of information at such values; more and more, Internet and multimedia configure themselves as being part of this group of services. Internet access via PDC/PHS *i-mode* [PDC00], [PHS00], and GSM WAP [GSM00] are certainly in this direction.

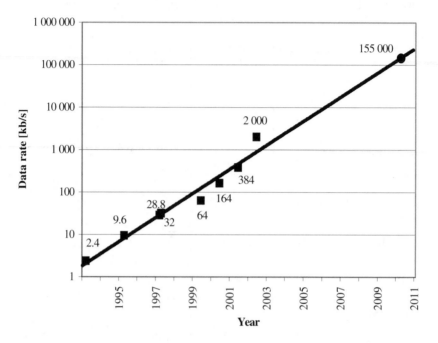

Figure 1.1.1: Evolution of data rates in cellular mobile communication systems.

In the very near future, mobile multimedia will be available via UMTS [UMTS00], which will enable users to exchange information at a data rate up to 2 Mb/s. Many people involved in the development of UMTS today hope that mobile multimedia will have the same success (and consequently expansion rate) experienced by mobile phones. It is not at all clear that this will happen, for reasons that will be mentioned later. It has after been suggested that this will depend on the definition of one or several 'killer applications' that will boost wireless multimedia services, thus creating conditions for that success. This is especially true for WBCS, due to the very high data rates involved (by today's standards), and the argument arises that there is no need to develop systems for non-existent applications. However, again taking the example of the computer industry, one can present arguments that contradict that conservative perspective: until some years ago, operating systems were based on line commands (e.g. VMS, DOS, Unix, and so on), which meant that the user had to memorise a lot of these instructions in order to use the computer; this situation was a barrier to the expansion of computers as a mass market, since one of the basic rules for this (ease of use) was not applied. Then Apple launched the windows-based Mac operating system, which was followed by Microsoft, and the rest of the story is well known. It is easy to find quotations from some years ago, coming from very important people (like company chairmen), stating that the computer industry would never be a mass market, or that a few computers would be enough for the whole world; similar statements on the lack of a future for mobile communications can also be found. As far as WBCS are concerned, perhaps the applications requiring these high data rates are still to be found, or perhaps the natural evolution of data rates over mobile systems (resulting from the evolution of existing applications) will reach those values in the future, as in the case of software and hardware in computers.

Another important aspect in the evolution of mobile communications is that new players are coming into the business. The business of manufacturers is different from that of operators, although they depend on each other. The former make money by selling either equipment to operators or mobile phones to customers, while the latter get their profit by selling airtime. The introduction of new features and systems is a closed circle situation: operators need equipment from manufacturers to put those services into function; in turn, manufacturers need operators to launch these novelties to sell phones to customers (even to those that already have a phone, but who are willing to pay for extra features). Moreover, manufacturers will only make such equipment if they are sure that operators will buy it, i.e. that they are willing to implement these new features or systems. This situation, which is not specific to telecommunications, can pose some problems in the development of WBCS: almost everyone in the business must be really convinced that these systems have a future, in order to start putting some real effort into their development, and this is still not the case today. Of course, before getting to WBCS, one has still to see what will happen to wireless multimedia applications to be developed within the incoming third generation cellular systems.

However the telecommunications business is more and more becoming one in which other companies besides operators and manufacturers have a role to play. The success of WBCS will depend very much on the existence of mass market multimedia services that will enable an increasing number of people to be seduced by them. This is the role of the content and applications providers. One can foresee a scenario in the future where many services that are not even available today on the Internet will be in common use in the future on a wireless multimedia terminal. One can already find many videos presenting visionary views of wireless multimedia services; two of the first that were developed are mentioned in what follows. One, produced by Hewlett-Packard [HP00], shows a disaster scenario resulting from an earthquake, where wireless multimedia terminals are used by rescuing teams; video-telephony is used as a basic service of communication among people, and some other applications are shown, like the exchange of data (building plans, locations on street maps, and so on) between headquarters and rescuers. The other, coming from the UK's GSM operator Cellnet [Cell00], presents a more general use of multimedia services by people in general, by introducing a day in the life of a family and their friends; besides video-telephony (again used as a basic feature), they shop, change travel reservations and do many other things via their wireless multimedia terminals. For either case, there are contents, services and applications provided to users, and it is here that companies other than operators or manufacturers will play an important role.

Many of the multimedia services to be provided in the future will probably be based on the Internet. Initially, the Internet was about getting and exchanging information (i.e. files), but new features have been developed based on it, and today many things can already be done via the Internet. Every day new services are offered to users: besides shopping (which is becoming increasingly popular), one can already book a flight, telephone (via the Internet), listen to radio, and many other things. It is still difficult to book a restaurant or to buy a ticket for a movie, not because it is not technological possible, but because almost all these type of businesses are not yet motivated to be on the Internet; even if they are, they need companies to develop their applications (which again poses no technological problem today). Although the Internet still cannot be considered a mass market, because its penetration has not yet achieved high numbers (as mobile phones have, for instance), it will definitely grow to that status in the near future. Perhaps the growth rate will also increase if an increasing number of services can be obtained via the Internet. Nevertheless, the Internet is still seen today as something to be used by fixed users, and almost all the available services are focused in that direction; if services directed to mobile users are conceived (and that is beginning to happen, for example the provision of a list of restaurants near the place where one is using the phone at that moment), then perhaps the penetration of mobile multimedia will be faster and have an increasing importance.

The concept and realisation of multimedia terminals are also very important to the success of wireless multimedia in general, and to WBCS in particular. At present it seems that the concept of a multimedia terminal is well established: basically, it will enable users to have video, audio and data/file exchange. With these basic features, a whole world of services is possible, depending only on what can be made available. Of course, one will have various types of terminals, providing more or fewer features, as is already the case today for mobile phones or computers: for the former, one can have a simple phone, or a merger of phone, fax and computer (such as Nokia's GSM Communicator [Noki00]), or even a merger of car radio and phone (such as Blaupunkt's GSM RadioPhone [Blau00]); for the latter, one has the example of Laptop Personal Computers (with all the available capabilities), with their increased capability to perform tasks that were not considered to be of computation some years ago.

Maybe it is not too risky to say that future WBCS terminals will need only a few buttons (compared to today's mobile phones), or even no buttons at all, since users will be able to give voice commands to the terminal (this is already the case for Blaupunkt's GSM RadioPhone). As a consequence, it is foreseeable that the size (surface) of the terminal may be mostly imposed by the size of the screen, since video cameras can already be made today with a very small size, while other technological constraints, coming from radio frequency devices, antennas and batteries, will determine its volume. Assuming that technology will continue to evolve as it has up to now, miniaturisation will certainly enable spectacles to become multimedia terminals: the supporting structure will integrate all the circuitry and antennas, images will be projected onto the lenses, and commands will be given by voice, hence requiring no buttons. Then, multimedia spectacles will become as popular as wristwatches are today, and the challenge is to imagine what R&D in mobile/wireless communications will be dealing with by that time.

These or similar visions are not only shared by scientists and engineers. One can observe high expectations of the economic world on the evolution of mobile multimedia radio systems and its market potentials. This is expressed by the amount of money spent to get a UMTS license. During the weeks in which this book has been finalised, five UMTS licenses have been auctioned in the United Kingdom [Spec00] for a total amount of 38 Billion Euro!

Within these exciting years, where such high expectations have grown up, COST 259 [COST00a] has dealt with some aspects of the evolution from present mobile communication systems to WBCS, in cooperation with projects within other European frameworks, namely ACTS [ACTS00].

1.2. Description of COST 259

Founded in 1971, COST [COST00c], is an intergovernmental framework for European Co-operation in the Field of Scientific and Technical Research, allowing the co-ordination of *nationally funded* research on a European level. COST Actions cover basic and pre-competitive research as well as the activities of public utility. The goal of COST is to ensure that Europe holds a strong position in the field of scientific and technical research for peaceful purposes, by increasing European co-operation and interaction within this field. COST has clearly shown its strength in non-competitive research, in pre-normative co-operation and in solving environmental and cross-border problems and problems of public utility. It has been successfully used to maximise European synergy, and has added value in research co-operation and it is a useful tool to further European integration, especially concerning Eastern and Central European countries. Ease of access for institutions from non-member countries also makes COST a very interesting and successful tool for tackling topics of a *truly global nature*.

To emphasise that the initiative came from the scientists and technical experts themselves, and from those with a direct interest in furthering international collaboration, the founding fathers of COST opted for a *flexible* and *pragmatic* approach. COST activities have in the past paved the way for community activities and its flexibility allows COST Actions to be used as a testing and exploratory field for emerging topics. The member countries participate on an 'à la carte' principle and activities are launched on a 'bottom-up' approach. One of its main features is its built-in flexibility. This concept clearly meets a *growing demand* and, in addition, it *complements* the European Community programmes. COST has a geographical scope beyond the European Union and most of the Central and Eastern European countries are members. COST also welcomes the participation of interested institutions from non-COST member states without any geographical restriction.

COST has developed into one of the largest frameworks for research co-operation in Europe and is a valuable mechanism for co-ordinating national research activities in Europe. Today it has almost 200 Actions in various areas (besides Telecommunications, it involves Agriculture & Biotechnology, Transportation, and Environment, among others), and involves nearly 30 000 scientists from 32 European member countries and more than 50 participating institutions from 11 non-member countries. In particular, COST Telecom [COST00b] involves over 2 000 scientists from network operators, research institutes, universities and manufacturers, dealing with: Optical Communications; User Requirements, Including Special Needs; Speech Technology; Multimedia Communications; Broadband Networking; Space and Satellite Networks; Antennas, Radiowaves Propagation and System Aspects; Mobile/Wireless Communications; Telecommunication Software and User Interfaces; and Electromagnetic Impact.

As mentioned before, mobile communications are evolving from second generation towards third generation systems, capable of accommodating a variety of services (from voice and data to video and multimedia) tailored to the customer's needs, in different environments (from macro-cells in rural areas to pico-cells, typical of in-building coverage), at different bit rates, according to the acknowledged concept of bandwidth on demand. Therefore, *personal* is becoming increasingly *personalised*, through the unique opportunities offered by the radio communication emerging technologies and the related advanced service provision capabilities. Such novel applications face demands for radio system aspects, network aspects, propagation issues, diversity countermeasures to the impairments experienced by the transmission channel, channel allocation strategies and planning tools, different to those addressed in the previous, very successful, COST Actions 207 [COST89] and 231 [COST99] that significantly contributed to the development of GSM, DECT and UMTS specifications, as well as to the deployment of the former.

The main objective of COST 259 was to increase the knowledge of radio system aspects for flexible personalised communications, capable of delivering different services, exploiting different bandwidths, and to develop new modelling techniques and related planning tools, in order to guarantee the continuity (and quality) of services delivered by networks of widely different capabilities and structures, across a number of different environments. Its title tries to summarise these objectives, according to the following rationale:

- **wireless**, to account for both *access mobility* (not only related to mobile communications, but also in terms of Wireless Local Loop (WLL), in an increasing perspective of convergence between mobile and WLL technologies), and *terminal mobility*, already provided by current digital cellular systems;
- **flexible**, to account for the progressive migration towards third generation systems (UMTS) and their merging into the fourth generation one, in a vision that encompasses a number of services, provided at different bit rates (according to the user's demand) in different bandwidths;
- **personalized**, to stress the concept of *personal mobility*, which is the main feature of Universal Personal Communications (UPC), in a multi-service perspective, to provide a wide range of user applications.

Close cooperation was maintained with other EU funded projects, the ACTS projects [ACTS00] accounting for the greatest exchange of information. It was done by taking into account the short–medium term goals of ACTS Projects on the one hand, and on the other the spirit of COST Actions that, although committed to medium–long term research perspectives, have to maintain the necessary (and fruitful) relationships with the activities currently in progress within other bodies. At the same time, all potential contributions to standardisation bodies, at European (ETSI) and international (ITU) levels have been examined.

COST 259 was organised into 3 Working Groups (WGs), and 3 Sub-Working Groups (SWGs), Figure 1.1.2. The WGs addressed the broad topics of:

- WG 1 – Radio Systems Aspects
- WG 2 – Propagation and Antennas
- WG 3 – Network Aspects

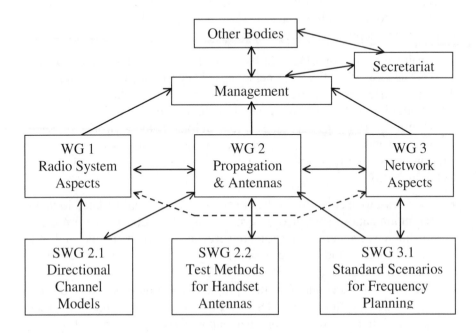

Figure 1.2.1: Structure of COST 259.

Much of the work was conducted within each WG, but some was also carried out in close co–operation among them. As a consequence, although each chapter reports on the results of each WG, in fact there is a lot of joint work in each one, i.e. there is not a full correspondence between the chapters and the WGs. The SWGs were set up to deal with very specific topics, by a smaller group of people, their results being reported in sections of the corresponding chapters.

- SWG 2.1 – Directional Channel Models
- SWG 2.2 – Test Methods for Handset Antennas
- SWG 3.1 – Standard Scenarios for Frequency Planning

More than 250 people have participated, belonging to more than 90 institutions from 19 countries; these numbers show the large effort involved, as well as the interest of the participating institutions in the project. Participants have contributed by presenting Temporary Documents (TDs) to be discussed in the meetings; some have volunteered to summarise the results into the sections of this book, which are listed in Annex I. A good balance was obtained between academia and industry (55 % to 45 %), which shows the interest of both in this type of close cooperation; Annex II lists the participating institutions and countries. There was definitely much to gain by participating in COST 259, since it enabled an exchange of information and cooperation among its participants that would not have been possible in another way.

As mentioned above, each chapter corresponds more or less to the work developed in each WG. Chapter 2 addresses radio systems aspects of future wireless communication systems, focusing primarily on physical layer issues. Section 2.1 deals with OFDM bearing in mind the transmission of high data rates, while CDMA is addressed in Section 2.2. Section 2.3 covers some more general aspects related to modulation and coding, and Section 2.4 presents work on optimum sampling associated with DECT receivers.

Antennas and propagation are dealt with in Chapter 3. Section 3.1 includes the part of propagation modelling below 5GHz, in both the deterministic and statistical perspectives. Section 3.2 addresses a framework for the spatial channel models, extending the famous delay models of COST 207 to the directional domain, while Section 3.3 is about smart antennas, being devoted to the engineering exploitation of the spatial nature of the radio channel. Propagation modelling at millimetre waves and associated antennas is presented in Section 3.4. Finally, the issue of the influence of the head in the antenna behaviour of handsets is addressed in Section 3.5, mainly by reporting on experimental results.

Chapter 4 is devoted to topics on network aspects. Section 4.1 deals with compatibility and spectrum efficiency issues, in systems ranging from GSM up to MBS, including UMTS, of course. Channel allocation strategies, namely aspects dealing with automated frequency planning, are the subject of Section 4.2. Section 4.3 is about cellular topics, i.e. cell modelling, tele–traffic engineering, and mobility modelling, while Section 4.4 addresses network optimisation, by describing techniques to be used by operators in order to increase or to trade-off the quality and capacity of their networks. Section 4.5 presents topics related to planning tools, discussing the geographical data, as well as techniques for automatic planning and some initial aspects of UMTS.

1.3. References⊗

[ACTS00]   http://www.infowin.org/ACTS
[Blau00]   http://www.blaupunkt.de
[Blue00]   http://www.bluetooth.com
[CDMA00]   http://www.cdg.org
[Cell00]   http://www.cellnet.co.uk
[CoPr97]   Correia,L.M. and Prasad,R., "An Overview of Wireless
                Broadband Communications", *IEEE Commun. Mag.*, Vol.
                35, No. 1, Jan. 1997, pp. 28–33.
[COST00a]   http://www.lx.it.pt/cost259
[COST00b]   http://www.cordis.lu/cost/src/telecom.htm
[COST00c]   http://www.belspo.be/cost
[COST89]   COST 207, *Digital land mobile radio communications*, Final
                Report, COST Telecom Secretariat, Brussels, Belgium,
                1989.
[COST99]   COST 231, *Digital mobile radio towards future generation
                systems*, Final Report, COST Telecom Secretariat, Brussels,
                Belgium, 1999.
[CTIA00]   http://www.wow-com.com
[GSM00]   http://www.gsmworld.com
[HP00]   http://www.hp.com
[IMT00]   http://www.itu.ch/imt
[Noki00]   http://www.nokia.com
[PDC00]   http://www.nttdocomo.com
[PHS00]   http://www.phsmou.or.jp
[RACE00]   http://www.analysys.com/RACE
[SAMB00]   http://hostria.cet.pt/samba
[Spec00]   http://www.spectrumauctions.gov.uk
[UMTS00]   http://www.umts-forum.org

⊗ In the future, referencing will probably be done exclusively to sources in
electronic format as opposed to paper, i.e. based on web sites and not on books and
journals. A problem may arise from this: books and journals can be kept on shelves
in many places all around the world, it thus being unlikely that one will never have
access to them, while many web sites may be stored in a single computer, and the
corresponding files can be erased at any time by its owner.

2

Radio Systems Aspects

Alister G. Burr

In this chapter, the radio systems aspects of future wireless communication systems are considered, focussing primarily on physical layer issues. Issues relevant to both third generation mobile systems (UMTS/IMT2000) and to future broadband wireless systems are covered. The chapter describes the research undertaken in Working Group 1 (WG1) of the Action. Participants were drawn from a range of European companies, universities and other research institutions, as listed in Annex II.

A substantial part of the work considered the two topics of OFDM and CDMA, particularly relevant to future broadband systems and to third generation systems respectively, and therefore these two topics are covered in the first two sections of the chapter. The topic of OFDM has been broadened to include other related techniques such as frequency domain equalisation. Section 2.3 covers some more general aspects related to Modulation and Coding, including both linear and non-linear modulation techniques, FEC coding, other types of equalisation, and adaptive modulation and coding techniques. One interesting area of work developed from the design of optimum receivers for the DECT standard, which has led to more generally applicable conclusions on optimum sampling in the presence of multipath. This is covered in Section 2.4.

Much of the work of WG1 has been carried out jointly with the other two Working Groups, and is therefore reported in other chapters of this book. The possibility of such interdisciplinary work has been one of the strengths of this Action. Thus, the group has collaborated with Working Group 2 on aspects of antenna arrays and smart antennas, where signal processing aspects of these systems, including space-time codes, have been the particular responsibility of WG1. This work is described in Sections 3.3.3.3 and 3.3.3.4. Two general areas have been considered in collaboration with Working Group 3: capacity enhancement of cellular systems, covered in Section 4.4 (and in Section 2.3.5 below), and protocols (Section 4.6).

2.1. OFDM and Frequency Domain Techniques

Andreas Czylwik and António Rodrigues

2.1.1. Introduction

Currently, third generation systems are being developed, which, compared to GSM, will support much higher data rates. Wireless local area networks are another development where new radio communications systems are being considered, also with the goal of transmitting high data rates. The applications supported by these systems will also include data transfer, video conferencing and mobile Internet access.

A broadband radio channel is characterised both by time-variant behaviour, caused by a moving receiver, RX, or transmitter, TX, and by frequency-selective fading, which is caused by multipath propagation. Therefore, if a conventional single carrier system is used for this purpose, channel equalisation can be very complex. If an OFDM system [AlLa87] is used instead, the radio channel is divided into many narrowband subchannels, which appear to be frequency non-selective. Therefore, the task of channel equalisation is reduced in a coherent RX to estimating a single complex factor (channel transfer factor) for each subcarrier.

Alternatively, differential modulation can be applied. In this case, no channel estimation is necessary and the computation complexity is very much reduced in each RX. A common M-DPSK scheme, which is a pure phase modulation, has been extended [EnRo95] for higher level differential amplitude and phase modulation (DAPSK) in order to achieve high bandwidth efficiency and still rather good performance.

The simplification of channel estimation and the high bandwidth efficiency and flexibility are the main motivations for using OFDM, which is always a preferred alternative if a high data rate is to be transmitted over a multipath channel with large maximum delay (see Figure 2.1.1).

Channel coding is important in OFDM systems. Due to the narrowband subcarriers, OFDM systems suffer from flat fading. In this situation, efficient channel coding leads to a very high coding gain, especially if soft decision decoding is applied. For this reason OFDM systems will always have to make use of channel coding [AlLa87]. Furthermore, OFDM allows multiple access techniques to certain time and/or frequency regions of the channel in a simple way [RoGr96].

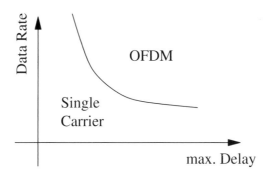

Figure 2.1.1: Influence of channel and data rate on the decision between a
single carrrier system and OFDM.

Finally, OFDM signals have a large peak-to-mean power ratio due to the
superposition of all subcarrier signals. In each TX, the power amplifier will
therefore limit the OFDM signal by its maximal output power. This also
disturbs the orthogonality between subcarriers, leading both to ICI and out-
of-band interference which is unacceptable [NeLo94], [MaRo98], [LaRo99].

The basic principles of OFDM have already been proposed as early as 1961
[FrLa61]. However, this idea could not be realised efficiently, since
powerful semiconductor devices were not available at that time. In
[AlLa87], OFDM has been proposed for broadcast applications and mobile
reception.

Meanwhile, the OFDM transmission technique has been standardised in the
DAB (Digital Audio Broadcasting) and DVB-T (Terrestrial Digital Video
Broadcasting) systems. Today, even relatively complex OFDM transmission
systems with high data rates are technically feasible, and such systems can
be taken advantage of in frequency-selective radio channels

An OFDM signal consists of N subcarriers spaced by the frequency distance
Δf. Thus, the total system bandwidth B is divided into N equidistant
subchannels. On each subcarrier, the symbol duration $T_S = 1/\Delta f$ is N times as
large as in the case of a single carrier transmission system covering the same
bandwidth. Additionally, each subcarrier signal is extended by a cyclic
prefix (called guard interval) with the length T_G. All subcarriers are
mutually orthogonal within the symbol duration T_S. The k-th subcarrier
signal is described analytically by the function $g_k(t)$, $k = 0, \ldots, N - 1$. For
each subcarrier a rectangular pulse shaping is applied.

$$g_k(t) = \begin{cases} e^{j2\pi k \Delta ft} & \forall t \in [-T_G, T_S] \\ 0 & \forall t \notin [-T_G, T_S] \end{cases} \tag{2.1.1}$$

The guard interval is added to the subcarrier signal in order to avoid Inter-Symbol Interference (ISI), which occurs in multipath channels. At each RX the guard interval is removed and only the time interval $[0, T_S]$ is evaluated. The total OFDM block duration is $T = T_S + T_G$. Each subcarrier can be modulated independently with the complex modulation symbol $S_{n,k}$, where the subscript n refers to the time and k to the subcarrier index inside the considered OFDM block. Thus, within the symbol duration T the following signal of the n-th OFDM block is formed:

$$s_n(t) = \frac{1}{\sqrt{N}} \sum_{k=0}^{N-1} S_{n,k} g_k(t - nT) \tag{2.1.2}$$

Due to the rectangular pulse shaping of the signal, the spectra of the subcarriers are *sinc* functions. The spectra of the subcarriers overlap, but the subcarrier signals are mutually orthogonal and the modulation symbols $S_{n,k}$ can be recovered by a simple correlation:

$$S_{n,k} = \frac{\sqrt{N}}{T_s} \left\langle s_n(t), \overline{g_k(t-nT)} \right\rangle \tag{2.1.3}$$

In a practical application, the OFDM signal $s_n(t)$ is generated in a first step as a discrete-time signal in the digital signal processing part of the TX. As the bandwidth of an OFDM system is $B = N \Delta f$, the signal must be sampled with sampling time $\Delta t = 1/B = 1/N \Delta f$. The samples of the signal are written as $s_{n,i}$, $i = 0, 1, ..., N - 1$, and can be calculated by an Inverse Discrete Fourier Transform (IDFT) which is typically implemented as an Inverse Fast Fourier Transform (IFFT).

$$s_{n,i} = \frac{1}{\sqrt{N}} \sum_{k=0}^{N-1} S_{n,k} e^{j\frac{2\pi ik}{N}} \tag{2.1.4}$$

The subcarrier orthogonality is not affected at the output of a frequency selective radio channel. Therefore, the received signal $r_n(t)$ can be separated into the orthogonal subcarrier signals by a correlation technique according to (2.1.3). Alternatively, the correlation at the RX can be done as a discrete Fourier transform (DFT) or an FFT (Fast Fourier Transform) respectively:

$$R_{n,k} = \frac{1}{\sqrt{N}} \sum_{i=0}^{N-1} r_{n,i} e^{-j\frac{2\pi ik}{N}} \tag{2.1.5}$$

where $r_{n,i}(t)$ is the i-th sample of the received signal $r_n(t)$ and $R_{n,k}$ is the received complex symbol of the k-th subcarrier.

If the subcarrier spacing Δf is chosen to be much smaller than the coherence bandwidth, and the symbol duration T much smaller than the coherence time of the channel, then the transfer function of the radio channel $H(f,t)$ can be considered constant within the bandwidth Δf of each subcarrier and the duration of each modulation symbol $S_{n,k}$. In this case, the effect of the radio channel is only a multiplication of each subcarrier signal $g_k(t)$ by a complex transfer factor $H_{n,k} = H(k\Delta f, nT)$. As a result, the received complex symbol $R_{n,k}$ after the FFT is

$$R_{n,k} = H_{n,k} S_{n,k} + N_{n,k} \qquad (2.1.6)$$

where $N_{n,k}$ is additive noise of the channel. A block diagram of an OFDM transmission system with convolutional channel coding is depicted in Figure 2.1.2.

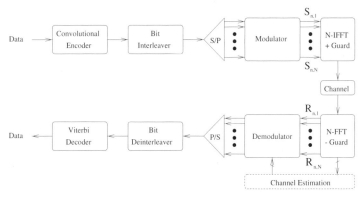

Figure 2.1.2: Block diagram of an OFDM transmission system with convolutional encoding.

2.1.2. Alternative forms

In general, multicarrier and related transmission methods are used for the transmission via wideband frequency-selective radio channels. For such channels equalisation in the time domain is a severe problem for single carrier transmission. For large time dispersion and high data rate, ISI smears the individual symbols over a long period of time. For example, a time dispersion of 5 µs at a symbol rate of 20 MHz causes ISI over 100 symbols.

For sufficient ISI reduction, time domain equalisers with 200 to 300 taps would be needed. Adaptive time domain equalisers for mobile radio systems with such complexity cannot be realised today for high data rates.

2.1.2.1. OFDM with adaptive modulation

Uncoded OFDM transmission with fixed modulation schemes for all subcarriers shows poor performance, because the error probability is dominated by the subcarriers with the smallest Signal-to-Noise Ratios (SNR). Therefore, the error probability decreases very slowly with increasing signal power.

The performance of OFDM can be improved significantly if different modulation schemes are used for the individual subcarriers. The modulation schemes have to be adapted to the prevailing channel transfer function. Of course, this method works only in a bi-directional transmission system where the channel transfer function can be estimated in the RX and communicated back to the TX.

In an OFDM system with adaptive modulation, the modulator adapts the signal alphabets of the individual OFDM subcarriers on the basis of the channel transfer function and SNR of the individual subcarrier. The block diagram of such a system is the same as for conventional OFDM except that an adaptive modulator is used at the TX and an adaptive demodulator at the RX.

It has to be assumed that the instantaneous transfer function of the radio channel varies slowly with time so that it can be estimated at the RX and communicated back to the TX via signalling channels. Therefore, only small velocities of the mobile (walking speed) can be accommodated. The temporal variation of the transfer function of the radio channel makes it necessary to adapt the modulation schemes of the transmitted subcarriers instantaneously. The adaptive modulator and demodulator have to be synchronised via a signalling channel, which is ignored in the following.

The input signal of the adaptive modulator is processed blockwise because of the blockwise signal processing of the FFT. The time-variant behaviour of a radio channel results in fluctuations of the channel capacity (in the sense of information theory). For a real transmission method, there is the following choice:

- keep the data rate constant and let the transmission quality (error probability) fluctuate, or
- keep the transmission quality constant and let the data rate fluctuate.

For most applications and services of broadband radio systems, a constant data rate is required. Therefore, only adaptive modulators that transmit with a constant data rate are considered.

This means that, with each FFT block, the same number of bits M is transmitted. The adaptive modulator adapts the distribution of bits on the individual subcarriers to the transfer function of the radio channel.

The adaptive modulator selects one of several different QAM modulation formats: no modulation, 2-PSK, 4-PSK, 8-QAM, 16-QAM, 32-QAM, 64-QAM, 128-QAM, and 256-QAM. This means that 0, 1, 2, 3, ... 8 bit per subcarrier and FFT block can be transmitted. In order to minimise the overall error probability, the error probabilities for all modulated subcarriers should be approximately equal.

The required SNRs for the above-mentioned modulation schemes are displayed in Figure 2.1.3 for given symbol error probabilities. With Gray coding, the bit error probability is approximately equal to the symbol error probability. The figure shows the bandwidth efficiency of different modulation schemes as a function of required SNR in comparison with channel capacity.

The adaptive modulator distributes the bits so that the overall error probability is minimum. The bit distribution (loading) algorithm maximises the minimum (with respect to all subcarriers) SNR margin (difference between actual and desired SNR for a given error probability). The result of the optimisation process is shown in Figure 2.1.4. For comparison, the upper diagram gives the absolute value of the transfer function. The lower diagram shows the optimised bit distribution vs. subcarrier frequency. A very similar shape of both curves can be observed: at frequency ranges with large power transfer factor, a high-level modulation scheme with high bandwidth efficiency is used. At frequency ranges with a small power transfer factor, a robust modulation scheme with a low bandwidth efficiency is used. The adaptive modulator can even decide not to utilise some subcarriers with very bad SNR.

Not only the bit distribution but also the power distribution on the individual subcarriers can be optimised. However, simulation results show that only a very small gain (0 to 0.5 dB) can be obtained from an optimised power distribution. Therefore, it does not make sense to use the additional optimisation of power distribution, since a lot of transmission overhead results from signalling of the quasi-analogue values of the transmit power to the RX.

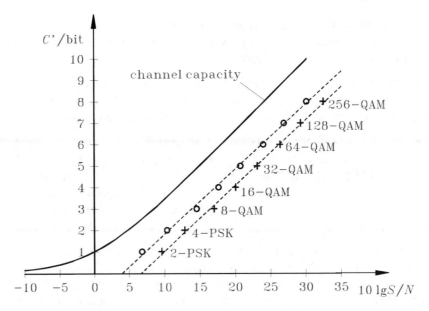

Figure 2.1.3: Required SNR for different QAM schemes: circles – symbol error probability $=10^{-3}$; crosses – symbol error probability $=10^{-5}$.

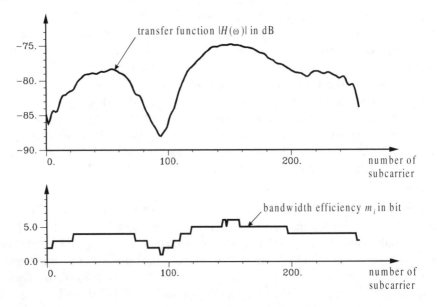

Figure 2.1.4: Example for the magnitude of the transfer function and optimised bit distribution.

Dramatic gain (5 to 14 dB at a bit error probability of 10^{-3}) from adaptive modulation has been reported from simulations of the transmission via measured and stored impulse responses [Czyl96a], [Czyl96b], [Czyl97]. An example is shown in Figure 2.1.5. The following uncoded modulation schemes are compared:

- system 1: single carrier transmission with a linear MMSE (Minimum Mean Square Error) equaliser in the frequency domain,
- system 2: conventional OFDM using the same modulation scheme on each subcarrier,
- system 3: OFDM with adaptive modulation using optimised modulation schemes on the subcarriers.

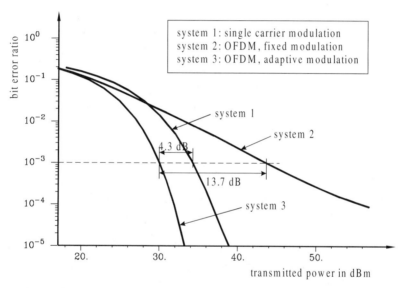

Figure 2.1.5: Comparison between modulation schemes: BER vs. SNR (example with high gain because of severe frequency selectivity).

Highest gain is obtained for the transmission via non-line-of-sight channels with severe time dispersion. Since NLoS (Non-Line-of-Sight) radio channels usually exhibit higher attenuation, this property is of particular advantage. For a frequency-flat channel conventional OFDM and OFDM with adaptive modulation use the same modulation alphabet. Therefore, no gain can be obtained in this case.

Adaptive OFDM also exhibits some disadvantages: the calculation of the distribution of modulation schemes requires a high computational effort.

Additionally, the channel must not vary too quickly because of the required channel estimation. A rapidly varying channel also requires increased signalling information with the effect that the data rate for communications decreases. Furthermore, an OFDM signal approximately exhibits a complex Gaussian distribution with a very high crest factor so that linear power amplifiers with high power consumption have to be used.

The main idea of using optimised waveforms is to obtain an equivalent receive filter, which is as narrow as possible so that the transmission method is less sensitive to narrowband interferers than OFDM. Also, the bandwidth requirements are improved with optimised waveforms: the guard band between different OFDM channels can be narrower. Waveforms are optimised in [KoMB98].

2.1.2.2. Single carrier transmission with frequency domain equalisation

There are several advantages of single carrier transmission compared with OFDM. In case of single carrier modulation, the energy of an individual symbol is distributed over the whole frequency spectrum. For this reason, severe narrowband notches in the channel transfer function caused by frequency-selective fading have only a small impact on error probability. Furthermore, the output signal of a single carrier TX exhibits only a small crest factor whereas an OFDM signal exhibits approximately a complex Gaussian distribution.

A solution to the complexity problem of equalisation in time domain for single carrier modulation is a Frequency Domain Equaliser (FDE). The block diagram of the considered single carrier transmission system with FDE is sketched in the lower part of Figure 2.1.6. The figure shows very clearly that single carrier modulation with frequency domain equalisation is very similar to OFDM transmission. The same blocks are needed; the main difference is that the block *inverse FFT* is moved from the TX to the RX [SaKJ94a], [SaKJ94b]. Therefore, single carrier modulation and OFDM without adaptation exhibit the same complexity.

Since the FFT algorithm is used in single carrier modulation with FDE, a blockwise signal transmission has to be carried out. As in an OFDM system, a cyclic prefix (guard interval) is required in order to mitigate interblock interference. For single carrier modulation a fixed symbol alphabet is used to realise a constant bit rate transmission.

There is, however, a basic difference between the single and multicarrier modulation schemes. For the single carrier system, the decision is carried out in the time domain. In the case of the multicarrier system, the decision is carried out in the frequency domain. In the single carrier system, an inverse

FFT operation is required between equalisation and decision. This inverse FFT operation spreads the noise contributions of the individual subcarriers over all samples in the time domain. Since the noise contributions of highly attenuated subcarriers can be rather large, a zero-forcing equaliser gives poor noise performance. For this reason, a MMSE equaliser is appropriate for the single carrier system. It minimises the sum of ISI and noise power.

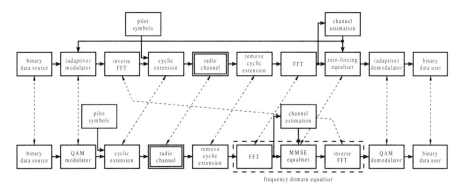

Figure 2.1.6: Comparison between block diagrams, upper part: OFDM transmission, lower part: single carrier transmission with frequency domain equalisation.

With single carrier modulation, a significantly better performance is obtained than with OFDM for fixed modulation schemes. But adaptive OFDM outperforms single carrier modulation by 3 to 5 dB (compare Figure 2.1.5).

The better performance of adaptive OFDM compared to single carrier modulation results from the capability of adaptive OFDM to adapt the modulation schemes to subchannels with very different SNRs in an optimum way. In order to improve the performance of single carrier modulation, antenna diversity with Maximum Ratio Combining (MRC) in the frequency domain can be used [Kade97]. If antenna diversity with MRC is used for both adaptive OFDM and single carrier modulation, simulations have shown that the performance advantage of adaptive OFDM is reduced by several dB. This is due to the fact that the equivalent transfer function resulting from maximum ratio combining shows much smaller fluctuations than the transfer function without antenna diversity.

2.1.3. Modulation and demodulation

As shown in (2.1.2), each subcarrier can be modulated independently by mapping a sequence of m bits to a specific point $S_{n,k}$ in the constellation diagram followed by an optional differential encoding. If no differential encoding is applied, data bits to be transmitted are directly mapped to modulation symbols $S_{n,k}$. Examples of these are PSK and QAM modulations.

Differential encoding means that the data bits to be transmitted are not directly mapped to the modulation symbols $S_{n,k}$, but rather to the quotient $B_{n,k}$ of two successive modulation symbols. This technique can be applied either in time or in frequency direction as described by the following equations:

$$S_{n,k} = S_{n-1,k} \cdot B_{n,k} \quad \text{or} \quad S_{n,k} = S_{n,k-1} \cdot B_{n,k} \tag{2.1.7}$$

In the case of M-DPSK for example, $B_{n,k} \in \left\{ e^{j2\pi i/M} \mid i = 0,...,M-1 \right\}$. However, M-DPSK has a poor performance if M is large. In this case a combined differential amplitude and phase modulation should be applied which is described below. Depending on the channel and on the design of the OFDM system, differential coding in the time or frequency domain may be more suitable.

For both types of modulation (differential or non-differential) there are different ways to demodulate the received complex sequence. In the case of non-differential modulation, coherent demodulation has to be applied. For this process, the RX needs to perform a channel estimation, which provides estimates $\hat{H}_{n,k}$ for the channel transfer factors. The demodulator decision is based on the quotient

$$D_{n,k}^c = \frac{R_{n,k}}{\hat{H}_{n,k}} = S_{n,k} + \frac{N_{n,k}}{\hat{H}_{n,k}} \Rightarrow \hat{S}_{n,k} = dec\left\{ D_{n,k}^c \right\} \tag{2.1.8}$$

The RX takes a hard decision according to given thresholds applied to $D_{n,k}^c$. The channel estimation can be performed on the basis of known symbols (pilot symbols), which are included in the OFDM transmit-signal.

If differential encoding is used in the TX, demodulation can be performed either non-coherently or quasi-coherently. With non-coherent demodulation the decision is based on the quotient of two successive symbols

$$D_{n,k}^{nc} = \frac{R_{n,k}}{R_{n-1,k}} = \frac{S_{n-1,k} \cdot B_{n,k} \cdot H_{n,k} + N_{n,k}}{S_{n-1,k} \cdot H_{n-1,k} + N_{n-1,k}} \quad \text{and} \quad \hat{B}_{n,k} = dec\left\{ D_{n,k}^{nc} \right\} \tag{2.1.9}$$

In general, successive channel transfer factors are strongly correlated so that $H_{n,k} \approx H_{n-1,k}$ and therefore cancel down in (2.1.9) (if the noise influence is neglected). Unfortunately, $D_{n,k}^{nc}$ is affected by twice the noise power of $D_{n,k}^{c}$ leading to a higher bit error rate than for coherent demodulation, if perfect channel estimation is assumed.

A quasi-coherent technique is an alternative to non-coherent demodulation [RoBr97]. Similar to the coherent case, the channel influence is first removed before differential decoding takes place. Due to differential encoding, there is no need to determine the channel factors $H_{n,k}$ unambiguously, but only in the phase interval of $2\pi / N_p$ rad (N_p being the number of phases). In this case, no pilot symbols are required to estimate the unknown channel transfer factors [Vite83].

Apart from differential decoding the processing in a quasi-coherent RX is similar to that in a coherent one, i.e.,

$$\hat{B}_{n,k} = \frac{\hat{S}_{n,k}}{\hat{S}_{n-1,k}} = \frac{dec(R_{n,k} / \tilde{H}_{n,k})}{dec(R_{n-1,k} / \tilde{H}_{n-1,k})} = \frac{dec(D_{n,k}^{c})}{dec(D_{n-1,k}^{c})} \tag{2.1.10}$$

In case of an incorrect decision this error influences two successive symbols due to differential encoding.

In order to increase the bandwidth efficiency, the well known M-DPSK can be extended to a differential amplitude and phase modulation (M-DAPSK), which shows a substantial performance improvement over M-DPSK for $M \geq 16$ [EnRo95]. DAPSK can be described as a differentially encoded APSK with the signal set

$$\Psi = \left\{ a^A \cdot e^{j\Delta\varphi \cdot P} \middle| A \in \{0,...,N_a - 1\}, P \in \{0,...,N_p - 1\} \right\} \tag{2.1.11}$$

As an example, a 64-APSK signal space diagram with $N_a = 4$ and $N_P = 16$ is depicted in Figure 2.1.7. Note that the amplitudes are spaced by a factor a. Mapping of the m input bits is done separately for amplitude and phase using m_a and m_P bits, respectively. To minimise the bit error rate, Gray mapping is used.

The number of amplitude circles $N_a = 2^{m_a}$, the number of phases per amplitude circle $N_p = 2^{m_p}$ and the amplitude ratio a are free parameters that have to be optimised depending on of the number of signal states $M = N_a \cdot N_p = 2^m$ and the demodulation method (quasi-coherent or non-coherent).

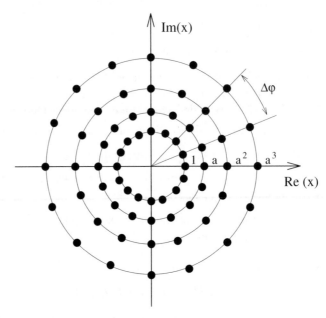

Figure 2.1.7: DAPSK signal space diagram.

The amplitude mapping can be developed so that differential encoding can again be described by a multiplication as in (2.1.7):

$$S_{n,k} = B_{n,k} \cdot S_{n-1,k}$$ (2.1.12)

where

$$B_{n,k} \in \Psi' = \left\{ a^{A'} \cdot e^{j\Delta\varphi P} \left| \begin{array}{l} A' \in \{-N_a + 1,..., N_a - 1\} \\ P \in \{0,..., N_p - 1\} \end{array} \right. \right\}$$ (2.1.13)

2.1.4. Channel coding

As explained above, the radio channel attenuates each OFDM subcarrier by a complex transfer factor $H_{n,k}$. If the channel is a multipath channel with many propagation paths and without a Line-of-Sight (LOS) path, then the amplitude of the transfer factors is Rayleigh distributed according to the central limit theorem. This means that even at a very large average SNR there are always some subcarriers that suffer from deep fading and produce many bit errors.

Without channel coding the typical flat fading curve is obtained for the BER. With channel coding, large SNR gains are achieved, especially if soft decision is applied. Soft output demodulation, which is required for soft decision decoding, has been considered for coherent, non-coherent and quasi-coherent techniques [MaRE98].

For comparison, an OFDM system according to Figure 2.1.2 with a total bandwidth of 7.16 MHz has been considered. The OFDM signal is transmitted over a frequency selective WSSUS radio channel described in [Hohe92] with the Power Delay Profile (PDP) Typical Urban (TU) from COST 207 [COST89], not time-varying.

For coherent demodulation ideal channel estimation is performed in the simulation program. Realistic channel estimation procedures need the transmission of pilot symbols and would cause an SNR loss of approximately 1.5 dB.

If differential modulation is applied, pilot signals are not necessary. For a system comparison convolutional codes are considered with different code rates. For example, if 6 % of all subcarriers are needed for the transmission of pilot signals in the coherent case and the code rate $R = 4/5$ is used, then a code rate of $R = 3/4$ is considered in the non-coherent demodulation of M-DAPSK. This means that the system comparison is based on the same net data rate, but different signal overhead.

Simulation results are shown in Figure 2.1.8 for differential modulation, code rate $R = 3/4$, with non-coherent and quasi-coherent and for non-differential modulation, $R = 4/5$. In the curves for non-differential modulation a loss of 1.5 dB due to the channel estimation has been assumed. The net data rates are 13.5 and 18 Mbit/s for 8-PSK/DPSK and 16-QAM/DAPSK, respectively [RoMa97], [RoMa98].

Expressions were derived in [Burr97a] for the interference between subcarriers (ICI) and intersymbol (ISI) due to the effects of dispersion in excess of the guard period. Frequency shift errors were also considered. Exponential PDPs such as the COST 207 Rural Area (RA) and Typical Urban (TU) channel models were used to give numerical examples.

Those results were used to estimate the BER performance of coded OFDM. The results show [Burr97a] that both guard period and FEC coding are essential to achieve optimum performance. An example is the combination of a rate 3/4 convolutional code with 33 % guard period to achieve a bandwidth efficiency of 1 bit/s/Hz. With 64 subcarriers this scheme can achieve 6.4 Mbit/s on a TU channel and 5.1 Mbit/s on a RA channel respectively.

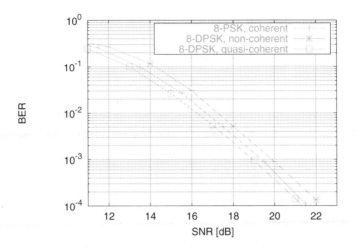

(a) 8-DPSK and 8-PSK

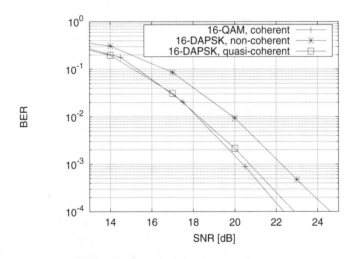

(b) 16-DAPSK and 16-QAM

Figure 2.1.8: Comparison of 8-DPSK, 8-PSK, 16-DAPSK and 16-QAM.

In [DaTr97] and [DaTr99] a coded OFDM system has been proposed for indoor broadband communications at 60 GHz taking into account actual environment propagation conditions. It was shown that with a 32 subcarrier system, code rate 1/2, sectored antennas (6 sectors), 10 dBm of transmitted power, it is possible to support packet transmission up to 155 Mbit/s with a coverage of 84 %.

2.1.5. Multiple access schemes

For OFDM based communication systems various multiple access schemes can be used. In the downlink, multiple access techniques are simpler. Obviously, the Base Station (BS) can arbitrarily assign subcarriers and/or OFDM blocks to specific users. OFDM-CDMA can also be used as a multiple access scheme [MuBR96], [RoGr97].

A pure TDMA scheme can be organised so that a number of OFDM blocks is assigned to each user. In this case, each user can make use of the diversity supplied by the total bandwidth. However, for each user, at least one entire OFDM block must be used for each transmission.

An FDMA scheme where a number of subcarriers is assigned to each user seems to be more flexible, since the data rate per user can be quantised more smoothly. A real advantage can be obtained in a TDD system where the BS can estimate the channel transfer factors for all users during the uplink. As the channel is unchanged in the subsequent downlink (provided that the channel is only slowly time-varying), the BS can now assign the best subcarriers to each user [RoGr96].

In the uplink, the TDMA approach where the users subsequently transmit a number of OFDM blocks is still relatively easy to implement. The FDMA approach, however, would mean that all users transmit several subcarriers at the same time. In this case, carrier synchronisation could be more difficult.

A combination of these techniques can be used and has been proposed for UMTS by Telia. A description of this technique can be found in [Rohl98].

2.1.6. Amplitude limitation

OFDM signals have a very large peak-to-average power ratio. In the TX, the maximal output power of the amplifier limits the amplitude of the signal to an amplitude threshold A_O, and this effect will produce interference both within the OFDM band and in adjacent frequency bands.

Two kinds of approach are investigated which assure that the transmitted OFDM signal does not exceed the amplitude A_O. The first method makes use of redundancy in such a way that any data sequence leads to a 'fitting' OFDM signal or that at least the probability of higher amplitude peaks is greatly reduced [JoWi96]. These approaches do not result in interference of the OFDM signal.

In the second type of approach, the OFDM signal is manipulated by a correcting function that eliminates the amplitude peaks. The out-of-band interference caused by the correcting function is zero or negligible.

However, interference of the OFDM signal itself is tolerated to a certain extent [MaRo98].

2.1.7. Intercell interference

A Multi-Carrier Code Division Multiple Access (MC-CDMA) scheme was proposed [NSRP98] to be used in a multi-cellular environment in order to reduce interference present in typical mobile systems. The conventional approach to limit the interference of users in adjacent cells is Frequency Reuse (FR). A new scheme – Code Reuse (CR) – was proposed in order to achieve a higher order of diversity. In this method, each cell in the cluster uses all the available carriers to transmit the users data signal.

The data signals of different users are distinguished by using orthogonal Walsh-Hadamard codes for each user. A comparison between the FR and CR schemes was done as a function of the environment path loss, number of carriers, frequency gap between adjacent carriers, number of users and cluster size. Results obtained for a cluster size equal to four show that, by increasing the number of carriers and the frequency gap between adjacent channels, the performance of both schemes is improved.

When the number of carriers is increased, the best improvement is achieved for the CR scheme, while for the frequency gap increase it is the FR scheme that improves the most. When a cluster size equal to seven is considered, CR achieves a better performance than FR since, in this case, the co-channel interference is clearly reduced due to the larger distance between cells using the same codes.

In conclusion, there are several ways to improve the performance in both schemes: increasing the number of carriers (increasing the diversity); increasing the frequency gap (less correlation in the carriers); reducing the number of users (less interference) or finally, by increasing the cluster size (reduce the interference). However, one should bear in mind that there is a limitation in the frequency spectrum and the enlargement of the cluster size is also a compromise against needed capacity.

The advantage of CR is that it allows a more effective and fast adaptation of the model of transmission to the conditions of the traffic. For instance, in the case of one cell having more traffic than the others, the proposed system needs to give more codes to this BS than the others. CR achieves better performance for low transmitted power (less then 14 dB), because there is a larger diversity (longer codes) in the transmitted signal.

The disadvantages of CR are that because of the loss of orthogonality of the codes, the performance of CR will be damaged when one is transmitting with higher power. The interference of signals transmitted by BSs of adjacent cells will severely damage the performance of the CR scheme. There is also loss of orthogonality in case of FR, but this loss will be higher for the case of CR.

2.1.8. Synchronisation

In general, for a digital transmission system the following parameters have to be synchronised: carrier frequency, carrier phase, clock frequency, clock phase, as well as the temporal position of the transmission frame. Additionally, for OFDM and related modulation schemes that use the FFT algorithm, a synchronisation of the temporal position of the FFT window is required. *FFT window* denotes the interval of samples that are processed within one FFT operation. Furthermore, for equalisation an estimate of the channel transfer function is required in many cases. It is needed for the equalisation in frequency domain, as well as the synchronisation of carrier and clock phase.

In both cases, OFDM and single carrier transmission with FDE, the synchronisation of carrier frequency is necessary. Although single carrier transmission is less sensitive to carrier phase noise than OFDM, an accurate synchronisation of carrier frequency is required. The synchronisation of carrier phase is necessary for coherent detection and is carried out by the channel estimation and equalisation.

In many cases, the precision and stability of clock oscillators is sufficiently high so that no clock frequency synchronisation is needed. Furthermore, clock and carrier frequency may be coupled so that by synchronisation of the carrier frequency, the clock frequency is also synchronised.

The clock phase need not be synchronised – its effect on system performance is compensated by channel estimation and equalisation. Only a coarse timing synchronisation is needed which optimises the position of the FFT window. The synchronisation of the FFT window and the frame synchronisation can be combined.

From the above considerations, it is concluded that in many cases only the following parameters have to be synchronised: carrier frequency and coarse timing of FFT symbols and transmission frame. Because of the similarity of single carrier and OFDM transmission, similar requirements have to be met for the synchronisation.

In the following, mainly pilot-aided synchronisation methods for OFDM and single carrier modulation schemes with FDE are discussed. In the case of

OFDM, pilot symbols can be distributed individually in the time and frequency domains.

Unlike OFDM systems, for single carrier systems it is not possible to distribute pilot symbols individually in time and frequency domains, since single carrier signals require the whole available frequency spectrum. Therefore, it is only possible to multiplex pilot symbols into the data stream in the time domain. Multiplexing of single pilot symbols within an FFT symbol does not make sense because the ISI of the transmission channel spreads the pilot symbols over a large number of symbols. Because of this reason, only complete (possibly shortened) FFT symbols are used as pilot symbols for channel estimation and frequency synchronisation.

In the following, a transmission frame with periodical synchronisation symbols is investigated which can be used for OFDM as well as for single carrier modulation with FDE. The synchronisation symbols can be utilised for all synchronisation tasks as well as channel estimation [Czyl98]. The transmission frame is displayed in Figure 2.1.9.

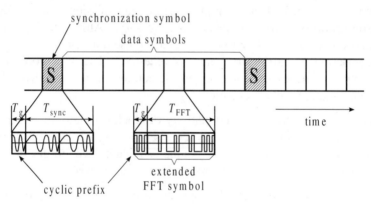

Figure 2.1.9: Transmission frame with periodical synchronisation symbols.

Each block consists of $N_{FFT} = 2^m$ complex data symbols and will be called FFT symbol. The temporal duration of an FFT symbol will be denoted by T_{FFT} and the guard time by T_g. Interblock interference is avoided if the guard interval is larger than the time dispersion of the radio channel.

The structure of a possible synchronisation symbol is displayed in Figure 2.1.9. It consists of two periods of a constant envelope signal (chirp signal). In the following, signals are described by their samples of the complex equivalent baseband signal. The periodicity of the synchronisation symbol can be expressed in time domain:

$$x_{S,i+N_p} = x_{S,i} \tag{2.1.14}$$

where the period within the synchronisation symbol equals $N_p = N_{FFT}/2$. Similar pilot symbols for synchronisation of OFDM systems have been proposed in [ScCo96].

For a corresponding OFDM system $\Delta f = 1/T_{FFT} = 1/(N_{FFT} \cdot \Delta T)$ equals the subcarrier spacing. ΔT denotes the duration of a data symbol. For the transmission frame displayed in Figure 2.1.9 the duration of the synchronisation symbol T_{sync} equals the duration of data FFT symbols T_{FFT}.

The FFT window has to be positioned to that time interval where the transients of the channel impulse response have died out and where a quasi-periodical signal is found. The periodicity within the synchronisation symbol is utilised for coarse timing synchronisation, since the periodicity is preserved even if the signal is transmitted via a time-dispersive channel.

Denoting the complex envelope samples of the received signal with r_i and the frequency offset with δf, the following correspondence holds during the periodic signal interval (noise is neglected):

$$r_{i+N_p} = r_i \cdot e^{j2\pi \, \delta f \, N_p \Delta T} \tag{2.1.15}$$

A periodical signal interval can be searched by minimising the periodicity metric [ChMU87], which calculates the squared average distance between the two considered signal periods:

$$M_p(k,\vartheta) = \frac{1}{N_\Sigma} \sum_{i=1}^{N_\Sigma} \left| r_{i+k+N_p} - r_{i+k} \cdot e^{j\vartheta} \right|^2 \tag{2.1.16}$$

The task of the phase term $e^{j\vartheta}$ is to compensate the effect of the carrier frequency offset. N_Σ denotes the number of terms in the sum ($N_\Sigma = N_p$) and k the time shift for coarse timing synchronisation. The optimum choice of ϑ is the negative phase $\vartheta = -\varphi(K)$ of the autocorrelation sum

$$K = \sum_{i=1}^{N_\Sigma} r_{i+k+N_p} \cdot r_{i+k}^* \tag{2.1.17}$$

Inserting the optimum phase into (2.1.16) yields [ChMU87]:

$$M_p(k) = \frac{1}{N_\Sigma} E - \frac{2}{N_\Sigma} |K| \tag{2.1.18}$$

where E denotes the signal energy of both periods:

$$E = \sum_{i=1}^{N_\Sigma} \left(\left| r_{i+k+N_p} \right|^2 + \left| r_{i+k} \right|^2 \right) \tag{2.1.19}$$

At time intervals where no signal is present (e.g. in a TDD transmission frame), a false lock-in can be avoided using the normalised periodicity metric:

$$M_{pn}(k) = \frac{E - 2|K|}{E} \tag{2.1.20}$$

The timing jitter can be reduced by lowpass filtering of the periodicity metric with respect to the position k of the FFT window. Additionally, the timing jitter is further reduced by averaging the position of the FFT window over several transmission frames.

For carrier frequency synchronisation it is assumed that the stability of the radio frequency oscillators is sufficiently high so that only a fine frequency offset synchronisation has to be carried out. This means that the frequency offset δf is smaller than half of the spacing between the samples in the frequency domain Δf.

The method for fine frequency estimation is based on the observation of the phase shift from the first to the second period of the synchronisation symbol. A maximum likelihood estimator for the frequency offset has been proposed [ClMS93], [Moos94], which calculates the frequency offset from the autocorrelation K. The estimated frequency offset is obtained from:

$$\hat{\delta f} = \frac{1}{2\pi N_p \Delta T} \cdot \varphi(K) \tag{2.1.21}$$

where $\varphi(K)$ is the phase of K. Due to the periodicity of the phase φ, the maximum detectable frequency deviation equals $\pm 1/(2 N_p \Delta T)$.

In the following, a system that reduces the effect of fading of a mobile radio channel by antenna diversity is considered. In the RX, MRC and equalisation are carried out simultaneously in the frequency domain. For single carrier transmission this concept has been proposed [Kade97]. In the following, the synchronisation of such an RX is analysed [Czyl99].

For an RX with two branches (antenna diversity), there are several
possibilities to synchronise by means of both received signals. In the
following several possibilities are discussed. A first possibility for the
simultaneous use of both signals is linear combining – similar to MRC.
After combining, the same synchronisation algorithms can be used as
discussed before. The block diagram of this approach is shown in Figure
2.1.10. The transmit signal $x(t)$ is simultaneously received via both channel
1 and channel 2. Unlike the case of MRC of the data signal, the transfer
functions $H_1(\omega)$ and $H_2(\omega)$ are unknown at the beginning of
synchronisation. For this reason, only frequency-independent coefficients
k_1 and k_2 may be used for weighting. The coefficients k_1 and k_2 have to
be chosen in such a way that the SNR of the output signal $z(t)$ becomes
maximum. The SNR is:

$$\left.\frac{S}{N}\right|_z = \frac{S_x \cdot \overline{\left|H_1(\omega)\right|^2} \cdot k_1^2 + S_x \cdot \overline{\left|H_2(\omega)\right|^2} \cdot k_2^2}{S_n \cdot k_1^2 + S_n \cdot k_2^2} \qquad (2.1.22)$$

It has been assumed that $H_1(\omega)$ and $H_2(\omega)$ are uncorrelated. S_x and S_n
denote the white power spectral densities of the transmit signal x and the
additive noise n_1 und n_2, respectively.

An analysis of equation (2.1.22) shows that there is no local optimum for the
coefficients. This means that only one of the branches has to be selected.
The branch with higher signal power has to be chosen. This method is
called *Selection Combining* (SC).

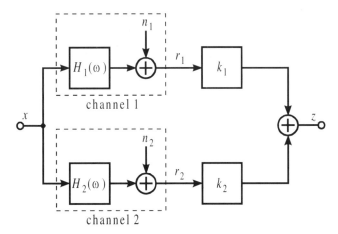

Figure 2.1.10: Linear combining of the received signals for synchronisation.

If the received signals $r_1(t)$ and $r_2(t)$ are simply added (equal gain combining – EGC) and one of the branches dominates, the output SNR is about 3 dB less than in case of selection combining.

The optimum linear combining method is SC, since the received signals $r_1(t)$ and $r_2(t)$ are uncorrelated. However, if SC is used, the information from the non-selected channel is lost. In order to take advantage from both channels, the first step should therefore be a *non-linear* operation on both signals so that correlation is generated. In a second step the correlated signals can be combined.

Therefore, the second approach to utilise the information of both branches is to combine the signals after a non-linear operation. A corresponding non-linear signal processing with subsequent linear combination for timing synchronisation is shown in Figure 2.1.11. It is a heuristic approach that is based on the periodicity metric shown in (2.1.20). The advantage of the displayed circuit is that in any case information from both branches is used simultaneously. The non-linear operation consists of calculation of the energy and correlation terms in (2.1.18). After the non-linear operation the signals are combined with a linear network. Finally, the linear combination of the energy and correlation terms is processed corresponding to the evaluation of a single channel.

The described synchronisation concept will be called *Equal Gain Combining (EGC) of energy and correlation terms*, since energy and correlation terms are simply added. Clearly, if one of the channels exhibits high attenuation, only noise is added from this branch. Therefore, an approach that weights the branches corresponding to their power yields better results. Such an approach is shown in Figure 2.1.12. It is called *Maximum Ratio Combining (MRC) of energy and correlation terms*. In this case the signals E_1 and $|K_1|$ are weighted with E_1. E_2 and $|K_2|$ are weighted with E_2.

The estimation error of timing synchronisation can be reduced by averaging the periodicity metric over N_F samples, if the channel does not exhibit severe time dispersion. The estimation error can be further reduced by filtering the estimated optimum position $\hat{k}$ over several transmission frames. Values that deviate severely from the mean can be omitted.

The same combining techniques as for timing synchronisation can also be used for frequency synchronisation. Figure 2.1.13 shows a circuit with MRC after non-linear signal processing. For MRC of the non-linear terms, the correlation results K_1 and K_2 are weighted with the energies E_1 and E_2, respectively.

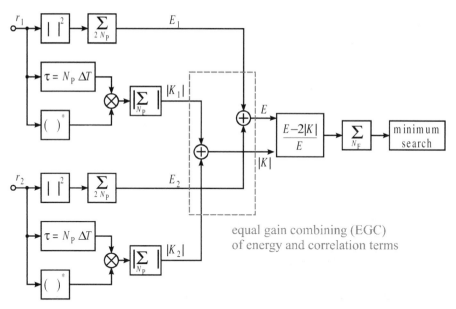

Figure 2.1.11: Timing synchronisation with antenna diversity: heuristic approach with equal gain combining of energy and correlation terms.

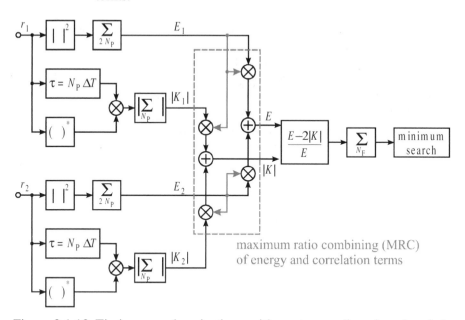

Figure 2.1.12: Timing synchronisation with antenna diversity: heuristic approach with maximum ratio combining of energy and correlation terms.

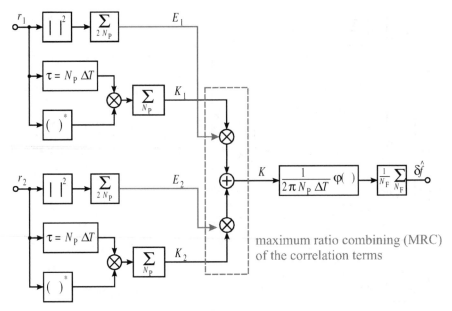

Figure 2.1.13: Frequency synchronisation with antenna diversity: heuristic approach with maximum ratio combining of the correlation terms.

After these heuristic non-linear approaches the question arises: what structure the optimum synchroniser should have? In case of a single RX branch, the Maximum Likelihood (ML) estimators for timing and frequency offset have been derived [BeSB96], [BeSB97]. It is shown that timing and frequency offset estimation can be carried out separately. The likelihood function of the received signal shows that the following metric has to be minimised to obtain the ML estimate of the timing offset:

$$M_{ML}(k) = \frac{\sigma_x^2}{\sigma_x^2 + \sigma_n^2} \cdot \frac{1}{N_\Sigma} E - \frac{2}{N_\Sigma} |K| \tag{2.1.23}$$

where σ_x^2 denotes the signal and σ_n^2 the noise variance. Obviously, for large SNR ($\sigma_x^2 \gg \sigma_n^2$) the ML metric turns into the periodicity metric. The ML estimate of the frequency offset is also given by (2.1.21). Also, in the case of two RX branches ML estimators can be derived. The block diagram of a corresponding timing offset estimator is shown in Figure 2.1.14.

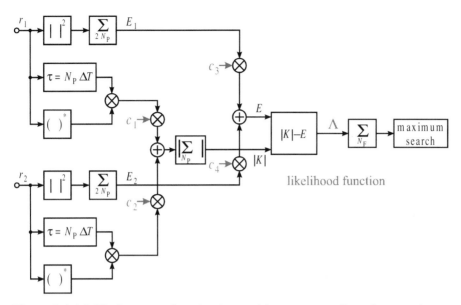

Figure 2.1.14: Timing synchronisation with antenna diversity: optimum maximum likelihood approach.

The coefficients $c_1 \ldots c_4$ depend on signal and noise variances and are given by:

$$c_1 = \frac{2\sigma_{x1}^2}{\sigma_n^2(2\sigma_{x1}^2 + \sigma_n^2)} \tag{2.1.24}$$

$$c_2 = \frac{2\sigma_{x2}^2}{\sigma_n^2(2\sigma_{x2}^2 + \sigma_n^2)} \tag{2.1.25}$$

$$c_3 = \frac{2\sigma_{x1}^4}{\sigma_n^2(2\sigma_{x1}^2 + \sigma_n^2)(\sigma_{x1}^2 + \sigma_n^2)} \tag{2.1.26}$$

$$c_4 = \frac{2\sigma_{x2}^4}{\sigma_n^2(2\sigma_{x2}^2 + \sigma_n^2)(\sigma_{x2}^2 + \sigma_n^2)} \tag{2.1.27}$$

The corresponding circuit for ML frequency offset estimation is displayed in Figure 2.1.15.

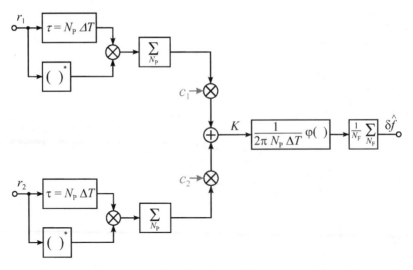

Figure 2.1.15: Frequency synchronisation with antenna diversity: optimum maximum likelihood approach.

In the following, the different methods for synchronisation in a system with antenna diversity are compared by means of simulation. The FFT length and the length of the guard interval are $N_{FFT} = 128$ and $N_g = 32$ samples, respectively. For the simulation, time-invariant channels with significant time dispersion were chosen. The PDPs are displayed in Figure 2.1.16. Both channels exhibit a direct path and an echo path attenuated by 10 dB. The delay between the direct path and the echo is 20 samples in both cases. The impulse response of channel 2 is delayed by one sample with respect to channel 1. The following two cases are considered: (a) channels 1 and 2 exhibit the same attenuation (ATT = 0); and (b) channel 2 is attenuated by 20 dB with respect to channel 1 (ATT = 20 dB). Figure 2.1.16 shows that the guard interval is sufficiently long for the channels considered.

The simulation results for timing synchronisation are displayed in Figures 2.1.17 and 2.1.18. The following combining techniques for the non-linear components E_1 and E_2 as well as $|K_1|$ and $|K_2|$ are compared: SC, EGC of non-linear terms, MRC of non-linear terms, and the optimum ML concept. No averaging filter that follows the metric calculation is used ($N_F = 1$). Figure 2.1.17 shows the performance when both channels exhibit the same attenuation. Clearly, SC yields the largest timing noise, since it does not utilise the information from both channels. All other synchronisers show almost the same performance. At high SNRs (≥ 14 dB) a small error floor is caused by the quantisation of the time axis (sampling).

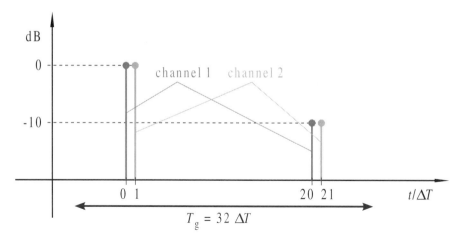

Figure 2.1.16: PDP of the impulse responses of both radio channels for simulations with antenna diversity.

Figure 2.1.18 shows that if channel 2 is attenuated by 20 dB, EGC of the non-linear terms yields the largest timing noise, since from channel 2 mainly noise and almost no information is added. Concerning both Figures 2.1.17 and 2.1.18, of course the optimum ML concept shows the best performance. But, there is almost no performance difference between the ML concept and the heuristic concept of MRC of the non-linear terms. On the other hand, the block diagrams of both methods show also almost the same computational complexity.

The proposed synchronisation algorithms can be extended easily to multi-antenna diversity RXs.

In principle, the synchronisation techniques discussed in this section can also be used if the inherent periodicity of OFDM and related signals (created from the cyclic prefix) is utilised. The main disadvantage of a synchronisation concept based on the cyclic prefix is that the synchronisation performance depends on the time dispersion of the channel. Especially, if the time dispersion is severe, the synchronisation performance will degrade significantly. ML synchronisers utilising the cyclic prefix are derived [BBBL97] for different signal models. The suitability of these synchronisers has been shown for the TDMA-OFDM UMTS proposal [BBBL98].

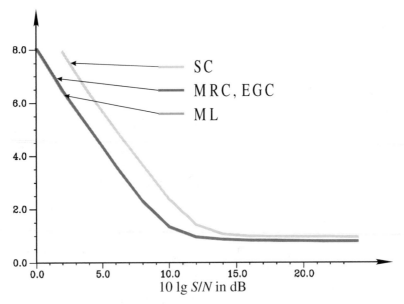

Figure 2.1.17: Standard deviation of the optimised FFT window position vs.
SNR. Both channels exhibit the same attenuation.

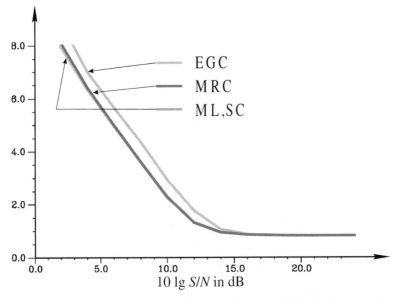

Figure 2.1.18: Standard deviation of the optimised FFT window position vs.
SNR. Channel 2 is attenuated by 20 dB with respect to
channel 1.

2.2. CDMA

Kimmo Kansanen and Laurent Schumacher

2.2.1. Introduction

The first commercial spread-spectrum-based multiple access scheme was implemented in a second generation cellular radio system in the American scene under the IS-95 banner. Simultaneously, European research efforts in both ACTS and COST 231 [COST99] frameworks paved the way for the implementation of CDMA schemes in third generation systems. These efforts led, in January 1998, to the selection of two air interfaces for UTRA (UMTS Terrestrial Radio Access) by ETSI. The first option is based on WCDMA (Wideband CDMA) for paired bands whereas the second solution relies on TCDMA (Time Code Division Multiple Access) for unpaired bands. The former is an FDD solution, while the latter employs TDD methods. The COST 259 community has been given strategic insight in the performance to be expected from these air interfaces.

Since the new air interfaces are to supplement and improve on current technologies, one of the basic services to be available is traditional speech communication. The provision of speech service by the WCDMA interface is studied in [MaMR98]. Required E_b/N_0 ratios to ensure a BER of 10^{-3} are computed in different transmission scenarios, including indoor and outdoor propagation and different terminal velocities. Similarly to other multiple access schemes, increasing mobility raises the required E_b/N_0 levels.

One significant feature of CDMA is its flexibility to offer different rate communications according to user needs and channel occupancy, making it attractive for variable rate wireless data communications. Chip-level simulations of different load scenarios (number of simultaneous connections, bit rates and spreading factor) are performed in [OlRu00]. The results are compared with those from the Gaussian approximation, which is found to provide a pessimistic error floor. An adaptive S-Aloha DS-CDMA scheme for packet data connections is described in [SaAg98a], where each mobile terminal is allowed to adapt its transmission rate with respect to the current load level of the system without additional signalling. This scheme is shown in [SaAg98b] to compare favourably with PRMA++ (Section 4.6.2) in cellular environments in terms of both performance and flexibility.

As far as TCDMA is concerned, the uplink spectrum efficiency and capacity are evaluated in [PaHW99], whereas the downlink spectrum efficiency is computed in [ScPa98]. In both cases, a multi-step procedure separating slow- and fast fading is applied. The distributions of inter-cell and intra-cell interference for a given scenario (reuse factor, number of users, propagation environment, velocity, beamforming and power control) are created. These are then used in extensive simulations to derive the spectrum efficiency and capacity under a provision of Quality of Service (QoS). The results show that the load of the system, the number of array elements at the BS and the reuse factor influence the spectrum efficiency. It also appears that the uplink has better spectrum efficiency than the downlink, mainly due to the availability of antenna and multipath combining and the diversity gains from these.

Most of COST 259 contributions deal with single-carrier CDMA, mainly due to the standardisation process. MC-CDMA is one of the most interesting schemes proposed outside of the scope of the European standardisation effort. The basic methods of network planning, code and frequency re-use, are introduced and analysed in [NSRP98]. Performance evaluation and investigation of the influence of these and other parameters (number of carriers, frequency gap, cluster size, path loss and system load) are also presented.

The section is organised according to results presented in COST 259, besides the general review on CDMA properties given here. Detector design has received much attention, from the RAKE RX (Section 2.2.2) to Interference Cancellation (IC) structures (Sections 2.2.3). The benefits of coding (Section 2.2.4) and of spatial diversity (Section 2.2.5) have also been evaluated. On the other hand, parameter estimation issues have been tackled (Section 2.2.6) and some hardware demonstrators have been presented (Section 2.2.7).

2.2.2. RAKE receivers

A RAKE RX is, in general terms, a single-user detector matched to one user's received signal. The RX combines separable multipath propagated signals, thus utilising multipath diversity. Usually the multipath combining is performed by a MRC, which provides the best possible SNR (after combining) in a single-user case. The bandwidth and the transmitted waveform define the time resolution of paths available for a RAKE, and the potential multipath diversity. The order of diversity available in a particular propagation environment is then defined by the delay profile of the environment. Increasing transmission bandwidth does not always improve the received power after multipath combining, since coherent combining requires accurate channel estimates. However, with a constant transmitted power, increasing multipath resolution decreases the power available for channel estimation per path, and thus the estimation accuracy.

In [AsAB98], the fading statistics of measured channels for 1.25 MHz (similar to IS-95) and 5 MHz (similar to UMTS) bandwidth transmissions are compared. The fading distributions are studied as a function of the number of RAKE fingers in the RX and also by analysing the excess transmit power required to achieve constant received power. The gain obtained by using a larger bandwidth is determined, as is the number of RAKE fingers that must be used. A similar approach is used [Alay98] for an analytical channel model with an exponential PDP. The fading statistics are analysed for a RAKE RX with ideal channel estimation and synchronisation.

2.2.3. Linear and non-linear interference cancellation

Linear interference cancellation schemes have mainly been proposed for CDMA uplink detection due to the availability of spreading sequences of all users, which are required in cancellation. In [DaJa99], an interesting linear cancellation scheme is proposed for the UTRA downlink. The scheme is based on taking advantage of the minimum spreading factor (4 in WCDMA) of the transmission. Linear IC is performed for the four parallel partial correlations prior to combining them for detection. The scheme is then applied to channel gain estimation in the RX. Figure 2.2.1 illustrates the impressive gains obtained with this IC scheme, especially in the high SNRs, where the BER floor of the RX is lower than for the conventional RX.

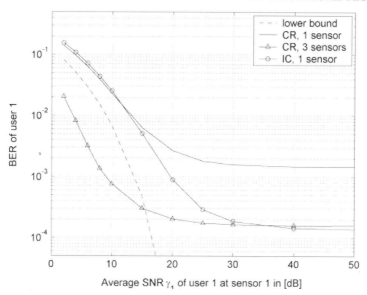

Figure 2.2.1: Bit-error rates in a 4-user UTRA downlink with channelisation code length of 16. The linear cancellation scheme outperforms the single-sensor correlation RX (© 1999 IEEE).

An uplink HD-PIC (Hard-Decision Parallel Interference Cancellation) is studied in [JLKK98], where effects of finite accuracy parameter estimation are evaluated. Realistic channel gain estimators and delay locked loops are applied in the RX. The RX is found to require accurate delay estimates to maintain performance close to that with perfect estimates. However, channel gain estimation remains the greatest influence on cancellation accuracy. The performance of channel coding with a HD-PIC RX is studied in [JuKa98], where a simple detector-decoder chain is compared to a joint detecting HD-PIC, which performs re-encoding of decoded data prior to cancellation. The performance comparison is shown in Figure 2.2.2. The potential gains of a HD-PIC in future systems are studied in [KaLJ99] by applying UTRA uplink signal structures in transmission and detection.

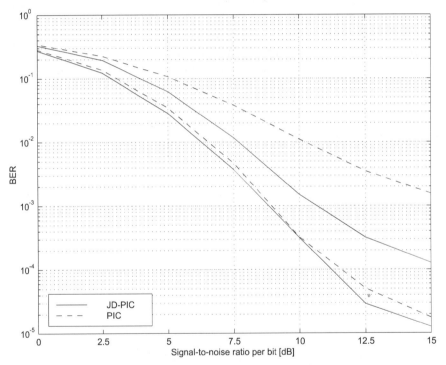

Figure 2.2.2:Bit-error rates of PIC and HD-PIC RXs in channel coded CDMA (code rate ½, constraint length 3) with channel estimation errors. Lower curves are related to an 8-user system, upper curves, to a 20-user system (© 1999 IEEE).

A cancellation scheme similar to that in [JuKa98] is proposed for TCDMA systems in [WWBO99]. Users are divided into two groups, which are then iteratively detected using a FEC decoder, re-encoded, and cancelled from the received signal. Multi-antenna reception is employed to increase performance of detection by utilising spatial separation of users and characteristics of the coloured inter-cell interference.

A method of implementing iterative Multi User Detection (MUD) with a RNN (Recurrent Neural Network) is presented in [Teic98]. Various iteration strategies are presented along with references to their counterparts in IC. Simulations are performed and the proposed detector is compared to the Maximum Likelihood Sequence Estimation (MLSE) detector, RAKE and the single-user bound.

2.2.4. Coding and modulation

The performance of conventional CDMA systems (single-user reception) depends highly on their ability to limit the MAI plaguing the transmission. It is well known that the spreading code sequences should be chosen wisely, such that their correlation properties keep the MAI level as low as possible. TCH (Tomlinson, Cercas, Hughes) sequences are low rate channel codes, which have been shown [AnCe97], [CoCe97] to be efficient waveforms in that respect. They exhibit simultaneously good auto- and cross-correlation properties. A TCH sequence-based CDMA system appears to be additive noise limited. Unfortunately, such a system has strict synchronisation requirements.

Coding schemes with iterative decoding have also been proposed for CDMA. Parallel concatenated recursive codes, better known as turbo-codes, are to be used in UTRA. In [HeSH97] the authors propose the use of iterative decoding in detection of users in the IS-95 uplink. The transmission can be interpreted as a serial concatenation of a systematic block code and a convolution code divided by an interleaver. Decoding methods are given for coherent and non-coherent detection along with simulation results for the Additive White Gaussian Noise (AWGN), static multipath, Figure 2.2.3, and time-variant multipath channels.

Turbo codes are introduced in the IS-95 downlink in [JHSH97] by using multiple code-channels in parallel. The increased resource usage is returned with impressive gains in required E_b/N_0 for a given BER. Gains can be as much as 2 dB in static multipath channels, Figure 2.2.4, and 1 dB even in time-variant channels. In a Frame Erasure Rate (FER) sense, however, gains remain large even in the latter case.

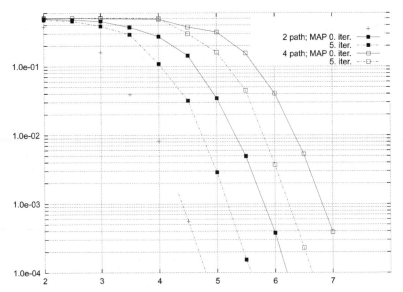

Figure 2.2.3:Bit-error-rates for a non-coherent RAKE RX (1, 2 and 4 paths
recombined) in time-invariant multipath channels. Iterative
decoding outperforms MAP decoding (© 1997 IEEE).

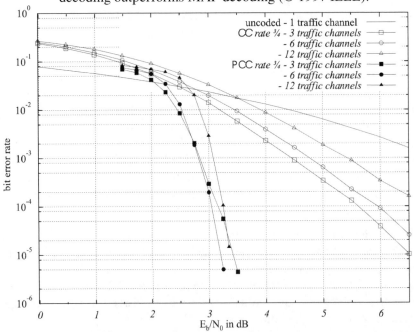

Figure 2.2.4:Bit error rates with code rate ¾ in time-invariant multipath
channel. PCC exhibit a gain of several dB with respect to CC
according to the number of traffic channels (© 1997 VDE).

Linear Joint Detection (JD) is known to cause noise enhancement. An optimal code rate, which compromises between noise enhancement and coding gain, is sought in [Burr97b], [BuWh99b]. For the conventional detector the optimal coding rate is the minimum that can be supported, but there always is an optimal rate for LDD (Linear Decorrelating Detector) due to the noise enhancement at lower rates. The linear Minimum Mean Square Error (MMSE) RX has an optimal rate at higher spectral efficiencies than 0.5, but below that no optimal rate can be found. This is intuitively satisfying, since the linear MMSE is approaching the RAKE at low system loads and the LDD at high system loads. The results are obtained by comparing analytical results utilising the Shannon bound and simulation of turbo and convolutional codes. Differences between the detectors are studied with different coding schemes and spectral efficiencies.

The search for the most efficient modulation and coding methods has finally led to the consideration of non-linear modulation schemes like Multi-h Continuous Phase Modulation (CPM). Such modulated signals exhibit a constant envelope, which is ideal when power amplification issues are to be taken into account. Moreover, when Multi-h CPM is combined with spectrum spreading, an appropriate choice of the spreading sequences increases the minimum Euclidean distance between two signals. The codes to be used should fulfil some requirements detailed in [PiRo97]. It is also illustrated that (h_1=4/8, h_2=3/8) is the most efficient 2-h scheme, as being the one with the best trade-off between implementation and power gain (2 dB at BER equal to 10^{-3} in AWGN channels).

2.2.5. Smart antennas

Many forms of diversity have been studied in conjunction with CDMA. Smart antennas, to which Section 3.3 is dedicated, are a promising concept in providing spatial diversity for CDMA. They enable an increase in capacity and/or decrease in transmission power and cluster size without changing the multiple access scheme. In addition to this, increased diversity stabilises the received power, relaxing power control dynamical requirements. Due to the flexibility of CDMA, smart antennas are relatively easy to utilise and benefit from. Indeed, such structures exploit the directional inhomogeneity of the channel with the help of beamforming and direction estimating algorithms.

A review of some smart antenna concepts (principal component MMSE, blind IC, ML and SAGE schemes) is presented in [JaDC99] for both the uplink and downlink of the WCDMA air interface. Modelling MAI as Gaussian noise prior to transmission and applying a ML criterion appears to be the most efficient scheme in the downlink with perfect knowledge of channel parameters. On the other hand, IC schemes (BIC in the downlink, SAGE in the uplink) exhibit the best performance for estimated channel parameters.

In contrast to [JaDC99], which does not carry out Direction of Arrival (DoA) estimation, the enhanced Zero Forcing-Block Linear Equaliser (ZF-BLE) detector and the MVU (Minimum Variance Unbiased) channel estimator introduced in [PFBB99] include DoA information in their processing. Depending on the chosen array structure, performance improvements up to 8 dB at BER equal to 10^{-3} are obtained. Moreover, the proposed ZF-BLE detector has been shown [WePH99] to be able to suppress inter-cell interference. This enhancement is obtained by taking the spatial covariance matrix of the interference into account.

2.2.6. Parameter estimation

Usually detectors rely on a preliminary and accurate estimation of the parameters of the received signal. For instance, the near-far resistant LDD requires the knowledge of the timing of the users. Timing recovery for the LDD is performed in [MiGB97], which presents a near-far resistant acquisition scheme suitable for fully asynchronous and bandlimited CDMA communications. Special, user-independent sequences are used for synchronisation purposes, creating an equivalent dedicated access channel. The timing information is derived from these acquisition sequences using spectral analysis techniques. This scheme is shown to ensure the acquisition of incoming users, in spite of the potential power imbalance. On the other hand, when combined with the one-shot LDD, it cancels the interference of a new user with respect to the already active ones.

Beside the afore-mentioned MVU (Minimum Variance Unbiased) channel estimator [PFBB99], phase recovery has also received attention. Since phase recovery algorithms are usually not standardised, the implementation and performance of these devices is a constructor specific issue. A phase recovery algorithm for WCDMA links is described [MaMR98] as a part of simulation of speech service in UTRA. A non-ambiguous absolute phase estimate is also derived along with the timing estimate in [MiGB97]. Issues of phase recovery are handled in a deeper and more general way in [ScVa99], where the benefits of extending MUD principles to phase estimation in CDMA communication systems are demonstrated.

2.2.7. Demonstrators

Most of the results mentioned in this section have been obtained through extensive Monte-Carlo simulations to prove the theoretical validity of the proposed schemes. Some contributors have also designed hardware implementations on which some of the schemes have been tested.

Four ASICs (Application Specific Integrated Circuit) are used in [Pavi99] to build a fully integrated digital RX (detection, synchronisation and channel estimation). This RX takes advantage of the path diversity in both indoor and outdoor communications.

On the other hand, Digital Signal Processors (DSPs) are part of the RX implementation in [AnBa97]. There, a combination of a one-shot LDD detector and the acquisition scheme proposed in [MaMR98] is realised. Near-far protection margins are so huge (90 dB) that the acquisition sequences can be transmitted by the incoming user at maximum power, without interference to the already active users. DSPs are also used in [DaHP97] to implement a 2-stage RAKE RX for a 2-user wireless communication system operating in the ISM band (2.4 GHz). Flat BER and DER (Delay Error Rate) curves with respect to power ratio between users illustrate near-far resistance of both detector and code synchroniser.

2.2.8. Summary and conclusions

Several contributions have demonstrated the validity of CDMA as a basis of future air interfaces. The flexibility of CDMA allows the use of several advanced transmission and RX technologies to enhance the link performance. The link-enhancing technologies are directly related to interference levels and the system behaviour and capacity. The future will show which of the advanced channel encoding, transmission and detection methods will be utilised in future systems.

2.3. Modulation and Coding

Alister Burr and Gorazd Kandus

Modulation and coding clearly remain central to the implementation of any wireless communication system, and development in the field is therefore continuing very rapidly. Indeed, FEC coding in particular has become increasingly important with the development of OFDM and CDMA techniques, both of which rely on FEC coding to achieve their performance. The development of third generation and broadband systems has also led to further developments of modulation schemes. This is firstly because there is increased requirement for spectral efficiency (especially with the introduction of spectrum charging in many countries), and secondly, because it has prompted a move to higher frequencies where unused spectrum is more readily available. Spectral efficiency pressures require more efficient modulation schemes than GMSK (Gaussian Minimum Shift Keying), which dominated second generation standards in Europe, leading to multilevel modulation schemes, both linear and non-linear. The move to higher frequencies (such as the LMDS or Local Multipoint Distribution System bands) has given rise to technological problems such as High Power Amplifier (HPA) non-linearity, which may be addressed by changes to the modulation techniques.

2.3.1. Linear modulation

These more spectrally efficient multilevel linear modulation schemes require linear amplification, and linear amplifiers are more difficult to implement at higher frequencies. [Burr99] considers two types of linear modulation with 16 point constellations, and evaluates their performance on non-linear channels. Conventional 16-QAM (whose constellation lies on a square grid) is compared with StarQAM (or 16-APSK), whose 16 constellation points lie on two concentric circles. On a linear channel the former is more power efficient by about 1.5 dB. Their performance on a non-linear channel is investigated using travelling wave tube (TWT) non-linearity as an example. The mean input power is backed off from saturation by 3 dB. In both schemes the non-linearity gives rise to constellation distortion due to AM-AM and AM-PM conversion (amplitude modulation distortion and amplitude to phase modulation conversion), self-interference (a random scatter of the constellation points even in the absence of thermal noise), and spectral re-growth. Spectral re-growth is the same in both cases, resulting in a re-grown sidelobe about 20 dB below the main lobe. However, the constellation distortion is less severe in StarQAM, and when compensated in the RX actually results in a slight improvement in the BER performance, because the AM-PM conversion increases the minimum constellation distance. 16-QAM performance, however, is degraded by more than 5 dB.

StarQAM or APSK is also compared with QAM in [EnRo95] (also [RoMa98]), this time in the context of OFDM on a fading channel. The former scheme lends itself to differential or quasi-coherent demodulation, which avoids the need for pilot symbols. The result is a reduction in performance, but in the quasi-coherent case, this is mainly accounted for by the difference in constellation shapes.

2.3.2. Non-linear modulation

Mobile communication systems operate in a bandwidth and power limited environment. Improvements in bandwidth and power efficiency are two important issues of future system development. Continuous phase modulation is a modulation technique that is widely used in mobile communications due to resistance to channel non-linearity and fading.

Multilevel CPM techniques can further improve spectral efficiency and increase data transmission rate. There are several possibilities to obtain multilevel CPM signals. One possibility is to vary the modulation index h from interval to interval. In this case, a scheme is called multi-h CPM. Another possibility is to superimpose several CPM signals with different amplitudes and transmit them simultaneously. In this case, multi-amplitude CPM is obtained.

Binary multi-h CPM signal with rectangular frequency-pulses can be written as a modulated waveform in the i-th symbol interval $(i-1)T \leq t \leq iT$ in the following form:

$$s(t,\alpha_i) = \sqrt{\frac{2E}{T}} \cos\left\{ 2\pi\left[f_c t + \frac{1}{2} h_i \alpha_i \left(\frac{t}{T} - (i-1) \right) \right] + \Phi_i \right\} \qquad (2.3.1)$$

where E is the symbol energy, T is the symbol duration, f_c is the carrier frequency, h_i is the modulation index and α_i is the symbol value in the i-th symbol interval. The value Φ_i is the excess phase at the symbol transition time $(i-1)T$.

It was determined that multi-h schemes when combined with spread-spectrum, present a better error performance for a convenient choice of spreading codes. In this case, the Euclidean distance is increased with the spreading of the signal.

The multi-amplitude or N-MSK signal is defined as a sum of N MSK signals with different amplitudes

$$s_{N-MSK}(t, \alpha_1, \alpha_2, ..., \alpha_N) = \frac{1}{\sqrt{\sum\limits_{j=1}^{N} 2^{2(j-1)}}} \sum_{j=1}^{N} s(t, \alpha_j) \qquad (2.3.2)$$

where $s(t, \alpha_j)$ is the j-th MSK component of the N-MSK signal

$$s(t, \alpha_j) = 2^{j-1} \sqrt{\frac{2E}{T}} \cos\left(2\pi f_c t + \frac{\alpha_j \pi t}{2T} + \Phi_{ij} \right) \qquad (2.3.3)$$

Multi-amplitude MSK schemes are more spectrally efficient than MSK, but power efficiency is reduced.

Varying the number of MSK components in response to mobile channel conditions results in variable rate transmission and system performance improvement. In the case of high CNR, all MSK signal components are transmitted, while in a deep fade situation only the greater one is transmitted.

In an N-MSK variable rate system, duplex transmission is required because information is needed about the quality of the link as perceived by the RX. The TX can respond to this information by adapting the number of transmitted bits per symbol interval according to the channel quality. In the case of adaptive N-MSK, the eye closure approach turned out to be the most effective criterion for switching between modulation schemes.

Figure 2.3.1 shows a comparison between the adaptive 4-MSK and constant rate N-MSK signals in slow varying Rayleigh fading channel.

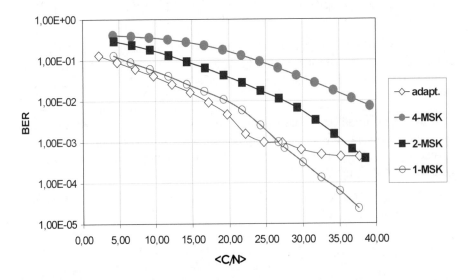

Figure 2.3.1:Performance of 4-MSK adaptive modulation technique with regard to constant rate 1-MSK, 2-MSK and 4-MSK.

2.3.3. FEC coding techniques

The work carried out on FEC coding falls into two categories. The first is the development of coding schemes themselves, which has been concerned with two types of code: Turbo-codes [BeGT93] and TCH codes [Cerc97]. The second is the application of codes in systems employing other techniques, such as OFDM and CDMA. Some aspects of this have already been covered in the previous two sections: here we will consider aspects such as comparison of the performance of different codes in these applications.

Turbo-codes (otherwise known as parallel-concatenated recursive systematic convolutional codes) were first introduced by Berrou, Glavieux and Thitimajshima in 1993. They attracted a great deal of attention because they represented the first close approach to the Shannon bound on channel capacity first discovered in 1948 [Shan48]. There are two aspects to Turbo-codes that result in their unique combination of powerful error-control properties and computationally feasible decoding.   The first is the combination of two (or more) very simple component codes, linked by a large pseudo-random interleaver, which results in a very complex composite code with particularly useful distance properties. The second is the iterative decoding algorithm, which enables simple decoders for the component codes to decode the full composite code, and whose structure leads to the name 'Turbo-codes'.

One disadvantage of Turbo-codes, as described in [BeGT93], is that a very long interleaver is used (64 kbit in that case) resulting in a very large latency and hence delay. Such codes cannot be used for delay-limited services such as speech, nor can they readily be adapted for short packet systems, and in particular ATM, which uses cells only 424 bits in length. Thus *short-frame* Turbo-codes are of interest, in which the interleaver size is restricted to a few hundred bits. However, the implicit restriction in block length from information theoretic considerations reduces the coding gain available [Burr97a].

However, it was shown that very significant advantages are available from the use of Turbo-codes even with such short interleavers. Figure 2.3.2 shows the BER performance of a rate 1/3 Turbo-code with interleaver size 256. The component codes here are 4-state and a pseudo-random interleaver as in [BeGT93] is used. Simulation results are given for eight iterations of the decoder. Note that performance is around 1.5 dB better at BER = 10^{-6} than the 64-state convolutional code of the same rate, and with such simple component codes decoding is in fact less computationally complex. Figure 2.3.2 also compares the simulation result with a union bound based on a

Monte-Carlo estimate of the weight distribution of the code, and with the uniform interleaver bound of Benedetto and Montorsi [BeMo96], showing that the latter over-estimates the so-called 'error floor' of the code by around an order of magnitude. The fact that the simulation result is above the union bound (which is an upper bound on BER for ML decoding) for lower signal to noise ratios also shows that the iterative decoder does not necessarily achieve ML decoding. Punctured rate 1/3 codes (giving rate 1/2) have also been considered, resulting in a similar comparison between the simulation result and the union bound. However, if the data sequence is punctured rather than the two code sequences only, the performance of the iterative decoder degrades further at low SNR, because the rate of the codes presented to the component decoders increases. Full details of this work are given in [BuWh99a].

TCH codes [Cerc97] are non-linear cyclic codes based on polynomials on the field GF(2) (i.e. 0, 1), derived from the field GF(p), where p is a Fermat prime number (a prime number that is one greater than a power of 2). The relevant Fermat numbers are 3, 5, 17, 257, and 65537, which means that the most useful TCH codes are of length 16 or 256. The codewords are cyclic shifts of one codeword derived from this basic polynomial, plus the sums of this codeword with certain cyclic shifts of itself chosen to maintain good autocorrelation properties. This means that the code may be decoded by determining the autocorrelation function of a received word and selecting the maximum, which can be implemented very efficiently using the FFT. Their construction means that TCH codes tend to be quite low rate. Typical examples are (16, 4), (16, 6) and (256, 16), although a (16, 8) code can also be obtained. However this is not a disadvantage when they are used in CDMA, as described in [CoCe97], since performance is best when as much of the spreading as possible is provided by the FEC code. This paper shows that TCH codes with rate 1/16 significantly outperform the standard convolutional codes in a CDMA system subject to both multipath and MAI.

Similarly, [JHSH97] (also [ScHJ98]) compares the performance of turbo-codes and convolutional codes in an IS-95-like CDMA system. The turbo-codes here use 16 state component codes and an interleaver length of 384 bits. The Turbo-codes have an advantage of some 3 dB at BER = 10^{-4} over the convolutional codes on a fading multipath channel. As mentioned above, the convolutional coding and M-ary orthogonal modulation used in IS-95 can also be treated as a (serially) concatenated code, and [HeSH97] (also [ScHe97]) shows that iterative (turbo) decoding can result in a performance improvement of around 0.6 dB.

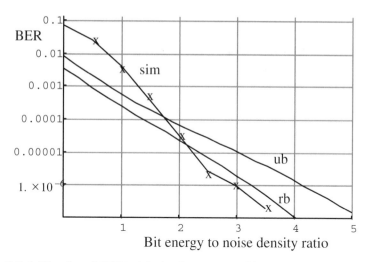

Figure 2.3.2:Simulated BER (sim) of rate one third length 256 turbo-code
compared with uniform interleaver bound (ub) and union bound
obtained from pseudo-random interleaver (rb) (© 1999 GET).

A comparison of Turbo- and convolutional codes, this time in the context of
OFDM, is reported in [BuWh99c]. The context here is digital broadcasting,
which is much less delay-limited than mobile systems, and hence a length
8192 interleaver can be used. Component codes are again 16 state, whereas
the convolutional codes are 64-state. Figure 2.3.3 shows the performance of
these codes on a Rayleigh fading channel, assuming independent fading of
code symbols. In the OFDM context this can be obtained by adequate
interleaving of the codeword over the multiplex. Note that a large part of the
advantage of the Turbo-codes on such a channel is due to the increased
diversity provided by the greater Hamming distance of the code. The
resulting advantage is some 8 dB at BER = 10^{-6}.

Relevant work on soft-decision decoding of coded multilevel modulation
symbol is also reported [MaRE98], [RoMa98], again in the context of
OFDM. Soft output coherent and non-coherent demodulation is described
for M-QAM and M-DAPSK modulation schemes (in M-DAPSK, also known
as Star-QAM, the constellation points are distributed on a series of
concentric circles). Optimum metric computation is described in each case.
Non-coherent demodulation of M-DAPSK is about 4 dB poorer than
coherent M-QAM, while quasi-coherent demodulation (which also does not
require pilot symbols for channel state estimation) is only 2 dB poorer than
coherent, a difference that is largely accounted for by the difference between
the QAM and the APSK constellations.

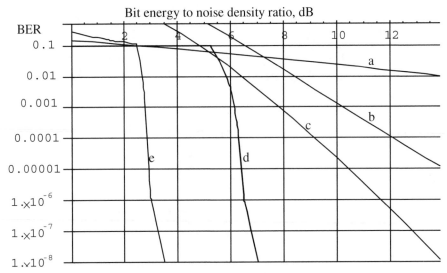

Figure 2.3.3. BER performance on Rayleigh fading channel: (a) uncoded; (b) rate 3/4 convolutional code; (c) rate 1/2 convolutional code; (d) rate 3/4 turbo-code; (e) rate 1/2 turbo-code (© 1999 IEE).

2.3.4. Equalisation

The environment in which wireless systems operate gives rise to multipath propagation, which in turn causes dispersion as well as fading. In all but the most narrowband systems (where the symbol rate is much less than the coherence bandwidth), the dispersion causes unacceptable intersymbol interference, which must be overcome by means of *equalisation* (Section 2.4 deals with one type of system in which the dispersion can be overcome by other means).

The conventional equalisation techniques are well known: namely linear equalisation, Decision Feedback Equalisation (DFE), and Maximum-Likelihood Sequence Estimation (MLSE): see (among other texts) [Proa95, Chapter 10]. Some more advanced schemes are described in [COST99, pp. 300–304].

Tomlinson–Harashima Precoding (THP) is investigated in [GFMR97] for application in a low-cost fixed wireless access (or wireless local loop) system. THP (introduced independently by Tomlinson [Toml71] and Harashima and Miyakawa [HaMi72]) shapes the signal at the TX in such a way that the signal may be recovered at the RX using only a modulo-N operation. This transposes the equalisation function to the TX side of the link, thereby reducing the complexity required in the RX. Adaptations to the original THP scheme for use on the wireless channel are described. The scheme has been implemented for typical rural and urban path delay profiles, demonstrating both effective equalisation and implementation feasibility.

[TJGB97] considers a rather different problem, related to the Wireless Local Area Network (WLAN) standard HIPERLAN I. This standard requires terminals to operate with a frequency offset up to 50 kHz (due to the use of low-cost oscillators), and at speeds up to 1.4 ms^{-1}. The symbol rate of the HIPERLAN system means that an equaliser is required. The mobility requirement is such that the channel impulse response could be regarded as quasi-static during one packet period, were it not for the frequency offset, which may give rise to total phase changes of 100π radian during one packet. It is shown that the use of a conventional adaptive equaliser to continuously update the tap weights is computationally infeasible, but joint optimisation of carrier phase and equaliser tap weights is both effective and computationally efficient. Results are obtained using both first and second order Phase-Locked Loops (PLL), and both linear equalisers and DFE. The second order PLL performs better, since a frequency offset does not give rise to a steady-state error. The DFE is both more robust on frequency selective channels, and more computationally efficient. The resulting system can tolerate frequency offsets greater than the HIPERLAN specification requires with no performance degradation.

Another form of equalisation is *Frequency Domain Equalisation* (FDE). This has many similarities with OFDM, and therefore is described in detail in Section 2.1. The principle is to perform equalisation in the frequency domain, using a FFT in the RX operating on a blockwise basis. Note that this in essence performs the same function as a linear equaliser (although it has advantages in implementation), and therefore is prone to the same noise enhancement effect as linear equalisation on a frequency-selective channel.

2.3.5. Adaptive coding and modulation

Work undertaken in COST 231 [COST99, p. 306] showed the potential advantages of adaptive coding and modulation in cellular systems, especially to optimise capacity in the presence of co-channel interference from neighbouring cells and of log-normal fading due to shadowing. This has been developed in [PeBT98] to include the effects of Rayleigh fading, and to determine the optimum form of power control in systems employing adaptive modulation.

[PeBT98] considers the effects of Rayleigh fading due to multipath in a single cell system. The work is aimed at Fixed Wireless Access (FWA) or Wireless Local Loop (WLL) systems in which fading can be assumed to be low enough for the link to adapt to it. It compares adaptive modulation and coding, both alone and joint with other techniques, against other strategies such as frequency hopping and 'frequency jumping'. The latter term refers to a form of adaptive channel allocation in a hybrid FDMA/TDMA system in which a link chooses the least faded frequency channel available, if necessary 'jumping' to another frequency channel as fade characteristics change, provided there is a TDMA slot available for it there.

Figure 2.3.4 shows the relative performance of the schemes considered in terms of the power margin required to ensure less than 1% packet loss. A hypothetical speech system is assumed, using six carrier frequencies in each cell, with six slots per frame. Voice activity is exploited, assuming that each speaker is active 41 % of the time. The 'baseline' scheme employs QPSK, relying on an increased power margin to overcome fading, while the adaptive modulation scheme uses a range of schemes from uncoded 16-QAM to rate 1/2 convolutionally coded QPSK. The frequency-hopping scheme is quite simple, using a single parity code with a single redundant symbol per frame so that the system is robust to the fading of one frequency channel. 'Controlled jumping' refers to a 'frequency jumping' scheme which is under the control of the BS, and can therefore choose the least faded frequency channel on which TDMA slots are available.

The results show that adaptive modulation and coding has advantages over most of the other schemes, and when combined with frequency jumping provides the optimum scheme in terms of minimising the power margin required. Note that in a cellular system the power margin requirement, which determines the required average SNR, translates into a minimum re-use distance requirement. Note also that the schemes that involve adaptive modulation and coding allow a trade-off between the number of conversations against the power margin. This means that the overall capacity per cell of a cellular system is less sensitive to the re-use pattern employed: if the cluster size is small a low rate coded scheme is required; if larger, a higher order modulation scheme is possible, compensating for the reduced re-use factor.

The previous work referred to above has assumed ideal power-balancing power control. [PeBT98] considers optimum power control schemes in both conventional TDMA and adaptive modulation and coding schemes. The comparison is made on the basis of the Shannon bound [Shan48], and thus is of use primarily to compare schemes rather than to predict attainable capacities. Two types of power control are described: the first is a generalised loss compensation scheme, in which the transmit power is varied according to the path loss. Transmit power (in dB) is increased by some fraction α of the path loss (also in dB). Then $\alpha = 0$ corresponds to no power control, while $\alpha = 1$ is perfect power balancing. It is shown that in conventional TDMA systems $\alpha = 0.5$ is close to the optimum under most circumstances: this is the 'square root power control' scheme that has been suggested by a number of researchers. However, in adaptive modulation and coding no power control ($\alpha = 0$) provides greater capacity.

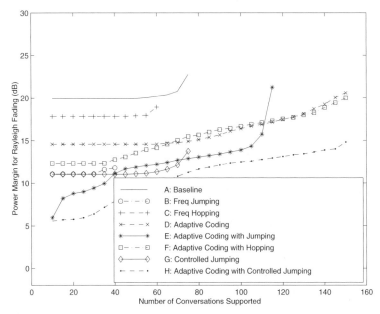

Figure 2.3.4: Power margin and supported conversations for 1% Packet Loss Rate in various systems employing adaptive modulation and coding (© 1998 IEEE).

The second power control scheme is Carrier-to-Interference Ratio (CIR) balancing. The system as a whole adapts in order to equalise the signal to interference ratio on all links. However, it has been shown that in an adaptive modulation and coding system the resulting capacity is much smaller than without power control. A scheme which maximises the overall capacity of an adaptive system by choice of transmit power on all links has also been developed, but in fact its capacity is not much greater than the scheme with no power control. This confirms the argument presented in [COST99, p. 306] that an adaptive modulation and coding scheme takes advantage of differences in signal to interference ratio to maximise overall capacity. The CIR balancing scheme, which averages out these differences, therefore reduces overall capacity.

The disadvantage of the adaptive modulation schemes is that they are 'unfair' to users who have a poorer communication channel, since they will be assigned a reduced information rate. This could be overcome by assigning a larger share of channel bandwidth to these users, although this will reduce the average overall capacity somewhat.

2.4. DECT and Adaptive Sampling

Andreas F. Molisch

2.4.1. Introduction and motivation

The DECT standard [ETSI91] is the pan-European standard for cordless telephones. It was originally designed in 1991 to replace the several analogue standards that existed at that time. The first commercial products went on sale in 1994, when about half a million terminals were sold. By the year 2000, some 30 million pieces will be in use. Designed to work in environments with extremely low time dispersion, it does not expect an equaliser to be used. However, nowadays it is used more and more for applications like WLL, private branch exchange systems, and low-mobility personal communications systems [PeRa96]. Due to these changes, the effect of time dispersion becomes increasingly important. COST 259 investigated both computation methods for the BER resulting from time dispersion effect, as well as methods for reducing those errors.

DECT is a TDMA/TDD system where 12 users share one frequency channel. Each time frame lasts 10ms, and is subdivided into 24 time slots. Each user is assigned two time slots – one for the uplink, one for the downlink. At the beginning of a timeslot, 32 known bits (A-field) are transmitted, followed by signalling information and 324 bits of user data. The bit duration is approximately 0.9 μs; the modulation format is Gaussian frequency shift keying with modulation index 0.5 and a normalised filter bandwidth of BT = 0.5. More details can be found in [ETSI91].

It is well known that a time-dispersive channel introduces errors into an unequalised TDMA system, because it causes ISI. This effect has been analysed since the early 1960s, leading to a wealth of computation methods. These include the analysis of quadratic forms of Gaussian random variables [Proa68], the determination of the Probability Density Function (PDF) of angle differences between Gaussian-distributed vectors [PaRR82], and the error region method [MoFP96]. The error floor, i.e. the BER due to ISI, turns out to be proportional to the square of the channel root mean square (RMS) delay spread (DS) normalised to the symbol period. For more details, see [Moli00] and [Moli98], where an overview of computation methods and results is given.

2.4.2. Principle of adaptive sampling

When we speak of fixed sampling, we mean that the RX samples the arriving signal at time instants $t_s' = kT + t_s$, where $k = 0, 1, 2, \ldots$ and t_s is a quantity that does not change during the interval $-\infty < t < \infty$; in the following, one does not usually distinguish between t_s' and t_s and calls them both 'the sampling time'. For adaptive sampling in DECT, the sampling time stays constant during each burst, but can change from burst to burst. As was shown in [MLPF97] (see also [MoNF99]), adaptive sampling can lead to a drastic reduction of the error floor in a DECT system. Intuitively, this can be explained by the fact that the position of the eye opening is shifted when the channel changes. The remarkable thing is, however, that the opening can shift by more than one symbol duration even if the DS is much smaller than a symbol duration, see Figure 2.4.1.

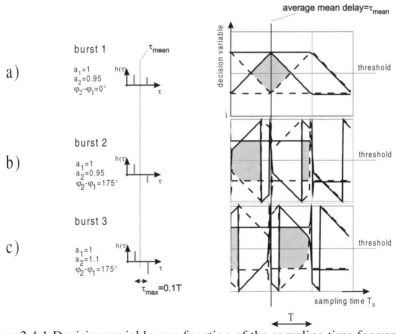

Figure 2.4.1:Decision variable as a function of the sampling time for various channel configurations. Solid lines: signals when '+1' is transmitted as the bit that is to be detected; dashed lines: transmitted '−1'.

For unfiltered MSK, adaptive sampling can lead to a complete elimination of the error floor provided only that the maximum excess delay is smaller than the symbol duration. In [MoBo98], it was shown that this is also true for most linear modulation formats: for coherent detection of BPSK, one requires that the basis pulse fulfils a symmetry condition in addition to the requirement of limited support of the pulse. For differential detection, no such symmetry requirement exists, because there is an inherent symmetry in the detection method. Quadrature-component interference prevents zero error floor for QPSK and any other linear modulation format, such as M-ary PSK and combined amplitude- and phase-shift keying, where the real and imaginary parts of the transmission symbols are uncorrelated.

2.4.3. Determination of optimum sampling time

In practice, the determination of the optimum sampling instant can be done as follows [PaMB00], [PaMo97], Figure 2.4.2: during the training sequence at the beginning of the burst, the signal is not sampled with frequency $1/T$, but oversampled with a factor N_{samp}. With these values, sequences are formed whose elements are spaced by T, indicated by the RAKE-like structures at the bottom of Figure 2.4.2.

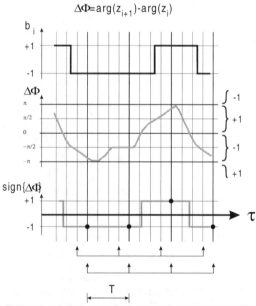

Figure 2.4.2: Determination of the optimum sampling instant: top figure: transmitted training sequence. Middle figure: decision variable of the arriving signal for a differential phase detector. Lower figure: output of a hard decision device.

The sequence is then selected which has the smallest deviation from the originally transmitted training sequence, i.e. minimises some metric for the transmission distortion. The sampling instant belonging to the best of all sequences is then used for the rest of the burst, assuming that the channel does not change significantly during one burst.

One possible metric is the bit errors of the received signal after hard decision (which is depicted in Figure 2.4.2). This is optimum in a noiseless channel, but sub-optimum when additive noise is present. An optimum metric can also be formulated for the noisy channel, but requires very complex computations. Use of a MMSE criterion can be shown to give almost optimum results, while keeping the hardware requirements low. Results of the BERs that can be achieved with this method can be found in [PaMB00].

2.4.4. Non-linear receiver structures

Another alternative to reduce errors in MSK (or generally in binary FSK) lies in the use of a receiver structure that consists of a frequency discriminator, a limiter, and a low-pass filter (usually, an integrate-and-dump filter) [MoPe00], [PeMo97]. This structure differs in one important respect from the usual Limiter-Discriminator-Integrator (LDI) structure: the limiter is placed *after* the frequency discriminator, and allows only instantaneous frequencies with $|fT| < h_{mod}/2$. It can be shown that in the usual LDI, time-dispersion errors are caused by bursts in the instantaneous frequency that occur in the bit transition regions. If these bursts are 'clipped', i.e. their magnitude is limited, the error floor is drastically reduced. The clipping also slightly enhances the effect of additive noise, but this effect is very small.

For unfiltered MSK and maximum excess delay smaller than half a symbol length, the error floor is completely eliminated. The error floor becomes finite when filtering is applied, because the filtering 'smears' the instantaneous-frequency bursts over a longer duration. Figure 2.4.3 shows the error floor for various filter bandwidths. The non-linear LDI can also be combined with adaptive sampling, further reducing the error floor.

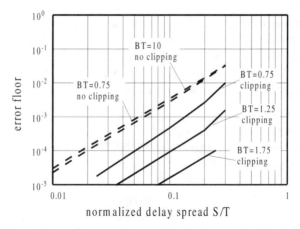

Figure 2.4.3:Error floor for non-linear discriminator (with fixed sampling) for various receiver filter widths.

2.4.5. Summary and conclusions

COST 259 has analysed the errors arising from time dispersion effects in DECT systems, and proposed remedies. Adaptive sampling and special receivers structures are suitable for strongly reducing the error floor, so that now DSs of approximately 300 ns can be tolerated for BER of 10^{-3}. Surprisingly, this is approximately the same value that can be tolerated by a PHS (Japanese cordless standard) system, even though PHS has a much longer symbol duration. The reason for this is that PHS uses DQPSK as its modulation format, where adaptive sampling does not lead to a significant improvement of the error floor.

2.5. References

[Alay98]   Alayon-Glazunov,A., "Considerations Regarding the Number of RAKE Fingers Required in CDMA RAKE-Receivers", in *Proc. of 1998 Nordic Radio Symposium*, Saltsjöbaden, Sweden, Oct. 1998 [also available as TD(97)064].

[AlLa87]   Alard,M. and Lassalle,R., "Principles of modulation and channel coding for digital broadcasting for mobile receivers", EBU Technical Review, No. 224, Aug. 1987, pp. 168–190.

[AnBa97]   Angulo,B. and Barton,S.K., "Hardware Implementation of a Near-Far Resistant Timing Acquisition Algorithm for CDMA", *COST 259*, TD(97)073, Lisbon, Portugal, Sep. 1997.

[AnCe97]   Antunes,L.C. and Cercas,F., "A simple CDMA-IC scheme based on TCH sequences", in *Proc. of 12th National URSI*

Symposium, Bilbao, Spain, Sep. 1997 [also available as TD(97)062].

[AsAB98]   Asplund,H., Alayon-Glazunov,A. and Berg,J.-E., "An Investigation of Measured and Simulated Wideband Channels with Applications to 1.25 MHz and 5 MHz CDMA Systems", in *Proc. of VTC'98 – IEEE 48th Vehicular Technology Conference*, Ottawa, Canada, May 1998.

[BBBL97]   van de Beek,J.-J., Börjesson,P.O., Boucheret,M.-L., Landström,D., Arenas,J.M. and Ödling,P., "On synchronisation in an OFDM-based proposal", *COST 259*, TD(97)063, Lisbon, Portugal, Sep. 1997.

[BBBL98]   van de Beek,J.-J., Börjesson,P.O., Boucheret,M.-L., Landström,D., Arenas,J.M., Ödling,P., Wilson,S.K., Östberg,C. and Wahlqvist,M., "Synchronisation of a TDMA-OFDM frequency hopping system", *COST 259*, TD(98)033, Bern, Switzerland, Feb. 1998.

[BeGT93]   Berrou,C., Glavieux,A. and Thitimajshima,P., "Near Shannon limit error-correcting coding and decoding: Turbo-codes", in *Proc. of ICC'93 – International Conference on Communications*, Geneva, Switzerland, May 1993.

[BeMo96]   Benedetto,S. and Montorsi,G., "Unveiling Turbo-codes: Some Results on Parallel Concatenated Coding Schemes", *IEEE Trans. Information Theory*, Vol. 42, No. 6, Mar. 1996, pp 409–428

[BeSB96]   van de Beek,J.-J., Sandell,M. and Börjesson,P.O., "*ML estimation of timing and frequency offset in multicarrier systems*", Research Report TULEA 1996:09, Lulea University of Technology, Lulea, Sweden, 1996.

[BeSB97]   van de Beek,J.-J., Sandell,M. and Börjesson,P.O, "Tracking of time and frequency offset in the uplink of an OFDM-based system for UMTS", *IEEE Trans. Signal Processing*, Vol. 45, No. 7, July 1997, pp. 1800–1805.

[Burr97a]   Burr,A.G., "Turbo-coded modulation", in *Proc. of International Seminar on Turbo-coding and Related Topics*, ENST Brest, Brest, France, Sep. 1997.

[Burr97b]   Burr,A.G., "Performance of linear separation of CDMA signals with FEC coding", in *Proc. of IEEE International Symposium on Information Theory*, Ulm, Germany, July 1997 [also available as TD(97)016].

[Burr99]   Burr,A.G., "Performance of 16QAM and STARQAM on a non-linear channel", *COST 259*, TD(99)102, Leidschendam, The Netherlands, Sep. 1999.

[BuWh99a] Burr,A.G. and White,G.P., "Comparison of iterative decoder performance with union bounds for short frame turbo-codes", *Annales des Telecommunications*, Vol. 54, No. 3–4, Mar.–Apr. 1999, pp 201–207 [also available as TD(99)003].

[BuWh99b] Burr,A.G. and White,G.P., "Combined performance of joint detection with FEC coding for CDMA mobile communication systems", in *Proc. of AMOS'99 – ACTS Mobile Communications Summit*, Sorrento, Italy, June 1999.

[BuWh99c] Burr,A.G. and White,G.P., "Performance of Turbo-coded OFDM", in *Proc. of IEE Colloquium on 'Turbo-coding in Digital Broadcasting – can it double capacity?'*, London, UK, Nov. 1999 [also available as TD(00)019].

[Cerc97] Cercas,F., "The New Class of TCH Codes", in *Proc. of I National Conference on Telecommunications*, Aveiro, Portugal, Apr. 1997.

[ChMU87] Chevillat,P.R., Maiwald,D. and Ungerboeck,G., "Rapid training of a voiceband data-modem receiver employing an equaliser with fractional-T spaced coefficients", *IEEE Trans. Communications*, Vol. COM-35, No. 9, Sep. 1987, pp. 869–876.

[ClMS93] Claßen,M., Meyr,H. and Sehier,P., "An all feedforward synchronisation unit for digital radio", in *Proc. of VTC'93 – 43rd IEEE Vehicular Technology Conference*, Secaucus, NJ, USA, May 1993.

[CoCe97] Correia,A.M.C. and Cercas,F.B., "On Channel Coding for CDMA over Multipath Rayleigh Fading Channels", *COST 259*, TD(97)057, Lisbon, Portugal, Sep. 1997.

[COST89] COST 207, *Digital Land Mobile Radio Communications*, Final Report, COST Telecom Secretariat, European Commission, Brussels, Belgium, 1989.

[COST99] COST 231, *Digital mobile radio towards future generation systems,* Final report, COST Telecom Secretariat, European Commission, Brussels, Belgium, 1999.

[Czyl96a] Czylwik,A., "Comparison of the channel capacity of wideband radio channels with achievable data rates using adaptive OFDM", in *Proc. of 5th European Conference on Fixed Radio Systems and Networks*, Bologna, Italy, May 1996.

[Czyl96b] Czylwik,A., "Adaptive OFDM for wideband radio channels", in *Proc. of GLOBECOM'96 – IEEE Global Telecommunications Conference*, London, UK, Nov. 1996.

[Czyl97] Czylwik,A., "Comparison between adaptive OFDM and single carrier modulation with frequency domain equalisation", *COST 259*, TD(97)029, Turin, Italy, May 1997.

[Czyl98] Czylwik,A., "Low overhead pilot-aided synchronisation for single carrier modulation with frequency domain equalisation", in *Proc. of GLOBECOM'98 – IEEE Global Telecommunications Conference*, Sydney, Australia, Nov. 1998.

[Czyl99] Czylwik,A., "Synchronisation and channel estimation for single carrier modulation with frequency domain equalisation and antenna diversity", *COST 259*, TD(99)014, Thessaloniki, Greece, Jan. 1999.

[DaHP97] Dalhaus,D., Heddergott,R. and Pesce,M., "A Concept of a Two-User CDMA Hardware Demonstrator for Wireless Outdoor Communications", in *Proc. of AMOS'97 – ACTS Mobile Communications Summit*, Aalborg, Denmark, Oct. 1997 [also available as TD(98)001].

[DaJa99] Dalhaus,D. and Jarosch,A., "Comparison of Conventional and Advanced Receiver Concepts for the UTRA Downlink", in *Proc. of UMTS Workshop*, Reisensburg, Germany, Nov. 1998 [also available as TD(99)001].

[DaTr97] Dardari,D. and Tralli,V., "Coded OFDM systems design for high-speed WLAN at millimeter waves", *COST 259*, TD(97)031, Turin, Italy, May 1997.

[DaTr99] Dardari,D. and Tralli,V., "High-speed indoor wireless communications at 60 GHz with coded OFDM", *IEEE Trans. Communications*, Vol. 47, No.11, Nov. 1999, pp. 1709–1721.

[EnRo95] Engels,V. and Rohling,H., "Multilevel differential modulation techniques (64-DAPSK) for multicarrier transmission systems", *European Transactions on Telecommunications*, Vol. 6, No. 6, Nov.–Dec. 1995, pp. 633–640.

[ETSI91] ETSI, "Radio Equipment and Systems Digital European Cordless Telecommunications Common Interface", *ETSI*, DECT- Specification, Part 1 to 3, Ver 02.01, 1991.

[FrLa61] Franco,G.A. and Lachs,G., "An orthogonal coding technique for communications", *IRE International Convention Record*, Vol. 9, Pt. 8, 1961, pp. 126–133.

[GFMR97] Gameiro,A., Fernandes,J., Matos,J.N., Ribau,P. and Silva,P., "Asymmetrical wireless radio system for local access", *COST 259*, TD(97)052, Lisbon, Portugal, Sep. 1997.

[HaMi72] Harashima,H. and Miyakawa,H., "Matched transmission technique for channels with intersymbol interference", *IEEE Trans. Communications*, Vol. 20, No. 4, Aug. 1972, pp. 774–780.

[HeSH97] Herzog,R., Schmidbauer,A. and Hagenauer,J., "Iterative Decoding and Despreading improves CDMA-Systems using M-ary Orthogonal Modulation and FEC", in *Proc. of ICC'97 – IEEE International Conference on Communications*, Montreal, Canada, May 1997.

[Hohe92]   Hoher,P., "A statistical discrete-time model for the WSSUS multipath channel", *IEEE Trans. Vehicular Technology*, Vol. 41, No. 4, Nov. 1992, pp. 461–468.

[JaDC99]   Jarosch,A., Dalhaus,D. and Cheng,Z., "Smart Antennas Concepts for the UMTS Terrestrial Radio Access", in *Proc. of Smart Antennas Workshop at the European Microwave Week*, Munich, Germany, Oct. 1999 [also available as TD(00)017].

[JHSH97]   Jungbauer,R., Herzog,R., Schmidbauer,A., Hagenauer,J. and Riedel,S., "Coding for a CDMA-System with Higher User Data Rates by Combining Several Traffic Channels", in *Proc. of EPMCC'97 – 2nd European Personal Mobile Communications Conference*, Bonn, Germany, Sep. 1997 [also available as TD(98)028].

[JLKK98]   Juntti,M.J., Latva-aho,M., Kansanen,K. and Kaurahalme,O.-P., "Performance of Parallel Interference Cancellation for CDMA in Estimated Fading Channels with Delay Mismatch", in *Proc. of ISSTA'98 – IEEE International Symposium on Spread Spectrum Techniques and Applications*, Sun City , South Africa, Sep. 1998 [also available as TD(98)038].

[JoWi96]   Jones,A.E. and Wilkinson,T.A., "Combined coding for error control and increased robustness to system non-linearities in OFDM", in *Proc. of VTC'96 – 46th IEEE Vehicular Technology Conference*, Atlanta, Georgia, USA, May 1996.

[JuKa98]   Juntti,M.J. and Kaurahalme,O.-P., "Performance of parallel interference cancellation for CDMA with channel coding", in *Proc. of VTC'99 Spring – IEEE 49th Vehicular Technology Conference*, Houston, Texas, USA, May 1999 [also available as TD(98)092].

[Kade97]   Kadel,G., "Diversity and equalisation in frequency domain – a robust and flexible receiver technology for broadband mobile communication systems", *COST 259*, TD(97)028, Turin, Italy, May 1997.

[KaLJ99]   Kansanen,K., Latva-aho,M. and Juntti,M., "Performance of Parallel Interference Cancellation in UTRA FDD Uplink", in *Proc. of AMOS'99 - ACTS Mobile Communications Summit*, Sorrento, Italy, June 1999 [also available as TD(99)044].

[KoMB98]   Kozek,W., Molisch,A.F. and Bonek,E., "Pulse design for robust multicarrier transmission over doubly-dispersive channels", *COST 259*, TD(98)078, Bradford, UK. Apr. 1998.

[LaRo99b] Lampe,M. and Rohling,H., "Reducing out-of-band emissions due to non-linearities in OFDM systems", in *Proc. of VTC'99 Spring – IEEE 49th Vehicular Technology Conference*, Houston, Texas, USA, May 1999 [also available as TD(99)048].

[MaMR98] Mastroforti,M., Melis,B. and Romano,G., "Preliminary performance evaluation of speech service for the WCDMA solution of UMTS", *COST 259*, TD(98)108, Duisburg, Germany, Sep. 1998.

[MaRE98] May,T., Rohling,H. and Engels,V., "Performance analysis of Viterbi decoding for 64-DAPSK and 64-QAM modulated OFDM signals", *IEEE Trans. Communications*, Vol. 46, No. 2, Feb. 1998, pp. 182–190.

[MaRo98] May,T. and Rohling,H., "Reducing the peak-to-average power ratio transmission systems", in *Proc. of VTC'98 – IEEE 48th Vehicular Technology Conference*, Ottawa, Canada, May 1998.

[MiGB97] Missiroli,M., Guo,Y.J. and Barton,S.K., "A Near-Far Resistant Acquisition Protocol for Joint-Detection CDMA", submitted to *IEEE Trans. Communications* [also available as TD(97)032 and TD(97)083].

[MLPF97] Molisch,A.F., Lopes,L.B., Paier,M., Fuhl,J. and Bonek,E., "Error floor of unequalized wireless personal communications systems with MSK modulation and training-sequence-based adaptive sampling", *IEEE Trans. Communications*, Vol. 45, No. 5, May 1997, pp. 554–562.

[MoBo98] Molisch,A.F. and Bölcskei,H., "Error floor of pulse amplitude modulation with adaptive sampling phase in time-dispersive fading channels", in *Proc. of PIMRC'98 – 9th IEEE International Symposium on Personal, Indoor and Mobile Radio Communications*, Boston, Mass., USA, Sep. 1998 [also available as TD(98)077].

[MoFP96] Molisch,A.F., Fuhl,J. and Proksch,P., "Error floor of MSK modulation in a mobile-radio channel with two independently fading paths", *IEEE Trans. Vehicular Technology*, Vol. 45, No. 2, May 1996, pp. 303–309.

[Moli00] Molisch,A.F. (ed.), *Digital Wireless Wideband Communications*, Prentice Hall, New Jersey, USA, 2000.

[Moli98] Molisch,A.F., "Bit error probability of cordless telephones in time-dispersive environments", *COST 259*, TD(98)046, Bradford, UK, Apr. 1998.

[MoNF99] Molisch,A.F., Novak,H. and Fuhl,J., *Basic Performance of DECT*, Sec. 5.7 in [COST99].

[Moos94]   Moose,P.H., "A technique for orthogonal frequency division multiplexing frequency offset correction", *IEEE Trans. Communications*, Vol. COM-42, No. 10, Oct. 1994, pp. 2908–2914.

[MoPe00]   Molisch,A.F. and Petrovic,R., "Reduction of the error floor of binary FSK by non-linear frequency discriminators", submitted to *IEEE J. Selected Areas Communications*, 2000.

[MuBR96]   Muller,T., Bruninghaus,K. and Rohling,H., "Performance of coherent OFDM-CDMA for broadband mobile communications", *Wireless Personal Communications*, Vol. 3, No. 2, 1996, pp. 295–305.

[NeLo94]   O'Neill,R. and Lopes,L.B., "Performance of amplitude limited multitone signals", in *Proc. of VTC'94 – IEEE 44th Vehicular Technology Conference*, Stockholm, Sweden, June 1994.

[NSRP98]   Nunes,M., Santos,J., Rodrigues,A., Punt,J., Nikokaar,H. and Prasad,R., "Effects of downlink intercell interference on MC-CDMA system performance", in *Proc. of PIMRC'98 – 9th IEEE International Symposium on Personal, Indoor and Mobile Radio Communications*, Boston, Mass., USA, Sep. 1998 [also available as TD(99)068].

[OlRu00]   Olmos,J.J. and Ruiz,S., "Chip Level Simulation of the Downlink in UTRA-FDD", *COST 259*, TD(00)023, Valencia, Spain, Jan. 2000.

[PaHW99]   Papathanassiou,A., Hartmann,C. and Weber,T., "Uplink Spectrum Efficiency and Capacity of TF-CDMA with Adaptive Antennas", *COST 259*, TD(99)113, Leidschendam, The Netherlands, Sep. 1999.

[PaMB00]   Paier,M., Molisch,A.F. and Bonek,E., "Training-sequence based determination of optimum sampling time in unequalized TDMA mobile radio systems", accepted in *IEEE Trans. Vehicular Technology*, 2000

[PaMo97]   Paier,M. and Molisch,A.F., "Determination of the optimum sampling time in DECT-like systems", *COST 259*, TD(97)012, Turin, Italy, May 1997.

[PaRR82]   Pawula,R.F., Rice,S.O. and Roberts,J.H., "Distribution of the phase angle between two vectors perturbed by Gaussian noise", *IEEE Trans. Communications*, Vol. 30, No. 8, Aug. 1982, pp. 1828–1841.

[Pavi99]   Paviot,O., *Study and design of digital spread spectrum receiver for indoor propagation channel* (in French), Ph.D. Thesis, INSA Rennes, France, June 1999

[PeBT98] Pearce,D.A.J, Burr,A.G. and Tozer,T., "Capacity of TDMA Cellular Systems with Slow Rayleigh Fading Counter-Measures", in *Proc. of PIMRC'98 – 9th IEEE International Symposium on Personal, Indoor and Mobile Radio Communications*, Boston, Mass., USA, Sep. 1998

[PeMo97] Petrovic,R. and Molisch,A.F., "Reduction of multipath effects for FSK with frequency-discriminator detection", *COST 259*, TD(97)013, Turin, Italy, May 1997.

[PeRa96] Pettersen,M. and Raekken,R., "DECT offering mobility on the local level - Experiences from a field trial in a multipath environment", in *Proc. of VTC'96 – 46th IEEE Vehicular Technology Conference*, Atlanta, Georgia, USA, May 1996.

[PFBB99] Papathanassiou,A., Furió,I., Blanz,J.J. and Baier,P.W., "Smart antennas with two-dimensional array configurations for performance enhancement of a joint detection CDMA mobile radio system", *Wireless Personal Communications*, Vol. 11, No. 1, 1999, pp. 89–108.

[PiRo97] Pinto,J.P. and Rodrigues,A., "Optimization of Spreading Sequences for Spread Spectrum Multi-h CPM", *COST 259*, TD(97)037, Turin, Italy, May 1997.

[Proa68] Proakis,J.G., "On the Probability of Error for Multichannel Reception of Binary Signals", *IEEE Trans. Communications*, Vol. 16, No. 2, Feb. 1968, pp. 68–71.

[Proa95] Proakis,J.G., *Digital Communications* (3rd Ed.), McGraw-Hill International, New York, NY, USA, 1995.

[RoBr97] Rohling,H. and Bruninghaus,K., "High rate OFDM-modem with quasi-coherent DAPSK", in *Proc. of VTC'97 – 47th IEEE Vehicular Technology Conference*, Phoenix, Ariz., USA, May 1997.

[RoGr96] Rohling,H. and Grunheid,R., "OFDM transmission technique with flexible subcarrier allocation", in *Proc. of ICT'96 – International Conference on Telecommunications*, Istanbul, Turkey, Apr. 1996.

[RoGr97] Rohling,H. and Grunheid,R., "Performance comparison of different multiple access schemes for the downlink of an OFDM communication system", in *Proc. of VTC'97 – 47th IEEE Vehicular Technology Conference*, Phoenix, Ariz., USA, May 1997 [also available as TD(97)022].

[Rohl98] Rohling,H., "OFDM transmission technique' in *Proc. of COST 252/259 Joint Workshop*, Bradford, UK, Apr. 1998.

[RoMa97] Rohling,H. and May,T., "Comparison of PSK and DPSK modulation in a coded OFDM system", in *Proc. of VTC'97 –*

47ᵗʰ IEEE Vehicular Technology Conference, Phoenix, Ariz., USA, May 1997 [also available as TD(97)023].

[RoMa98] Rohling,H. and May,T., "Soft decision decoding for OFDM radio transmission", COST 259, TD(98)004, Bern, Switzerland, Feb. 1998.

[SaAg98a] Sallent,O. and Agustí,R., "A Proposal for an Adaptive S-ALOHA Access System for a Mobile CDMA Environment", *IEEE Trans. Vehicular Technology*, Vol. 47, No. 3, Aug. 1998, pp. 977–986 [also available as TD(98)007].

[SaAg98b] Sallent,O. and Agustí,R., "An Efficient Data Transmission Policy in an Integrated Voice-Data DS-CDMA Network", in *Proc. of VTC'98 – IEEE 48ᵗʰ Vehicular Technology Conference*, Ottawa, Canada, May 1998 [also available as TD(98)008].

[SaKJ94a] Sari,H., Karam,G. and Jeanclaude,I., "An analysis of orthogonal frequency-division multiplexing for mobile radio applications", in *Proc. of VTC'94 – IEEE 44ᵗʰ Vehicular Technology Conference*, Stockholm, Sweden, June 1994.

[SaKJ94b] Sari,H., Karam,G. and Jeanclaude,I., "Frequency-domain equalisation of mobile radio and terrestrial broadcast channels", in *Proc. of GLOBECOM'94 – IEEE Global Telecommunications Conference*, San Francisco, Calif., USA, Nov. 1994.

[ScCo96] Schmidl,T.M. and Cox,D.C., "Low-overhead, low-complexity (burst) synchronisation for OFDM", in *Proc. of ICC'96 – IEEE International Conference on Communications*, Dallas, Texas, USA, May 1996.

[ScHe97] Schmidbauer,A. and Herzog,R., "Iterative decoding and despreading improves CDMA systems using FEC and M-ary orthogonal modulation", *COST 259*, TD(97)069, Lisbon, Portugal, Sep. 1997.

[ScHJ98] Schmidbauer,A., Herzog,R. and Jungbauer,R., "Coding for a CDMA-System with Higher User Data Rates by Combining Several Traffic Channels", *COST 259*, TD(98)028, Bern, Switzerland, Feb. 1998.

[ScPa98] Schmalenberger,R. and Papathanassiou,A., "Downlink Spectrum Efficiency of a JD-CDMA Mobile Radio System", *COST 259*, TD(98)094, Duisburg, Germany, Sep. 1998.

[ScVa99] Schumacher,L. and Vandendorpe,L., "MAI Mitigation in DA ML Carrier Phase Recovery Loops for DS-CDMA Systems", in *Proc. of VTC'99 Fall – 50ᵗʰ IEEE Vehicular Technology Conference*, Amsterdam, The Netherlands, Sep. 1999 [also available as TD(99)054].

[Shan48]   Shannon,C.E., "A mathematical theory of communication", *Bell System Technical Journal*, Vol. 27, July and Oct. 1948, pp. 379–423 and 623–656.

[Teic98]   Teich,W., "Multiuser Detection for DS-CDMA Communication Systems based on Recurrent Neural Network Structures", in *Proc. of ISSTA'98 – IEEE International Symposium on Spread Spectrum Techniques and Applications*, Sun City, South Africa, Sep. 1998.

[TJGB97]   Tellambura,C., Johnson,I.R., Guo,Y.J. and Barton,S.K., "Equalisation and frequency offset correction for HIPERLAN", in *Proc. of PIMRC'97 – 8th IEEE International Symposium on Personal, Indoor and Mobile Radio Communications*, Helsinki, Finland, Sep. 1997 [also available as TD(97)036].

[Toml71]   Tomlinson,M., "New automatic equaliser employing modulo arithmetic", *Electronics Letters*, Vol. 7, No. 5/6, Mar. 1971, pp 138–139.

[Vite83]   Viterbi,A.J., "Nonlinear Estimation of PSK-modulated carrier phase with application to burst digital transmission", *IEEE Trans. Communications*, Vol. IT-29, No. 4, July 1983, pp. 543–551.

[WePH99]   Weckerle,M., Papathanassiou,A. and Haardt,M., "Estimation and utilization of spatial intercell interference covariance matrices in multi-antenna TD-CDMA systems", in *Proc. of PIMRC'99 – 10th IEEE International Symposium on Personal, Indoor and Mobile Radio Communications*, Osaka, Japan, Sep. 1999.

[WWBO99]   Weckerle,M., Weber,T., Baier,P.W. and Oster,J., "Adaptive Array Processing for Time Division CDMA Utilising Multi-Step Joint Detection", *COST259*, TD(00)002, Valencia, Spain, Jan. 2000.

3

Antennas and Propagation

Ernst Bonek

Radio propagation is at the heart of mobile communications. Multipath, fading and shadowing set the ultimate limit of how much information can be transmitted and at what quality. Communications engineers might see the mobile radio channel as a problem, but this is too narrow a view. On the contrary, a truly thorough understanding of this channel opens vast possibilities for the exploitation of its subtle phenomena by smart engineers.

Propagation research has repeatedly been declared exhausted. This Action has proved the opposite, not least because of the vivid interaction between radio systems people, network operators, and propagation and antenna researchers. The following sections will present the details of this proof.

Section 3.1, which is about propagation modelling up to frequencies including the 5 GHz ISM (Industrial, Scientific, Medical) band, shows the steady progress achieved in both statistical and deterministic modelling of the mobile radio channel. Polarisation has gained importance in relation to path loss and fading.

Section 3.2 addresses a framework for spatial channel models, extending the famous delay models of COST 207 to the directional domain. This topic arose only very recently, but turned out to be so prolific that a dedicated Sub-Working Group was established.

Of course, it has been always known that the multipath components arrive at the receiver from different directions. However, the large body of new experimental evidence and the need for a widely accepted model of this propagation phenomenon necessitated co-ordinated research and effort. Both network planning and *ad hoc* radio networks will benefit.

Section 3.3, about smart antennas, is devoted to the engineering exploitation of the spatial nature of the radio channel. Smart antennas show great promise for boosting the capacity of mobile networks, both in terms of subscribers and data rate.

Emerging technologies for multi-Mbit/s data rates will require higher carrier frequencies than are in use today. Section 3.4 deals with propagation modelling for millimetre-wave bands and associated antennas.

Evidently, personal phones operate in close proximity to persons. So, why have cellular network planners traditionally inserted the *free space gain value of 0 dBi* in their link budgets for the terminal antenna? Section 3.5 reports on the experiments that shatter this erroneous belief and show realistic gain values in operational environments that are much smaller. The proper measurement techniques have been discussed in another dedicated Sub-Working Group.

Given the surprising and exciting new results of antenna and propagation research of our group in the last three years, I am confident that many phenomena still await their discovery and theoretical explanation.

3.1. General Aspects of Propagation

Ralf Kattenbach

Although current research in the area of propagation measurements and modelling largely focuses on directional channels and the use of higher frequencies in the millimetre-wave bands, which will be covered in other sections of the current report, there has still been quite some work done in COST 259 on the general aspects of propagation. One could object that these general aspects have already been exhaustively studied in the past; however, since new systems and services use other frequencies or bandwidths, there is a need to at least validate if well established propagation models are still applicable. It also has to be taken into account that measurement techniques have been improved during the last years, thus it seems to be reasonable to use these improvements for validating and improving the statistical and empirical models in particular. Due to restrictions in computational resources, previous models have often been developed to be simple and also with a certain application in mind. However, due to the enormous increase in computer-speed, the capacity now exists to develop and apply more universal and realistic models for improving existing systems as well as the development of new systems and services.

The term 'General' in the title of the current section already implies that the aspects covered herein form a kind of 'basis', and thus are valid or at least valuable for directional channel descriptions, and for propagation at higher frequencies. To avoid an overlap with the sections that are dedicated to these special research topics within COST 259, the current section mainly focuses on *non-directional* channels at frequencies below 6 GHz. The sub-section on statistical and empirical modelling describes the validation and improvement of existing models as well as some new approaches. These models are especially valuable for system designers, but also for network planning. The deterministic modelling, which is applicable mainly for network planning, since it is usually fixed to a certain environment, is described in another sub-section. Another general aspect of propagation are measurements and the equipment used for such measurements, since statistical and empirical models are based on the evaluation of measurement results, and both statistical and deterministic modelling approaches have to be validated by appropriate measurements to end up with realistic models. All measurement campaigns and channel sounding equipment used in COST 259 are thus tabulated in a sub-section of the current section.

3.1.1. Statistical and empirical modelling

Ralf Kattenbach

In contrast to deterministic models, which are based on a more or less detailed reproduction of the actual physical wave propagation process, the statistical and empirical models attempt to reproduce, either directly or by statistical means, certain characteristics observed from measurements of the mobile radio channel. Three fundamental characteristics covered in respective sub-sections are the path loss, the small-scale fading and the large-scale fading, which form a kind of hierarchical structure, since path loss values serve as an input for large-scale fading models whereas the latter determine the power values for small-scale fading modelling. Another sub-section is dedicated to statistical polarisation modelling. The term WSSUS is frequently used in context with statistical modelling, thus there is a special sub-section on the definition of WSSUS, its validity and the occasional mix-up with other statistical models. Finally, a sub-section on miscellaneous topics covers aspects that have impacts on several of the models in the other sub-sections.

3.1.1.1. Path loss models

Christian Bergljung and Peter Karlsson

This sub-section is concerned with statistical and semi-empirical path loss models. The former type is often derived using simple linear regression on measured data, whereas the latter is generally based on deterministic methods (ultimately governed by Maxwell's equations) that are modified according to experimental data. Obviously, the distinction between these two types is ambiguous, so the reader should also consult Section 3.1.3 (Deterministic Modelling). The greater part of this section deals with path loss models for urban areas, and the starting point is often the work carried out within the COST 231 project [COST99]. However, finding a robust model for the propagation over a row of buildings still seems to be an open problem. The frequencies of concern here are 900 MHz and 1800 MHz (or thereabouts).

Some indoor results are also presented with a focus on office environments and in frequency bands above UHF, notably in the 5 GHz band (a potential worldwide allocation for Radio-LANs).

3.1.1.1.1. Semi-Empirical Models for Urban Small-Cell Deployments

Path loss models for urban areas often comprise two components that correspond to the dominant mechanisms of propagation: an expression relevant for the propagation in the vertical plane (over rooftops) and another for the horizontal plane (along street canyons). The former will be, broadly speaking, more dominant far away from the Base Station (BS), whereas the latter can be expected to play the dominant role in the neighbourhood of the BS. There are, of course, cases in which both are important, which is dependent on the BS height relative to surrounding buildings.

Most of the models of rooftop propagation are based on various approximate solutions to the problem of diffraction by multiple (infinitely thin) screens. They may then be augmented by empirical correction factors derived from measurement data, the COST 231 Walfish-Ikegami (COST 231-WI) being one particular example. It is recalled that the total path loss is usually described by three terms: the free space loss L_0, the additional multiple-screen diffraction loss L_{msd} and the rooftop to street loss L_{rts} (see also [COST99]). Figure 3.1.1 shows the general geometry, the different path loss terms and some of the input parameters to the models.

In [AlRC98], five models of rooftop propagation are compared, namely Ikegami [IYTU84], Walfish-Bertoni (WB) [WaBe88], Maciel [MaBX93], Xia [XiaH96] and COST 231-WI [COST99]. Several combinations are also considered, where different path loss terms from different models are used. The models cannot be applied to all kinds of environments in a general way, since each one has its own characteristics. The free space term L_0 is always included, Ikegami's model then only accounts for L_{rts}, whereas Maciel's only adds L_{msd}. The remaining models include all of the terms listed above.

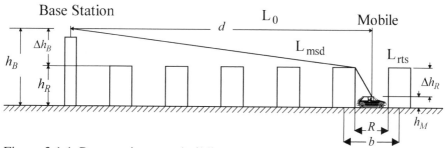

Figure 3.1.1: Propagation over buildings.

Some parameters have a stronger influence on the path loss values than others; the distance d between the BS and the mobile is of course one of them. The following parameters are also important (see [AlRC98] for a more comprehensive discussion):

- BS height h_b, particularly when the BS is located above the rooftops;
- The orientation of the street (where the mobile is located) with respect to the direct radio path between the BS and the mobile;
- The (average) distance between buildings b, which is mostly accounted for in the Walfish-Bertoni model.

Two test scenarios were selected in [AlRC98], and measurements were taken at 900 MHz. Both had a rectangular street pattern, but the building heights were more variable in one of them. Even though these scenarios are regular, some approximations with regard to the geometry must be made in order to apply the models. A vertical cross section (compare Figure 3.1.1) with buildings of different heights, for example, is approximated by a profile of buildings of a constant (average) height and spacing b. This procedure will of course have an impact on the final result. Figure 3.1.2 shows the path loss prediction supplied by the five models alone or combined for the scenario in which the building height was almost constant. The BS antenna was located above the rooftops.

Clearly, the models cannot predict the power peaks of the measured signal, which occurred when the mobile passed street junctions (crossroads). The overall results from both scenarios revealed that Xia's model and the Maciel-Xia's combination gave the best results (that is, smallest mean error and standard deviation) for the case in which the BS antenna is above the rooftops. The Xia-Ikegami combination yielded the best results for BS antenna locations below the rooftops. In the previous cases, the standard deviation was within the 4–5 dB range; in the latter cases the range was larger, 3–8 dB. It should be noted that other test scenarios might have given different results.

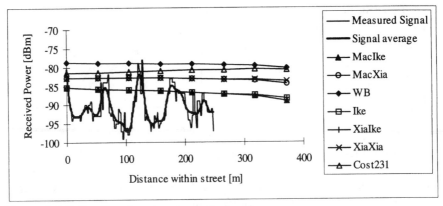

Figure 3.1.2: Predicted and measured signals.

More results for the case in which the BS antenna is located below rooftops are presented in [Ortg99], albeit with a carrier frequency twice as high. The models employed are COST 231-WI and the Uni-Valencia [CaMA95], and measurement data were taken in the city centre of Stuttgart at 1 980 MHz. A procedure similar to that of [AlRC98] was used to convert the actual terrain (or building) profile (obtained from a 3D database) into a regular counter-part. Using any of the two models under study, the order of magnitude as well as the trend of the path loss could be predicted. The mean prediction errors were of the order of ±2 dB with standard deviations in the range 2-8 dB. However, for COST 231-WI, the standard deviation was about 2 dB lower than that of the Uni-Valencia model.

The application of various semi-empirical (or rather semi-deterministic) models to a city with undulating streets and irregular building types is considered in [IHAP99]. For all models except one ([Vogl82]), the terrain profiles between the BS and the mobile were converted into regular profiles with a constant average building height and spacing, just as in [AlRC98], [Ortg99]. The best results were supplied by Xia's model [XiaH97], whereas the COST 231-WI gave the largest standard deviation of the error.

One of the most common problems associated with the use of purely statistical or semi-empirical models is that tuning with measured data is necessary in order to achieve a low mean and standard deviation of the prediction error. Ideally, one would like to choose the set of parameter values relevant to a particular model *a priori*, that is, before comparing measured data with the results supplied by the model. In [Basi99], a multiple-screen model based on a parabolic differential equation [Berg94] is used, together with a medium-high resolution database (25×25 m^2). Contrary to most of the models referred to above, this model allows a varying screen height along the terrain profile. An advanced statistical database is created by defining a number of clutter classes, each of which is associated with a mean building height and a building density (defined on a 25 × 25 m^2 pixel). Each class is then mapped on to a particular screen height.

The clutter classes defined in [Basi99] were obtained by resampling original 1 m resolution databases (to 25 and 50 m) for nine cities in Sweden and Italy. The classes were in fact mapped onto three different sets of screen heights, characterised as 'standard', 'low attenuation setting' (that is, lower screen heights) and 'very low attenuation setting'. The most interesting case is perhaps that in which the untuned model is used, that is, the multiple-screen model with the standard setting for each city. To elucidate the overall performance of the model, three different cases of settings were defined; one in which the prediction results are obtained considering the performance of the model with the default setting for each city (Case 1); another evaluated on the basis of the best setting for the screen height in each city (Case 2); the last one using the low attenuation for all cities (Case 3). Furthermore, the comparison between measured and predicted data was made in different ways: firstly, the global error was obtained as the mean value of the errors of the model for every city, and secondly, the mean error was obtained using all available data for all cities. The results are summarised in Table 3.1.1; the left one of the two sub-columns shows the relevant results for the first way of analysing the data. The results in [Basi99] suggest that it is possible to use a half-screen model without making an independent tuning for every city.

Table 3.1.1: Performance of untuned and tuned versions of the multiple-screen model.

	Setting	Mean error [dB]		Av. STD [dB]	
Case 1	Default	-2.0	-1,3	2.2	8.9
Case 2	Low	-0.2	0.0	0.9	8.6
Case 3	Very low	0.5	1.2	2.3	8.9

Now, turning to the propagation along street canyons (usually modelled only in the horizontal plane), [Ortg99] also evaluates various models of diffraction around a street corner with the BS antenna located below the rooftops. An empirical multiple-slope model (see [Ortg99], [EGTL92]) yielded the best performance, particularly far into the crossroads.

3.1.1.1.2. *Behaviour at Street Junctions*

The models of propagation over rooftops do not account for the influence of street junctions (or crossroads) on the received power. Normally, the signal level increases at the junctions due to the propagation along the streets, the local peaks in Figure 3.1.2 provide a particular example (BS antenna above rooftops). This increase may cause additional interference, which is important from a cellular planning viewpoint.

An attempt to remedy this is presented in [ClFC99]. The path loss L_p is then described by the following expression:

$$L_p = L_0 + L_{msd} + L_{rts} + L_{cross}, \qquad (3.1.1)$$

where L_{cross} accounts for the crossroads, while the remaining terms are defined as above. The relevant geometry is shown in Figure 3.1.3.

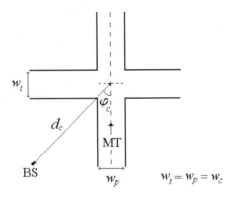

Figure 3.1.3: Geometry for the Gonçalves model.

The decrease in path loss as the mobile passes a crossroad is assumed to be of the following simple form (Gonçalves model):

$$L_{cross} = \begin{cases} 0 & \left| d_{street} \right| \geq \dfrac{\Delta d_{ext}}{2} \\[2ex] -A_M \exp\!\left[-18(d_s / \Delta d_{ext})^2\right] & \dfrac{\Delta d_{int}}{2} < \left| d_{street} \right| < \dfrac{\Delta d_{ext}}{2} \\[2ex] \Delta L_{int} & \left| d_{street} \right| \leq \dfrac{\Delta d_{int}}{2}. \end{cases} \qquad (3.1.2)$$

The propagation loss as a function of the distance from the mobile to the road junction d_{street} is depicted in Figure 3.1.4. The dependence of Δd_{int}, Δd_{ext}

and Δd_{street} on the geometrical parameters is rather involved; the details are given in [ClFC99].

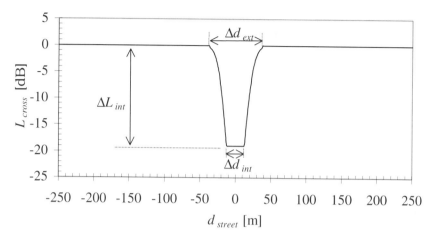

Figure 3.1.4: Variation of L_{cross} vs. d_{street} for $\varphi_c = 90°$.

For a regular 'Manhattan'-type of environment, the model (3.1.2) supplied reasonable results when $\varphi_c > 20°$ (the mean error and the standard deviation were in the order of 4 dB); for $\varphi_c < 20°$, the errors were much larger.

3.1.1.1.3. Frequency Dependence: 900 MHz and 1 800 MHz

The results described above have been obtained either at 900 MHz or 1 800–1 900 MHz. In [ChOC00], measurements taken at 900 MHz and 1 800 MHz in the same areas (urban small cells) are analysed and compared. The measurements were carried out in two distinct areas of Lisbon. The path loss is predicted according to (3.1.2). Hence Gonçalves' model for the crossroads is employed, while the multiple-screen loss is predicted by Bertoni's [XiBe92] and Ikegami's models [IYTU84].

As regards the difference between the measured path loss in the two bands, the mean absolute value of the difference was 7.8 dB and 9.3 dB for the two respective areas. One of the test scenarios was characterised by an irregular street and building structure, which resulted in large prediction errors; the mean error was about 15 dB (up to 40 dB). The worst results occurred when $\varphi_c < 20°$ (the street angle), as reported in [ClFC99]. Obviously, the treatment of a crossroad seems to be an open problem.

The next section contains some results pertaining to the frequency dependence of the path loss in indoor environments.

3.1.1.1.4. Path loss Models for Indoor Environments

This sub-section primarily considers the path loss within buildings, but the results can also be used to estimate the path loss between outdoor and indoor systems. In particular, a non-linear model for the path loss as a function of the number of walls traversed by the direct path is presented. This model has mainly been analysed at 5 GHz, but it is likely to be applicable in many other bands. Therefore, the frequency dependence is discussed first.

Multifrequency measurements for a single building are presented in [Nobl99]. The frequencies selected for this study were 2, 5, 17 and 60 GHz, and the measurements were taken on several floors of an office building. Using simple linear regression, the power decay inside a room with LoS propagation was found to be less than that of free space. In particular, the decay exponent n ranged from 1.3 to 1.7 for all frequencies for transmitter-receiver, TX-RX, separations from 1 to 10 m. A similar propagation effect, due to reflections from walls, floor and ceiling, was also found in an LoS corridor up to 30 m separation at 5.3 GHz [KiVa97]. This wave-guide effect has been noted in many studies, irrespective of carrier frequency used, see also [KaEM92], for example. This is not surprising.

The path loss in obstructed and NLoS environments is of course strongly dependent on furniture, the wall material and the building geometry. For light wall structures like plasterboard the average wall attenuation loss L_{wi} per wall, that is, the loss in excess of free space loss L_{FS}, was found to lie in the range from 2.5 dB to 4 dB [Nobl99], [KiZV99]. This result is obtained by assuming a linear behaviour of the total wall attenuation as a function of the number of penetrated walls k according to the Motley-Keenan model [KeMo90].

The excess loss L_{ex} in dB is given by

$$L_{ex} = L - L_{FS} = L - 20 \cdot \log_{10}(4\pi d / \lambda). \tag{3.1.3}$$

However, the relative power of the direct path is reduced when the attenuation of each wall increases. This is due to the fact that significant paths through doorways, windows and ventilation ducts, etc. exist. This behaviour can be accounted for by taking a non-linear approach according to

$$L_{ex}(k) = L_{wi} \cdot k^{((k+1.5)/(k+1)-b)} \quad \text{[dB]} \tag{3.1.4}$$

For an office building with a mixture of wall materials such as plasterboards, glass and concrete, the *average* excess loss per wall was $L_{wi} = 8.4$ dB and the non-linear parameter b was 0.4 at 5.8 GHz. In fact, the parameter b also depends on L_{wi}. If the wall loss is small, one can expect a more linear behaviour for small k, whereas for large wall losses, other propagation paths (not penetrating walls) quickly become dominant. Using estimated mean values of the excess loss data presented in [Nobl99], [KBTB99], [KiZV99], the following relationship between b and L_{wi} was found:

$$b = -0.064 + 0.0705L_{wi} - 0.0018L_{wi}^2 .$$ (3.1.5)

A plot of L_{ex} vs. L_{wi} and the number of penetrated walls k in the direct path traversing the walls is given in Figure 3.1.5. Note the nearly linear behaviour (in dB) of the loss for small wall losses, and that the slope of the curve decreases quickly for large L_{wi}.

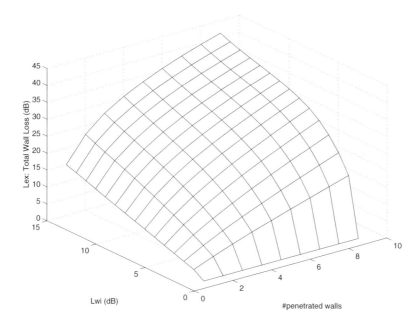

Figure 3.1.5: Average excess path loss in the 5 GHz band vs. the wall loss L_{wi} and the number of penetrated walls in the direct propagation path.

As an example of the spatial behaviour of the loss, the average excess loss values for 31 office rooms are plotted in Figure 3.1.6. The partitions of the rooms are generally made of wood and plasterboards (a few are concrete walls), whereas the outer walls, as well as the supporting walls in the middle of the office plan, are made of concrete. The TX is located just at the points where loss values are given, and the (fixed) receiver position is denoted RX.

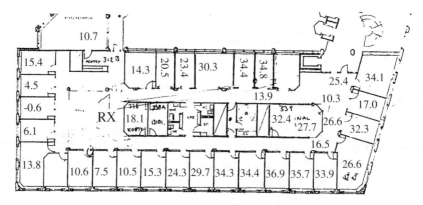

Figure 3.1.6:Average excess path loss in the 5 GHz band for 30 TX locations inside office rooms and five locations in the corridors.

A graph showing the average excess path loss as a function of k in the 5 GHz band for three different offices is drawn in Figure 3.1.7. The values referred to as 'PeK', 'PhN' and 'JaK' are extracted from [KBTB99], [Nobl99] and [KiZV99], respectively. Note that the increase of the loss with k is smaller for large k. If $b \approx 0.5$ (with L_{wi} about 8 dB) in (3.1.4), then the excess loss behaves asymptotically as $L_{ex} \propto L_{wi} \sqrt{k}$ for large k.

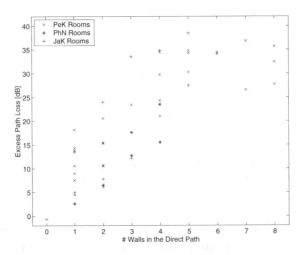

Figure 3.1.7: Average excess path loss in the 5 GHz band in three different offices vs. k.

It is also interesting to note the frequency dependence of the wall loss. In Table 3.1.2, results obtained from the multifrequency data (valid for an office environment) presented in [Nobl99] are given. For this particular example, the thin walls are represented by plasterboard subdividings between rooms, and the thick walls are (generally) made of reinforced concrete. It is important to note, however, that the value L_{wi} in (3.1.4) is an *average* value that represents all the different wall types within the particular prediction area (range of a BS, for example).

Table 3.1.2: Penetration loss as a function of frequency for two types of walls.

Frequency [GHz]	Loss for thin walls [dB]	Thick walls [dB]
2	3.3	10.9
5	3.4	11.8
17	3.5	16.1

3.1.1.1.5. Regions of Validity of Different Models, Methods for Comparison

Given a set of propagation models, being either statistical or semi-empirical, one may ask about the regions of their validity. An approach to determine this for both statistical and deterministic models is proposed in [FCGB98]. The statistics used for a particular prediction model is mainly the standard deviation of the error as a function of the distance from the BS. It is found that the Hata formula may not work at larger distances, within its supposed range of applicability. For the particular scenarios (small cell and micro-cellular), it is suggested that semi-empirical prediction models for rooftop propagation (Saunders and Bonar [SaBo94], in this case) can be replaced by statistical (power-law type) models beyond 1 km if the BS antenna height is at the same level or above the rooftops. It should be noted, however, that statistical models require tuning.

A methodology for comparison between propagation models is presented in [BaGi98]. Apart from the mean error of the predicted power and the corresponding standard deviation, the cumulative density function of the error is used as well as the figure-of-merit F suggested in [WaLL97]. We recall that $F = (N_{cc} + N_{cnc}) / N_{tot}$, where N_{cc} and N_{cnc}, respectively, denotes the number of locations where coverage and non-coverage is correctly predicted, while N_{tot} is the total number of locations. A comparison in terms of the cumulative distribution curves and that of the second order statistics does not show any significant difference, whereas the figure-of-merit appears to give some new information.

3.1.1.2. Large-scale fading

Ralf Kattenbach

The term large-scale fading denotes the variations of the mean amplitude or mean power of a received signal when a time- or space-variant multipath channel is used as the transmission medium. Usually these mean values remain approximately constant for a short time- or space-interval, and thus variations can be observed only on a large-scale basis. For time-variant channels the equivalent denotation long-term fading is frequently used. The major mechanism causing large-scale fading in context with mobile radio channels is the changing visibility or obstruction of multipath components (MPCs) when moving through a certain scenario. This causal reason has lead to the denotation shadow fading, which is sometimes also used for the large-scale fading. An accepted empirical bound up to which the mean amplitude or power values remain approximately constant has been set up by Parsons and Bajwa [PaBa82] to be movements over a few tens of a wavelength. Usually, the large-scale characteristics of a channel are evaluated on the basis of measurements; however, due to a lack of measurements some researchers [LiDe99], [DiKR00], [DiRe00] have evaluated the results from deterministic propagation prediction tools instead.

3.1.1.2.1. Narrowband Modelling

Large-scale fading was originally studied from measurements with a CW-signal (continuous wave). From such measurements it can be observed that the mean amplitude a of the received signal is log-normally distributed over time or space:

$$f_a(a) = \frac{1}{\sqrt{2\pi} \cdot \sigma \cdot a} \cdot e^{-\frac{(\ln a - \mu)^2}{2\sigma^2}} , \ a > 0 \quad , \tag{3.1.6}$$

i.e. power or path loss values in dB appear to be normally distributed. This result is an accepted narrowband model for the large-scale fading, and is frequently used for analytical calculations or system simulations (e.g. [RuCO98]). The mean value μ is determined from the path loss models covered in Section 3.1.1.1; for the standard deviation σ appropriate values can be determined from measurements for the respective environment.

A somewhat different behaviour for the narrowband large-scale fading in tunnels has been reported in [LiDe99]. The evaluation of results from ray-tracing, RT, simulations showed that there is no log-normally distributed large-scale fading; instead, the mean received power is identical to the value determined by path loss modelling. This result, which appears to be reasonable from the fact that in a (straight) tunnel there is no 'shadowing' like, for example in urban areas, has been confirmed by measurements in [LiBD00].

3.1.1.2.2. Wideband Modelling

Wideband (i.e. frequency-selective) channel modelling is usually based on the time- or space-variant Channel Impulse Response (ChIR). For a frequency-selective channel the ChIR shows several distinct components (which in the following will be denoted as 'paths') at certain delay times that can be related to certain propagation lengths, whereas for narrowband channels there is only one single path in the ChIR. Thus, it is quite common in channel modelling to treat each of the paths in a wideband ChIR like a narrowband channel, which means assuming a log-normal distribution for the large-scale fading of each path [ZwFW98], [Zwic00]. In context with directional channel modelling, the description by means of a ChIR has to be extended by the angle of incidence. For a means of classifying different channel types, this extended ChIR has been decomposed into parts resulting from different scattering areas in [FuMB98]. The part related to the LoS path in [FuMB98] is assumed to reveal log-normally distributed large-scale fading over time. An extension of the ChIR that includes information both on the angle of incidence and on polarisation is the Field Delay-Direction Spread Function (FDDSF) [COST99], [HeBF97]. It is assumed in [HeBF97] that all components of the FDDSF show log-normally distributed large-scale fading over time or space.

An important quantity in context with wideband channel modelling is the RMS Delay Spread (DS). Some work has been done within COST 259 to investigate the large-scale behaviour of the RMS DS. There is a distinct correlation between the received power vs. space and the RMS DS vs. space [KiVa97], [KiVa98a]. This coincides with results reported by other researchers, and thus it is proposed in [AsLB98], [AsBe99b] to model the large-scale behaviour of the RMS DS as described in [GEYC97] to be log-normally distributed with a correlation function for the logarithm of the DS that is identical to the correlation function of the large-scale fading.

For a realistic wideband channel modelling, not only the large-scale *fading* but also other large-scale *variations* should be taken into account. These are the appearance and disappearance of paths when the scenario changes [ZwFW98], [HeBF97], and also time- or space-variant changes in the delay time of the paths [HeBF97] due to changing propagation lengths. In context with geometry based stochastic models, which describe the channel properties on the basis of different scattering areas like those proposed in

[FuMB98], appearance and disappearance of clusters have to be taken into account [SACH99], [AsBe99b] as a large-scale effect.

3.1.1.3. Small-scale fading

Ralf Kattenbach

The term small-scale fading denotes rapid amplitude variations of a signal received via a time- or space-variant multipath channel. These rapid variations can occur for movements even in the order of a wavelength, and are caused by constructive or destructive superposition of MPCs that reveal different phase-evolution over time or space, respectively. For small-scale movements (according to the accepted bound introduced by Bajwa and Parsons up to some tens of a wavelength [PaBa82]), the mean amplitude remains approximately constant; thus, for large-scale movements the small-scale fading is superimposed on the large-scale fading. Since movements over small areas are related to a short time interval, in context with time-variant channels the small-scale fading is also frequently denoted as short-term fading. Due to the rapid variations, the small-scale fading is also sometimes denoted as fast fading. The investigation of small-scale fading is usually based on the evaluation of measurements; however, especially in recent years, the results from deterministic propagation prediction tools have also been used instead of measurements for studying small-scale fading effects in micro- and pico-cells. A comparative study at 1.8 and 2.4 GHz in an indoor environment [LTDV99] shows that, although the fading profiles themselves are different, at least the Ricean K-factor is comparable when derived from RT results and measurements.

3.1.1.3.1. Narrowband Modelling

The most common model for the small-scale fading is the assumption of a Rayleigh fading amplitude (e.g. [Lee82], [Pars92]). The Rayleigh distribution:

$$f_a(a) = \frac{a}{\sigma^2} \cdot e^{-\frac{a^2}{2\sigma^2}}, \ a \geq 0 \tag{3.1.7}$$

results for the amplitude distribution of the sum a of a large number of uncorrelated rotating vectors, each with equal amplitude and uniformly distributed phase. Thus, there is a physical reasoning for Rayleigh distributed amplitudes in mobile radio channels when assuming the received signal as the superposition of a large number of uncorrelated MPCs with approximately the same amplitude [BrDe91]. The phase of the resulting vector a will be uniformly distributed in the interval $[0, 2\pi]$. The term $2\sigma^2$ can be identified with the signal power, and can thus be determined from the results of a large-scale fading model. The Rayleigh distribution can be decomposed into uncorrelated zero-mean Gaussian distributions for the real and imaginary part of the resulting vector a, which is advantageous for simulation purposes.

Rayleigh fading amplitudes were originally observed from measurements with a CW-signal. The Rayleigh fading model, in the first instance, thus is a narrowband model for the small-scale fading, which in this form, due to its simplicity, is still frequently used for analytical calculations or system simulations (e.g. [RuCO98]). Investigations of measurements in [KiVa98a], [KiVa99] at 5.3 GHz in an indoor environment have confirmed the narrowband Rayleigh fading hypothesis both for LoS and NLoS. Also, from the investigations in tunnels at 2 GHz, the small-scale fading amplitudes appear to be Rayleigh distributed [LiDe99], but a dependency of the goodness-of-fit on the extent of the investigated space-interval was found [LiBD00].

From the investigations in [KaEn97], the time-variant transfer function at each single frequency can be regarded as a narrowband measurement of the small-scale fading. For these measurements, it turns out that for frequencies between 1.5 and 2.1 GHz in an indoor environment, the Rayleigh distribution is not always the most appropriate function. Much better results came out for the Rice distribution:

$$f_a(a) = \frac{a}{\sigma^2} \cdot I_0\left(\frac{a \cdot \rho}{\sigma^2}\right) \cdot e^{-\frac{a^2+\rho^2}{2\sigma^2}}, \ a \geq 0 \quad . \tag{3.1.8}$$

This distribution function describes the amplitude distribution of the resulting vector a from the sum of a dominant rotating vector and a large number of much smaller uncorrelated rotating vectors with equal amplitude and uniformly distributed phase. Ricean fading can thus be interpreted as the coherent superposition of a dominant component with a Rayleigh fading signal. In (3.1.8) I_0 is the zero-th order modified Bessel function of first kind. The parameter ρ^2 describes the power amplitude of the dominant component and $2\sigma^2$ describes the power of the Rayleigh distributed components. Thus, when there is no dominant component (i.e. $\rho^2 = 0$), the Rice distribution consequently turns into a Rayleigh distribution. The ratio between the power of the direct component and the power of the Rayleigh part is often expressed by the Ricean K-factor $K = \rho^2/(2\sigma^2)$, which thus describes the deviation from Rayleigh fading in a concise form. Due to its relation to the Rayleigh distribution, for simulation purposes Ricean fading can be created by simply adding the dominant component to the complex valued Gaussian noise sources that are used to simulate Rayleigh fading. The phase (e.g. see [BiDI91], [MaFN98]) will then no longer be uniformly distributed:

$$f_\varphi(\varphi) = \frac{1}{2\pi} e^{-\frac{\rho^2}{2\sigma^2}} \left[1 + \sqrt{\frac{\pi}{2}} \frac{\rho\cos\varphi}{\sigma} e^{\frac{\rho^2\cos^2\varphi}{2\sigma^2}} \left[1 + \mathrm{erf}\left(\frac{\rho\cos\varphi}{\sqrt{2}\sigma}\right) \right] \right], \qquad (3.1.9)$$

$$|\varphi| \le \pi \quad .$$

Besides [KaEn97], narrowband Ricean fading has also been assumed for the investigations in [LTDV99] at 1.8 and 2.4 GHz in an indoor environment. However, this choice was based on a more or less general reasoning.

For the aforementioned investigations with respect to the narrowband Rayleigh or Ricean fading hypothesis, different evaluation methods have been used. In [KiVa98a], [KiVa99] and [LiDe99] the Cumulative Distribution Function (CDF) has been determined from the measurements and has been evaluated by visual inspection in comparison to the theoretical curve. The investigation in [LiBD00] was based on the fact that both the mean and the standard deviation of the measured fading profile are related to the parameter σ of the Rayleigh distribution. Thus, the parameter σ has been determined both from the mean and the standard deviation of the measured data and the ratio of both results has been evaluated with respect to its deviation from 1. In [KaEn97] the CDF has been determined from measurements and the theoretical CDF has been fitted to the results from measurements by using an optimisation algorithm (modified simplex [NeMe64]). The remaining mean square error for a fit that has been found to be optimum can then be used as a quality criterion.

The discrepancy in the found suitability of the Rayleigh distribution from the different investigations may be explained in several ways. The investigations in [KiVa98a] and [KiVa99] were based on fading profiles of several tens of metres length, and those in [LiDe99] and [LiBD00] on profile lengths of up to a few hundred metres. In all cases, the large-scale fading has been removed before the data has been examined with respect to the small-scale fading. In [KaEn97] the investigated profile lengths were 1.8 m, and the large-scale fading has not been removed due to the short distance covered. In accordance with the result from [LiBD00], which found that the Rayleigh distribution tends to fit better with increasing profile length, the discrepancy may thus become reasonable. Another difference is that the investigations in [KaEn97] were based on linearly scaled amplitude values, whereas in [KiVa98a], [KiVa99], [LiDe99], and [LiBD00] logarithmically scaled values were used, and thus low amplitudes were weighted much stronger. A third difference is that these latter investigations are based on some profiles only (an absolute number is not given in the documents), whereas the investigations in [KaEn97] are based on a total number of 15 639 profiles, from which about 4 000 showed very good results, even for the Rayleigh distribution. Thus the suitability of a Rayleigh distribution also depends on the profile, which in turn depends on the environment.

3.1.1.3.2. Wideband Modelling

As already mentioned in Section 3.1.1.2 in the context of large-scale fading, wideband (i.e. frequency-selective) channel modelling is usually based on the ChIR by treating each path in the ChIR as a narrowband channel. The time-variant ChIR $h(t,\tau)$ can be expressed as a sum of weighted delta-functions:

$$h(t,\tau) = \sum_i a_i(t) \cdot e^{j\varphi_i(t)} \cdot \delta(\tau - \tau_i) \quad , \tag{3.1.10}$$

where i is the path number, τ_i is the delay of the ith path, $a_i(t)$ and $\varphi_i(t)$ are the time-dependent amplitude and phase of the respective path i. In general, the path delay may also be time-variant, i.e. $\tau_i \rightarrow \tau_i(t)$. Treating each path as a narrowband channel now leads to Rayleigh distributed amplitude values $a_i(t)$ and uniformly distributed phase values $\varphi_i(t)$ for a wideband Rayleigh channel model. For the wideband Ricean channel model, the amplitude values $a_i(t)$ are thus Ricean distributed, and the phase $\varphi_i(t)$ is distributed according to (3.1.9). It is quite common to assume a Rice distribution only for the first path, since this path usually includes the LoS-component as a dominant part, and assuming a Rayleigh distribution for all other paths. This can be achieved by values of $K \neq 0$ for the first path and $K = 0$ for all other paths in the wideband Ricean fading channel model. The frequency selective fading models described by (3.1.10) can be implemented for simulation purposes by a tapped delay-line, with taps at each τ_i and randomly varying complex valued tap-weights determined from a Rayleigh or Rice distribution with related random phase values.

The power of each single tap is assumed to reveal log-normally distributed large-scale fading, as mentioned before. In contrast to narrowband small-scale fading models, the mean power of each tap can no longer be determined by the (narrowband) path loss models. For frequency-selective channels, the power is spread over a certain range of delay times (time-dispersion). This distribution of power vs. delay time is described by the Delay Power Density Spectrum [Bell63], more commonly known as the

Power Delay Profile (PDP). From the PDP, values for the power of taps in a wideband model, and thus the parameters for the distribution functions, can be determined. The most common assumption for the shape of the PDP is an exponentially decaying function (as proposed, for example, by Lee [Lee82]):

$$P_h(\tau_e) = P_0 \cdot e^{-\frac{\tau_e}{S_\tau}}, \quad \tau_e \geq 0 \quad .$$
(3.1.11)

This PDP is dependent on the excess delay $\tau_e = \tau - \tau_0$, rather than the delay τ, with τ_0 being the delay of the strongest path. The power for this strongest path, and thus the amplitude value P_0 in (3.1.11), is usually derived from path loss models. The decay of the exponential PDP is determined by the RMS DS S_τ, which in general is defined as the square root of the second central moment of the PDP normalised to the total power:

$$S_\tau = \sigma_\tau = \sqrt{\frac{\int_{-\infty}^{+\infty} (\tau - m_\tau)^2 P_h(\tau) d\tau}{\int_{-\infty}^{+\infty} P_h(\tau) d\tau}} \quad ; \quad m_\tau = \frac{\int_{-\infty}^{+\infty} \tau P_h(\tau) d\tau}{\int_{-\infty}^{+\infty} P_h(\tau) d\tau} \quad .$$
(3.1.12)

The RMS DS thus is a measure for the 'width' of the PDP, and hence for the time-dispersiveness of the channel; m_τ is the mean delay, i.e. the first order moment of the normalised PDP.

The correlation between the random amplitude and phase values over time for each tap can be described by appropriate correlation functions. However, it is more common to describe the correlation by the related power density spectra in the frequency domain, i.e. by Doppler spectra. If MPCs of approximately the same amplitude arrive uniformly distributed from all directions, for a moving TX or RX this results in the so-called 'Classical Doppler Spectrum':

$$P_s(\nu) = \frac{1}{\pi} \frac{1/\nu_{max}}{\sqrt{1 - (\nu/\nu_{max})^2}} \quad ; \quad |\nu| \leq \nu_{max} \quad ,$$
(3.1.13)

which is also known as 'Jakes' Spectrum'. In (3.1.13) ν is the Doppler shift and ν_{max} is the maximum Doppler shift, which occurs for MPCs impinging from behind or in front of the mobile. For the case of Ricean fading, the dominant component arrives under a certain angle at the mobile, and thus has a certain Doppler shift. This is considered by a δ-pulse at the respective Doppler shift in the spectrum. Describing the correlation by Doppler spectra is advantageous, especially with respect to model implementations. The amplitude and phase values can first be generated as statistically independent random variables, which are thus uncorrelated and have a constant Doppler spectrum. Then the correlation can be induced by simply using a filter whose transfer function has the shape of the desired Doppler spectrum.

The number of taps N in the tapped delay-line is usually fixed, and the delay for each tap is determined by the reciprocal of the channel bandwidth B, i.e. $\Delta\tau = \tau_i - \tau_{i-1} = 1/B$, which is the path resolution. This will result in several tap weights with zero values for the cases when there is no path in the ChIR. To determine if and where there is a path unequal to zero, i.e. an actual path, the occurrence of paths is frequently modelled as a Poisson process [AlBe99b], as originally proposed by Turin *et al*. [TCJF72]. The number of actual paths is sometimes also modelled by statistical means, e.g. in [HeBF97] as a Poisson process.

The wideband channel model described up to now, i.e. a tapped delay-line with Rayleigh/Ricean fading tap weights, classical Doppler spectrum and exponential PDP, can be regarded as a kind of 'standard wideband model'. This is commonly accepted and frequently used (at least its fundamental features) for analytical calculations or simulative investigation of communication systems and their components (e.g. in [CoCe97], [MASA98], [AlBe99b]). Essential principles of the standard model have also been adopted for the directional channel models in [ZwFW98], [SACH99]. Considerable work has been done within COST 259 to validate or improve the standard model, to derive appropriate parameters for different situations and environments or to establish efficient methods for parameter derivation. Particular details of this work within COST 259 will be described in the following.

Measurements at 5.3 GHz with a bandwidth of 53.75 MHz in different indoor environments, both under LoS and NLoS have been investigated in [ZhKV00], and the parameters for a tapped delay-line model as derived from the measurements are given in tables. An arbitrarily fixed number of four taps has been chosen; the tap delays are determined taking into account the decay of the PDP and the dynamic range of the measurements. All taps are assumed to be Rayleigh fading, except for the first tap, which has been chosen to be Ricean fading in some cases. The Doppler spectra were determined to be classical, for Ricean fading taps additionally with a dominant component. The tap power has been extracted directly from the measurements rather than a PDP model. Correlation between the taps has also been investigated; in most cases there is only a slight correlation.

In [MASA98] tables are given for the model that has been decided by ETSI/BRAN for the use in HIPERLAN/2 simulations. In principle, the model is identical to the standard model, however, in order to reduce the number of taps, non-equidistant tap delays have been chosen with an increasing tap delay at higher excess delays. The total number of taps is fixed to 18. All taps are assumed to be Rayleigh fading with corresponding classical Doppler spectrum for all channel types, except for the first tap in the case of LoS in a large open space environment, which is Ricean with a K-factor of 10. The measurements in [MASA98] are intended to confirm these modelling assumptions by measurements at 5.22 GHz, with a 16 MHz bandwidth, in office and large open indoor environments. According to the authors of the document, their measurements indicate that the PDP is exponential with a spike at zero excess delay for LoS, the fading is Rayleigh except for the LoS spike, which has a K-factor between 5 an 15, and the Doppler spectrum is classical with a LoS spike.

Detailed investigations of measurements at 1.88 GHz, with a 5 MHz bandwidth, in urban and suburban environments have been performed in [AlBe99a]. A total of 71 measurements, each consisting of 1 430 instantaneous ChIRs, served as the basis. The small-scale fading of different paths has been investigated not only with respect to a Rayleigh and a Rice distribution, but also the Nakagami distribution (which yields an excellent physical explanation for the fading [Naka60], [BrDe91] – however it is only an approximate solution for the superposition of randomly rotating vectors), the Weibull distribution and the log-normal distribution have been taken into account. The comparative study of the goodness-of-fit for the different distribution functions shows (in principle) similar results as has been obtained for indoor environments in [Katt97a], [KaEn98]. From the comparative study in [AlBe99a], the Rice distribution turned out to be the best choice, Rayleigh appears to be inappropriate. Also, the (approximate) equivalence of the Rice and the Nakagami distribution has been addressed and investigated (compare [Katt97a], [KaEn98]). An efficient determination of the K-factor is done by the Fading Delay Profile (FDP), which displays the K-factor vs. delay. From FDPs averaged over all measurement locations it turned out in [AlBe99a] that a K-factor of about 6 can be used for the first (strongest) path, and a K-factor of about 1.6 is appropriate for all other paths. The globally averaged PDP was found to be appropriately modelled by an exponentially decaying function with a single peak at zero excess delay. The decay constant was derived from the measurements rather than from the DS (a comparison to the DS is not given in the document); values for the decay constant as well as for the relative power of the peak are given in a table in [AlBe99a].

Measurements in [PMFF98] at 1.8 GHz in urban and rural environment have confirmed the exponential PDP hypothesis with the DS as the decay constant. Doppler spectra turned out to match the classical spectrum in an urban environment and a two component truncated Gaussian spectrum in a rural environment, which coincides with a proposal from COST 207 [COST89].

The modelling approach in [MaFN98] is based on the idea of combining deterministic and statistical methods, since purely deterministic modelling cannot take into account all of the details of the propagation process. The approach consists of a tapped delay-line with Ricean fading taps and a phase distribution according to (3.1.9). However, no empirical PDP shape and no empirical K-factors are used; these values are determined by a deterministic model based on analytical calculations. With this deterministic model, all MPCs are first calculated directly during the simulation, and then for each time interval (according to a tap-length) the power of the strongest component and the mean value of the power of the other components are extracted to determine the parameters ρ and σ of the Rice distribution (3.1.8). A validation of the model is done, due to a lack of measurements, by comparing the CDF of amplitude values as well as the phase values vs. distance created by the model with those extracted from RT simulation results and a fairly good agreement has been found. For comparison, the deterministic phase increment model proposed in [Hash93] has also been implemented, and it turned out that a phase variation according to (3.1.9) seems to be more realistic than such a deterministic phase increment model.

The channel model in [AsLB98] has been proposed to be used for the evaluation of time based positioning methods in T1P1.5 standardisation of GSM in the US. The approach is based on the CODIT model [PéJi94], which has been slightly modified and especially extended by DS modelling and angle of arrival modelling. The CODIT model, in principle, is also based on the tapped delay line approach, but there is no assumption of an exponential PDP; the power values for the taps are given by appropriate Ω values for the Nakagami distribution, which are given in tables in [PéJi94]. The small-scale fading is modelled by the Nakagami distribution [Naka60]; appropriate m-values are also given in [PéJi94]. Doppler spectra are not classical in the CODIT model, but are characterised by several truncated Gaussian curves with mean values derived from the angles of arrival for certain distributions of scatterers in the environment. From these Doppler spectra, the time-dependent phase values can be determined directly. The DS model in [AsLB98] is based on [GEYC97], and describes the RMS DS as a log-normally distributed value with a mean that increases with distance; parameters for different channel types are given in [AsLB98]. The DSs are uncorrelated at different BSs. A matching of the DS model and the CODIT model is achieved by rescaling the time axis (for urban/suburban environment), or by adjusting the relative powers of the diffuse and the non-diffuse part (for mountainous environments), respectively.

For the channel model in [HeBF97] the small-scale fading is not explicitly modelled by distribution functions, but each single MPC is created with randomly chosen amplitude, delay, and incidence direction in such a way that the resulting power density spectra of the channel are close to empirical spectra, e.g. to an exponential PDP. Each MPC itself is not fading for small-scale movements, but the amplitude of the vectorial sum of MPCs within a certain delay interval will reveal small-scale fading due to different phase variations of the single MPCs. This method, which has become quite popular in recent years, especially in the context of directional geometry based stochastic models, thus reproduces the emergence of small-scale fading in a realistic way. However, the derivation of parameters for such models demands high resolution measurements, since in principle the stochastic properties of single MPCs have to be known.

3.1.1.3.3. *Power Delay Profile and Delay Spread Evaluations*

Quite a lot of work has been done on evaluations of the 'shape' of the PDP. Graphs of PDPs derived from measurements in different environments and situations can be found in numerous COST TDs [AsBe97], [PMFF98], [MASA98], [AlBe99a], [HRSS99], [PLNR99], [KaPa99], [LOAN99], [BeKB99], [EnKF99], [HNBK00]. The shape of these PDPs can in many cases, at least approximately or on average, be classified as exponentially decaying from visual inspection. Also, exponentially decaying clusters in the PDP resulting from different scattering areas, as proposed for the use in some models [FuMB98], [SACH99], can be observed from some measurements, especially for macro-cells. Certain distinct peaks in the measured PDPs can be related to dominant components, which then have to be modelled by larger K-factors, such as those proposed for the first paths in the models in [AlBe99a] or [MASA98]. It has to be noted that there are different types of PDPs, which are usually all termed by the general expression 'PDP'. The instantaneous PDP is simply the absolute squared value of the Impulse Response (IR) a certain instant or location, the averaged PDP is the average of the instantaneous PDP over a measurement run of certain length or time-interval in the same scenario, and the globally averaged PDP is the average over several measurements in different scenarios of usually the same type of environment. These differences have to be taken into account when comparing results in the aforementioned documents. It also has to be taken into account that the PDP can either be displayed in logarithmic amplitude scale (usually in dB) or in linear amplitude scale, which results in different shapes of the profile.

The investigations in [Burr97] are based on the channel description by different scattering areas in [FuMB98]. For the case of a non-directional description, a formulation for the cluster caused by the local scatterers in the PDP is derived analytically. Results are plotted for a uniform distribution of scatterers in the whole space, for a uniform distribution within a circle and for a Gaussian distribution around the mobile. The results do not match an exponential PDP, except for larger excess delays in the case of Gaussian distributed scatters; for small excess delays the curves follow an inverse cubic curve, approximately. In [Burr98] the work from [Burr97] is extended by modelling the arrival of MPCs within the PDP as a Poisson process. It turns out that the arrival rate for the Poisson process is not constant, which besides coincides with observations from measurements reported in [Katt97a], [Katt98]. Additionally, an inverse square/fourth power law with a breakpoint as a path loss model and Rayleigh fading is considered. An investigation of the analytically derived average PDP in [Burr98], taking into account path loss and arrivals of paths, shows that the result is not an exponentially decaying function, but follows an inverse power law.

In [EnKF99] PDPs derived from measurements at 5.2 GHz in indoor environments have been investigated with respect to their shape by fitting exponential curves to the measured PDPs using an optimisation algorithm (modified simplex). The remaining mean square error for a fit that is found to be optimum by this algorithm is then used as a quality criterion. One result is that there are different optimal curves depending on whether the PDP is logarithmically or linearly scaled. This has lead to a new approach, which divides the PDP into two parts with different exponential curves for each part. With this approach the PDP model can be equivalently used both in logarithmic and linear scale. The best results however turned out for matching an exponential curve to the PDP in logarithmic scale, which is incompatible with the usual assumption (i.e. linear curve in logarithmic scale). A further result from the investigations in [EnKF99] is that the decay factor of the exponential curve for a linearly scaled PDP in many cases corresponds very well with the RMS DS, as assumed for (3.1.11).

An analytical derivation of the PDP in a closed cubic room is given in [HNBK00]. The result is an exponentially decaying function with the decay factor depending on the room dimensions and the mean reflection-loss. For a cubic room with two open surfaces (i.e. a 'short tunnel') the result is the product of an exponential curve and an inverse linear curve. Thus for arbitrary openings the product of the exponential curve with an inverse power curve is proposed:

$$P_h(\tau_e) = \frac{\gamma}{(c\tau)^\delta} P_0 \cdot e^{-\frac{\tau_e}{S_\tau}} \quad , \tag{3.1.14}$$

with γ being some proportionality constant, c the velocity of light, S_τ being determined from the room dimensions and the mean reflection loss by formulations given in [HNBK00], $\delta = 1$ for the 'short tunnel' and $\delta \neq 1$ for arbitrary openings. Comparisons with results from RT simulations and also measurements in an RF-shielded chamber, as well as in realistic environments, have shown that the model performs very well, even in not exactly cubic rooms and also under NLoS.

Summarising the results from investigations of the PDP shape, it can be stated that the exponentially decaying PDP, with a decay factor determined by the RMS DS is in fact applicable as a first and easy to handle modelling approach. However, for more realistic modelling, the exponential PDP seems to be multiplicatively influenced by an inverse power curve.

It has become obvious from the descriptions up to now that the RMS DS (3.1.12) is an important quantity in context with the PDP. But even more so, the RMS DS can be used for direct investigations of Inter-Symbol Interference (ISI) [Chua87], and thus for a direct determination of Bit Error Rates (BER) [MoFP96]. When looking through the literature, it turns out that it has become quite popular to characterise different channels by (and often only by) measured RMS DSs. There have also been measurements of the RMS DS within COST 259; results are either displayed in tables [KiVa97] or by the CDF of the instantaneous RMS DS [KiVa97], [KiVa98a], [MASA98], [PLNR99], [KiVa99], [LOAN99]. The mean of this CDF can be used as a measure for the frequency-selectivity of the channel, whereas the standard deviation of the CDF can be used as a measure for the time-variant channel behaviour.

3.1.1.3.4. Approaches for Bandwidth Independent Wideband Models

The 'standard model' described and investigated up to now was based on equation (3.1.10). A closer inspection reveals that the δ-function in (3.1.10) implies an infinite bandwidth. The investigations described above and derivations of parameters were based on measurements with a finite bandwidth. Also, simulations based on the model can only be performed with certain bandwidth restrictions. A more realistic description will thus result if the δ-function in (3.1.10) is replaced by a $\sin(x)/x$-function [KaEn97], which for an infinite bandwidth turns into a δ-function. However, the problem still remains that the model parameters derived from measurements will be dependent on the bandwidth, since with increasing bandwidth more and more MPCs can be resolved and less components superimpose for one path in the IR. Thus, due to the changes in actually superimposing MPCs, the parameters describing the fading (e.g. the Ricean K-factor) will change. For the extreme case of an infinite bandwidth, each MPC can be resolved and there will thus no longer be any small-scale fading at all. Due to the decreasing superimposing components for high bandwidths, the modelling by statistical distribution functions becomes questionable, since a statistical description is usually based on the assumption of a large number of superimposing components.

There were two different approaches proposed in COST 259 to overcome the bandwidth dependence of the parameters for the small-scale fading. The approach in [Aspl00a] is based on the assumption of Ricean fading for the path at zero excess delay and Rayleigh fading for all other paths. The shape of the PDP is assumed to be exponentially decaying with a strong peak at zero excess delay. From measurements, it has been observed that narrowband K-values are log-normally distributed. It was further found from measurements that there is a linear dependence of the logarithmic K-factor (both for narrowband and wideband) on the excess path loss (i.e. the path loss in relation to free space path loss). Taking into account these assumptions and observations, a formula is derived in [Aspl00a] that describes the K-factor of the first path as a function of the narrowband K-factor K_0, the bandwidth B and the RMS DS S_τ:

$$K(B) = \frac{K_0}{1 - e^{-\frac{1}{B \cdot S_\tau}}} \quad . \tag{3.1.15}$$

However, it has to be borne in mind that the description is only valid for the assumed certain structure of the ChIR.

Another approach to overcome the bandwidth dependence of modelling parameters, and even to overcome the general problem that statistical modelling becomes questionable for high bandwidth, is described in [Katt97a], [KaEn97]. The basic idea is to use the Time-Variant Transfer Function (TVTF) instead of the Time-Variant Impulse Response, since according to Bello [Bell63], all system functions of time-variant channels are equivalently applicable for channel modelling. There are several strong arguments for preferring the TVTF for statistical modelling. The TVTF can be regarded as the sum of adjacent narrowband channels, thus a statistical modelling of this function can be regarded as a consistent extension from narrowband statistical modelling to wideband statistical modelling, and hence all known properties from narrowband models can be applied straightforwardly. Each value of the TVTF is determined by the superposition of all (i.e. the maximum number of available) MPCs, which permits a statistical modelling independently of bandwidth. Thus a graphically displayed TVTF always looks like a 'random function' [Katt97a], [KaEn97]. Another reason is the time-frequency duality already pointed out by Bello [Bell63], which implies that the frequency-selective fading can and should be modelled by the same means as the time-selective fading. There are some more, in fact less strong arguments, given in [Katt97a], [KaEn97]. Measurements have been evaluated in [KaEn97] with respect to the proposed approach, and it turned out that a Rice distribution, a Nakagami distribution and a Weibull distribution are almost equivalently appropriate to describe the fading. For simulation purposes, it is proposed creating the values as statistically independent (and thus uncorrelated) random numbers, and then inducing the correlation by filtering with appropriate scattering functions. In that context, a mixed statistical and deterministic approach is also discussed. Recently, the approach in [Katt97a], [KaEn97] has been extended for the use in directional channel models [Katt99].

3.1.1.3.5. Wideband Power Investigations

An investigation that is also concerned with the bandwidth dependence of small-scale fading, but in this case with the fading of the signal received via a multipath channel rather than the fading of paths in the ChIR, is described in [AsBe97], [AlBe99b]. The (wideband) power of a signal received via a frequency-selective channel can be determined as the power sum of all taps [AlBe99b], i.e. at different delay times; thus, the wideband power can be expressed mathematically as the integral of the (instantaneous) PDP over τ:

$$P_W(t) = \int_{-\infty}^{+\infty} |h(t,\tau)|^2 d\tau \quad . \tag{3.1.16}$$

Note that in contrast to (3.1.16) the narrowband power, i.e. the power that occurs for a non frequency-selective channel, would be:

$$P_N(t) = \left| \int_{-\infty}^{+\infty} h(t,\tau) d\tau \right|^2 \quad . \tag{3.1.17}$$

The small-scale fading of the wideband power (3.1.16) is examined for different bandwidths in [AsBe97] on the basis of measurements, and in [AlBe99b] by using simulation results from the 'standard model'. It can be observed that there is a less distinct time-selective fading with increasing bandwidth, i.e. the standard deviation of the CDF of the (instantaneous) wideband power becomes smaller with increasing bandwidth. The difference of the power for different bandwidths at a certain CDF level is called fading gain. A relation is found between the fading width (difference of signal power at 99 % and 1 % CDF levels) and the product of bandwidth and DS. Since the wideband power describes the power that can be obtained by maximum ratio combining with an ideal RAKE RX, the observations are used in [AsBe97], [AlBe99b] to evaluate the performance of ideal RAKE RXs at different bandwidths, and to determine the maximum number of RAKE fingers needed.

3.1.1.4. WSSUS

Ralf Kattenbach

The term WSSUS (Wide Sense Stationary Uncorrelated Scattering) was originally introduced by Bello in a frequently cited paper [Bell63]. In this paper, which was intended to derive different models for time-variant channels, Bello first describes the system functions, which are related to each other in a cyclic manner by Fourier transforms. Based on each of the system functions, a related correlation function can be defined by ensemble averages. These correlation functions are related to each other in a cyclic manner by two-fold Fourier transforms. The WSSUS assumption now leads to some simplifications for the input-output relations and also for the channel modelling [Bell63], [Katt97a]. WSSUS is equivalent to wide-sense stationarity with respect to the variables t and f of the TVTF $T(t,f)$:

$$R_T(\Delta t, \Delta f) = E\{T^*(t, f) T(t+\Delta t, f+\Delta f)\} \quad , \tag{3.1.18}$$

with R_T being the Time-Frequency Correlation Function and $E\{.\}$ denoting ensemble averaging. From (3.1.18), using the Fourier transform relationships between the functions, formulations for the other correlation functions of a WSSUS channel can be derived [Bell63], [Katt97a], i.e.:

$$R_h(\Delta t; \tau, \tau') = E\{h^*(t, \tau) h(t+\Delta t, \tau')\} = \delta(\tau' - \tau) \cdot P_h(\Delta t, \tau), \tag{3.1.19}$$

with R_h being the Time-Delay Correlation Function, $h(t,\tau)$ the Time-Variant Impulse Response and $P_h(\Delta t, \tau)$ the Delay Cross Power Spectral Density. Further:

$$R_H(v, v'; \Delta f) = E\{H^*(v, f) H(v', f+\Delta f)\} = \delta(v' - v) \cdot P_H(v, \Delta f), \tag{3.1.20}$$

with R_H being the Doppler-Frequency Correlation Function, $H(v,f)$ the Doppler-Resolved Transfer Function and $P_H(v,\Delta f)$ Doppler Cross Power Spectral Density. Still further:

$$R_s(v,v';\tau,\tau') = E\{s^*(v,\tau)\,s(v',\tau')\} = \delta(v'-v)\cdot\delta(\tau'-\tau)\cdot P_s(v,\tau)\,, \quad (3.1.21)$$

with R_s being the Doppler-Delay Correlation Function, $s(v,\tau)$ the Doppler-Resolved Impulse Response and $P_s(v,\tau)$ Scattering Function. Functions $R_T(\Delta t,\Delta f)$, $P_h(\Delta t,\tau)$, $P_H(v,\Delta f)$ and $P_s(v,\tau)$ turn out to be related in a cyclic manner by Fourier transforms. Due to the δ-functions in (3.1.19), (3.1.20) and (3.1.21) the functions P_h, P_H and P_s are sufficient to describe the correlation properties of a WSSUS channel, and thus they are often denoted as the 'correlation functions' of a WSSUS channel, although in strict sense P_h, P_H and P_s are spectral power densities, whereas R_T, R_h, R_H and R_s are the real correlation functions of a WSSUS channel. Anyhow, for the functions describing the correlation properties of a WSSUS channel (i.e. $R_T(\Delta t,\Delta f)$, $P_h(\Delta t,\tau)$, $P_H(v,\Delta f)$ and $P_s(v,\tau)$) there is a simplification, since these functions are dependent on two variables rather than four variables for the correlation functions in the general case. The origin of the denotation WSSUS can be seen from (3.1.19), since for this correlation function, which is a channel description completely in the time-domain, there is Wide Sense Stationarity (WSS) with respect to t and there is zero correlation if $\tau' \neq \tau$. Since different delay times τ are related to different scatterers, the latter behaviour can be reasonably denoted as Uncorrelated Scattering (US).

Difficulties occur when trying to apply the theory of WSSUS channels to measured system functions, because it is not possible to gather an ensemble of functions but only one realisation. Thus, it is not possible to determine ensemble averages, but only averages over the variables of the respective realisation, e.g. time averages. In the case of ergodicity, the averages over a variable of a realisation are identical to the ensemble average. Furthermore, ergodicity implies stationarity, and so the assumption of wide sense ergodicity with respect to t and f is a possibility to apply the theory of WSSUS channels to measured data, which can be regarded as a further specialised form of a WSSUS channel. A complete derivation of the entire correlation functions of a WSSUS channel with the additional assumption of ergodicity is given in [Katt97a].

The terms WSSUS or WSSUS-model have frequently been mixed up with the 'standard model' described in Section 3.1.1.3.2, since there are some similarities with the 'sampling model for the Time-Variant Impulse Response of a channel with input bandwidth restriction' described in Bello's paper [Bell63]. These similarities are the tapped delay-line structure, the uncorrelatedness of the tap coefficients and the fact that the PDP (in terms of Bello: the Delay Power Density Spectrum) results as a special case of the Delay Cross Power Spectral Density, which in turn is one of the functions describing the correlation properties of a WSSUS channel. But there are also certain differences between Bello's model and the 'standard model'. The first is that in Bello's model the taps have equidistant delays with a value equal to the reciprocal of the bandwidth, which is not always chosen for the 'standard model'. Bello's model is also not based on the sum of weighted δ-functions (3.1.10), but rather on a sum of $\sin(x)/x$-functions. Bello did not make any presumptions on the statistical distribution functions of the tap-weights of his model for the IR. Finally, the correlation properties of the tap-weights are determined by samples of the Delay Cross Power Spectral Density, thus, a description by PDPs and Doppler spectra will only be valid for independently time- and frequency-selective channels. Also, it has to be taken into account that WSSUS is even more general, and that Bello has described numerous models for WSSUS channels in the second part of his basic paper than only the model that has some similarities with the 'standard model' described in Section 3.1.1.3.2.

The WSSUS assumption is usually presumed to be valid in order to treat the tap-weights in the 'standard model' as uncorrelated random variables, but there are only few evaluations with respect to the correlation between tap-weights. In most cases, the validity of WSSUS is assumed according to the work of Cox [CoLe75], [Cox73] and Bajwa/Parsons [PaBa82], [BaPa82], who stated that WSS is valid for RX movements over some tens of a wavelength and that US is unrestrictedly valid. However, their arguments can be shown to be questionable or inconsistent, as explained in [Katt97a], [Katt97b]. The major inconsistencies (among others) are the fact that the investigations on WSS were based on time-averaging of only single realisations of the IR, which implies ergodicity and thus stationarity in advance. The investigations on US were based on the Scattering Function P_s, although, according to (3.1.21), US can only be seen from a δ-function in R_s and thus not from P_s. That does not mean, however, that all investigations up to now, which were based on the statements of Cox and Bajwa/Parsons, are invalid; rather they turn out to be based on the so-called 'hypothetical WSSUS assumption', which has been defined in [Katt97a], [Katt97b]. The hypothetical WSSUS assumption means that for each single realisation, as far as the other realisations are unknown, these other realisations can hypothetically be assumed to be such that the channel is stationary and ergodic with respect to t and f. Thus, each investigation that evaluates only single realisations is actually based on the hypothetical WSSUS assumption. The work of Cox and Bajwa/Parsons was also valuable in the context of investigations on the bound between large- and small-scale fading, but this should not be mixed up with the WSS assumption defined by Bello.

Now the question still remains of whether the channel is in real sense a WSSUS channel, or if it is even ergodic. The investigation of statistical stationarity and ergodicity requires the ensemble of realisations of a process. Since only one realisation can be recorded by measurements, additional theoretical considerations about the whole process are necessary, which permit statements on stationarity and ergodicity, or at least about other realisations such that ensemble averages can be calculated. Based on general considerations about the investigated channel, a proposal is made in [Katt97a], [Katt97b] on how to generate an ensemble from one realisation. From investigating such ensembles, it turns out that a time-variant channel with a moving RX in static environment is generally WSS as well as ergodic with respect to t, and that the validity of US can be evaluated from correlation functions based on averages over the samples along a measurement run in time or space. Quite by chance, this turns out to be similar to investigations made by others on the 'tap correlation' (e.g. [KiVa98a]), however these investigations were not based on any assumptions about the ensemble, which strictly speaking are necessary. The investigation of correlation functions for a moving RX in static indoor channels in [Katt97a], [Katt97b] reveals that, strictly speaking, US is never valid. However, US can be assumed to be at least approximately valid in many, but not all cases. In cases when US is (approximately) valid, however, ergodicity with respect to t and f is also (approximately) valid. Another important result from [Katt97a], [Katt97b] is that, strictly speaking, a differentiation between the mathematical US assumption ($R_h = 0$ for $\tau' \neq \tau$) and the physical US assumption (i.e. each path shows a correlation only with itself and not with other paths) is necessary, since both assumptions may not match if time-variant changes in the delay of paths occur.

In recent years, attention has mainly focussed on directional channel modelling. In this case, the IR is usually described by formally adding a dependence of the time-variant IR on the angle of incidence ϕ. An extension of WSSUS for directional channels has been described in [HRSS99], [SACH99], [Katt99], such that for directional WSSUS there is US with respect to ϕ or equivalently WSS with respect to the aperture. There are eight functions describing the correlation properties for a directional WSSUS channel. These functions are related to each other by Fourier transforms in a kind of three-dimensional cyclic manner, which can be visualised most expressively by arranging the functions on a cube, as in [HRSS99], [Katt99].

3.1.1.5. Stochastic polarisation modelling

Patrick C.F. Eggers

With the increased density of cells and BS locations to provide cellular capacity in urban areas, non-intrusive antenna systems have become highly valuable.   Here polarisation antenna systems offer compact diversity solutions. This subsection describes COST259 modelling work relevant for BS applications of such systems.

Applications of polarisation sensitive antennas at handsets together with 3D dual polarised indoor environment descriptions are found in Section 3.5.

3.1.1.5.1. Outdoor Macro-Cells

The polarisation modelling for macro-cells has been focused on single TX cases, i.e. single polarisation MS transmit and dual polarimetric BS receive (co- and cross-polarised components).

Substantial experimental investigations indicate that orthogonal (here assuming vertical and horizontal linear) polarisations at the BS are uncorrelated, i.e. $\rho_{V,H} \approx 0$.

Contrary to pure space diversity systems, polarisation diversity systems may exhibit pronounced branch power difference expressed via the Cross-Polar Discrimination (XPD).   With the aforementioned non-correlated co- and cross-polarised signals, the XPD is the determining factor in polarisation diversity systems relying on environment induced polarisation cross-coupling.   In case of MS handset terminals with random polarisation state and orientation, XPD ≈ 0 dB may occur.

In [COST99] work is presented which indicated that spatial auto-correlation (with spatial displacement Δd) for the two orthogonal polarisations is about equal in NLoS cases, i.e. $\rho_V(\Delta d) \approx \rho_H(\Delta d)$ even though XPD $\neq 0$ dB. This indicates similar angular spread of the two orthogonal polarisations.

This latter assumption has been verified experimentally in [PeMF99b] investigating 2D azimuth (φ) spread at 1 800 MHz (for MS location close to the BS bore sight).   Concluding from [PeMF99b], we may deduce polarisation behaviour in the angular domain as:

$$\overline{\varphi}_V \cong \overline{\varphi}_H \quad , \tag{3.1.22}$$

$$P_V(\varphi,\tau) \cong XPD \cdot P_H(\varphi,\tau) \cong XPD \cdot \frac{1}{\sigma_\tau} e^{\frac{-\tau}{\sigma_\tau}} \cdot \frac{1}{\sqrt{2}\sigma_\varphi} e^{\frac{-\sqrt{2}|\varphi|}{\sigma_\varphi}} \quad . \tag{3.1.23}$$

Here $P(\varphi,\tau)$ is the temporal (τ) dependant Power Angular Profile (PAP), i.e. the average azimuthal response $P(\varphi,\tau) = E\{|h(\varphi,\tau)|^2\} = P(\varphi)P(\tau)$. This seems to be modelled well with independent temporal and azimuthal distributions (PAP and PDP are Laplacian and Gaussian, respectively) [PeMF99b]. $h(\varphi,\tau)$ is the angular dependent ChIR and σ_τ, σ_φ are the delay and azimuth spread, respectively.

The instantaneous angular dependent ChIRs $h_V(\varphi,\tau)$ and $h_H(\varphi,\tau)$ are modelled as independent, due to $\rho_{V,H} \approx 0$.

3D angular investigations at a BS site have been performed in [LHTB99], [KLTH00]. The investigation shows street canyon dominated propagation. The over roof propagation present, mainly stems from high-rise buildings in the surrounding areas. It is found that the detailed source locations in azimuth and elevation do not fully coincide for the co- and cross-polarised components. This difference does not appear to be very pronounced if considering an elevation integration and averaging as assumed in the 2D case of (3.1.23).

Theoretically, the diversity potential is unchanged when rotating the BS polarisation system (around bore sight axis) with respect to the MS reference polarisation. The only effect of the BS rotation of for example ±45° systems, with a 90°(vertical) MS reference polarisation, is that the branch power levels are equalised (apparent XPD ≈ 0 dB), but traded in for a higher branch correlation. This effect is described in [COST99]. There may be practical limitations of non-ideal equipment that show up in differences between a co-aligned BS polarisation system as compared to a rotated one.

[AaAP98] found that BS polarisation antenna rotation aligned with the MS (V, H) polarisation instead of ±45° only show little better performance with higher power vehicle mounted terminals (which can be assumed, being more polarisation pure than handset terminals). For handset, like MS situations, the difference is marginal.

For the case of channel models used to develop Direction of Arrival (DoA) algorithms, [SeLS99] has found that the inclusion of polarisation information overcomes inconsistency problems with closely spaced sources, thus providing consistent estimation of both DoA mean value and angular spread.

3.1.1.5.2. Penetration Effects

The increased use of handset terminals has given a large user density inside buildings. Thus to supply coverage the network has to deal with the building penetration and the possible effects this has on the BS antenna diversity already applied for outdoor coverage.

Urban investigation in [EgKO98] shows that the polarisation effects (XPD and cross-polar correlation) are largely unaffected by building penetration and elevation (building floor). The only effect noticed is the penetration loss being higher at the ground floor level than top floor level (this height gain may just as well be due to higher illuminating power present at top floors (this was not tested). This indicates that the major polarisation cross-coupling effects due to the environment occur on the outdoor link between BS and the building the MS is located in.

Thus, building penetration can be modelled by equal loss (L) for both orthogonal polarisations:

$$L_{V,building} \cong L_{H,building} \quad , \tag{3.1.24}$$

thus preserving the XPD and the polarisation cross-correlation (from outdoors to inside the building):

$$XPD_{inside} \cong XPD_{outside} \; ; \rho_{VH,inside} \cong \rho_{VH,outside} \quad . \tag{3.1.25}$$

Consequently, the outdoor polarisation diversity potential is preserved for indoor coverage, and the network only needs to modify the link budget with respect to the penetration loss.

3.1.1.5.3. Outdoor and Indoor Pico-Cells

Full polarisation matrix (see Figure 3.1.8) sounding is presented in [NeEg99]. The indices VV, VH, HV, HH show the components of the polarisation matrix, i.e. co- and cross-polarised signal components. Consequently, XPDs can be defined to any reference co- and corresponding cross-polarisation.

Measurements and simulations in [NeEg99] show differences in case of symmetric environments (closed 'tunnel'-like indoor corridors) vs. unsymmetrical environments (open 'waveguide'-like street canyons).

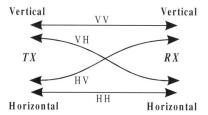

Figure 3.1.8: Schematic plot of the transmission link between TX and RX.

In LoS cases co- and cross-power levels (and XPDs) are function of overlapping footprints of RX and TX antennas in azimuth and elevation. In NLoS street canyon cases the RX XPD for horizontal TX signals is lower than for vertical TX signals. In NLoS indoor corridors the XPDs of both TX polarisation have a higher degree of similarity, due to the near symmetric conditions in elevation and azimuth.

Contrary to the open BS sites considered in macro-cells, pico-cell polarisation effect modelling is highly dependent on actual antenna footprint with respect to vertical and horizontal surfaces.

3.1.1.6. Miscellaneous

Ralf Kattenbach

An empirical attempt that may have impacts both on path loss, large-scale and small-scale fading is described in [AsBe99a]. A building database of an urban environment with a cell size of 500×500 m^2 and the BS in the centre above rooftop level (macro-cell) has been evaluated with respect to the probability of the occurrence of LoS as a function of distance d and BS height h_{BS}. The main trend in the results can be expressed by:

$$P_{LoS}(d, h_{BS}) = \begin{cases} \dfrac{(h_{BS} - h_B)}{h_{BS}} \cdot \dfrac{(d_{CO} - d)}{d_{CO}} & d < d_{CO}, h_{BS} > h_B \\ 0 & d \geq d_{CO}, h_{BS} \leq h_B \end{cases} \quad , \qquad (3.1.26)$$

with h_B being the average building height and d_{CO} the cut-off distance, e.g. in this case $d_{CO} = 500$ m. This model has been proposed in [AsBe99b] to be used in the 'standard model' with Ricean fading for the first tap and Rayleigh fading for all other taps.

In [Zwic00] a model for the probability of LoS has also been proposed, but for indoor environments rather than for macro-cells. In this case, an exponential decay of the probability as a function of distance has been found to yield a good description.

3.1.2. Propagation measurements and channel sounders

Jörg Pamp

The radio channel has an obvious and direct influence on the performance of any wireless communication system. Less obvious, but nonetheless significant, is the influence the radio channel can have on the economical implementation of a high-quality radio network.

As the radio channel evolves with the progression of the information society to ever-higher data rates, and thus wider transmission bandwidths, higher carrier frequencies, and more sophisticated modulation, the necessity to measure and characterise the radio channel still assumes great importance.

With the advent of adaptive antenna systems in wireless telecommunications over the last few years, the focus of both measurement systems and measurement campaigns shifted towards adding angular resolution capabilities on top of mere delay resolution. Obviously, this called for even more sophisticated hardware, then called 'vector' channel sounders. Some prominent examples of this new generation channel sounders were available within COST 259 and ground-breaking work on directional channel modelling has been performed based on measurement data obtained with these systems (cf. Sections 3.2 and 3.3).

This section provides an overview of the radio channel measurement campaigns performed during COST 259 (Table 3.1.4) and to give an inventory of the measurement equipment used to obtain experimental data (Table 3.1.5) from which parameters for an adequate description of the radio channel can be derived. Please note that some of the technical specifications are interdependent and represent maximum values that possibly could not be reached all together at the same time.

Complementary information on this subject can be found in [Lehn98] for current channel sounders 'outside' COST 259 and [COST99] for earlier models. An in-depth assessment of the results of some of the measurement campaigns is presented in Sections 3.1.1, 3.2 and 3.4.

Table 3.1.3: Specific acronyms used in Tables 3.1.4 and 3.1.5.

DSB	Double Side Band
HhT	Handheld Terminal
K	Ricean K-Factor
n.a.	Information not available
OCXO	Oven-Controlled Crystal Oscillator
PL	Path loss
PN	Pseudo-Noise
Rb	Rubidium Reference Oscillator
SpA	Spectrum Analyser
T/R	Transmitter/Receiver
UCRA	Uniform Cross Array
VNA	Vector Network Analyser
VSA	Vector Signal Analyser
XO	Crystal Oscillator

Table 3.1.4: Radio channel measurement campaigns reported in COST 259.

		0.45	0.9	0.9	0.9 / 1.8
Frequency [GHz]		0.45	0.9	0.9	0.9 / 1.8
Environment		rural	out- to indoor	Urban	urban
Measured Quantity		ChIR	signal power	signal power	signal power
Bandwidth [MHz]		1	0.2	0.2	0.2
Measurement System		10	TEMS	TEMS	TEMS
Tx Antenna	Type	omni	sector	Sector	sector
	Gain [dBi]	7	10 – 14	10 – 14	10 – 14
	Polarisation	vertical	vertical	Vertical	vertical
	Height [m]	21 / 120	n.a.	20 | 25	n.a.
	Mounting	mast	Above rooftop	below rooftop | above rooftop	above rooftop
Rx Antenna	Type	monopole	monopole	HhT internal	HhT internal
	Gain [dBi]	2	2	0	0
	Polarisation	vertical	vertical	Vertical	vertical
	Height [m]	2.6	1	1.7	1.7
	Mounting	car roof	trolley	handheld	handheld
T/R Separation [m]		several 1 000	n.a.	500	900
Sampling Interval		10 - 20 m / λ/10 - λ/20	480 ms	480 ms	480 ms
Analysis Results		APDP, DS statistics, CIR statistics	signal power	signal power	signal power, PL comparison
Institution		TR (S), Farsta	IST (P)	IST (P)	IST (P)
TD References		[LOAN99]	[ÂnNC98]	[AIRC98], [GoCo98], [CIFC99]	[ChOC00]

Table 3.1.4 (continued): Radio channel measurement campaigns reported in COST 259.

Frequency [GHz]	0.9 / 1.8 / 2.5 / 5.2	1.739	1.8		1.8 / 2.4
Environment	indoor	urban / suburban / indoor / outdoor	urban	Rural	indoor
Measured Quantity	signal power	ChIR	ChIR		signal power
Bandwidth [MHz]	0.001	44	8.2		CW
Measurement System	SpA	2	3		VSA
Tx Antenna Type	monopole	dipole / helix / patch	dipole		omni
Gain [dBi]	2	n.a.	2		n.a.
Polarisation	vertical	n.a.	vertical		vertical
Height [m]	2	1.7	2		2.1
Mounting	tripod	pedestal	car roof		stand
Rx Antenna Type	monopole	directional	ULA		omni
Gain [dBi]	2		n.a.		n.a.
Polarisation	vertical	dual, 45° slant	vertical		vertical
Height [m]	1.4	48 / 4	20 / 26 / 32	47 / 27	1.5
Mounting	rotating	above rooftop	above rooftop		rail
T/R Separation [m]	5 – 30	500 – 1 000	300 – 3 000	1 000 – 8 000	5 – 10
Sampling Interval	18° (9.4 cm)		5 ms (56 mm)		λ/8
Analysis Results	signal statistics	XPD, envelope fading CDF, cross-correlation	power delay-azimuth-Doppler spectra, Doppler spectra,		mean signal power, signal statistics
Institution	IHE (D)	CPK (DK)	CPK (DK), ERA (S)		Uni Cantabria (E)
TD References	[Zwic00]	[EgKO98]	[PeMF98, PMFF98, AsBe98]		[LTDV99]

Table 3.1.4 (continued): Radio channel measurement campaigns reported in COST 259.

		AaAP98	KaEn97	LaFr99b	RuSA99
Frequency [GHz]		1.8	1.8	1.8	1.8
Environment		urban / sub-urban / rural	indoor	indoor	out- to indoor
Measured Quantity		signal power	transfer function	transfer function	signal power
Bandwidth [MHz]		CW	600	600	CW
Measurement System		SpA	14	14	n.a.
Tx Antenna	Type	directional	discone	discone	monopole / directional
	Gain [dBi]	18.5	0.65 (sic)	0.65	2 / 10
	Polarisation	horizontal, vertical, 45° slant	vertical	vertical	n.a.
	Height [m]	2.5	1.3	1.4	1.5, uptilt
	Mounting	car roof	stand	stand	n.a.
Rx Antenna	Type	directional	discone	discone	monopole
	Gain [dBi]	16	0.65	0.65	n..a.
	Polarisation	vertical, circular, 45°	vertical	vertical	n.a.
	Height [m]	2.5	1.3	1.4	1.5
	Mounting	24 / 5 / 12	rail	rail	
T/R Separation [m]		≤ 500 / ≤ 1500 / ≤ 2000	5 – 22	6	15 – 30
Sampling Interval		167 ms (16.7 mm)	λ/8	λ/8	n.a.
Analysis Results		signal statistics	amplitude, phase statistics	MD, DS w/o a human present in room	penetration loss, PL
Institution		Telenor (N)	Uni Kassel (D)	Uni Kassel (D)	UPC (E)
TD References		[AaAP98]	[KaEn97]	[LaFr99b]	[RuSA99]

Table 3.1.4 (continued): Radio channel measurement campaigns reported in COST 259.

Frequency [GHz]	1.8	1.88	1.88	1.89	1.89
Environment	urban / suburban	urban / suburban	suburban	street canyon	indoor
Measured Quantity	ChIR	ChIR	ChIR	ChIR	
Bandwidth [MHz]	150	5	5	50	
Measurement System	11	4	4	2	
Tx Antenna — Type	monopole / dipole	directional	directional	4-ele patch	
Gain [dBi]	2	15.5	12.5	13	
Polarisation	vertical	vertical	vertical	dual horizontal / vertical	
Height [m]	1.8	25	25	2.05 / 1.6	
Mounting	car roof / artificial head	above rooftop	above rooftop	trolley	
Rx Antenna — Type	ULA	dipole	dipole	4-ele patch	
Gain [dBi]	n.a.	2	2	13	
Polarisation	dual X	vertical	vertical	dual horizontal / vertical	
Height [m]	20 / 45	2	2	2.6	1.9
Mounting	above rooftop	car roof	car roof	below rooftop	n.a.
T/R Separation [m]	n.a.	150 – 2 500	200 – 2 000	150	15
Sampling Interval	400 μs	109 μs	0.007 – 0.017 λ	24 mm	
Analysis Results	APS, ADPS	signal statistics	APDP / signal statistics	XPD, correlation signal power	
Institution	TR (S), Farsta	ERA (S)	ERA (S)	TUW (A) / CPK (DK)	
TD References	[Lars98]	[AlBe99a]	[AsBe97]	[NeEg99]	

Table 3.1.4 (continued): Radio channel measurement campaigns reported in COST 259.

Frequency [GHz]	1.96	2.1	2.1		2.154
Environment	urban	tunnel / indoor	urban	suburban	urban
Measured Quantity	ChIR	ChIR	ChIR		ChIR
Bandwidth [MHz]	120	50	50		30
Measurement System	9	16	8		6 (sliding correlator)
Tx Antenna — Type	discone	dipole	dipole		discone
Gain [dBi]	2	2	2		2
Polarisation	vertical	vertical	vertical		vertical
Height [m]	5 / 10	2	2.5		2
Mounting	n.a.	stand	car roof		below rooftop
Rx Antenna — Type	monopole	dipole	ULA		ULA
Gain [dBi]	2	2	n.a.		5 (single element)
Polarisation	vertical	vertical	vertical		vertical
Height [m]	2.1	2	25 / 2.3	7	4
Mounting	n.a.	car roof	above / below rooftop	below rooftop	car roof
T/R Separation [m]	≤ 350	≤ 500	120 – 380	270 – 510	100 – 200
Sampling Interval	0.2 m	1 cm	2 cm	1 m, 2 cm	static
Analysis Results	PL	PL, signal statistics, PDP, DS	APS, PDP, AS, DS		field direction-delay-spread function
Institution	T-Nova (D)	Uni Lille (F)	Telenor (N)		HUT (FIN)
TD References	[Ortg99]	[LiDe99], [LiBD00]	[PLNR99]		[HeKa97]

Table 3.1.4 (continued): Radio channel measurement campaigns reported in COST 259.

Frequency [GHz]		2.154	2.154	2.154	2.154
Environment		university campus	out- to indoor / suburban / highway	urban / out- to indoor	suburban / out- to indoor
Measured Quantity		ChIR	ChIR	ChIR	ChIR
Bandwidth [MHz]		30	30	30	30
Measurement System		6 (direct sequence)	6 (direct sequence)	6 (direct sequence)	6 (direct sequence)
Tx Antenna	Type	discone	discone	directional	discone
	Gain [dBi]	2	2	17	2
	Polarisation	vertical	vertical	vertical	vertical
	Height [m]	1.8	50	8	1.8
	Mounting	trolley	tower	below rooftop	rooftop
Rx Antenna	Type	ULA	spherical array	spherical array	4 different HhT antennas
	Gain [dBi]	5 (single element)	8 (single element)	8 (single element)	0 – 8
	Polarisation	vertical	dual	dual	dual
	Height [m]	11.5	1.65	1.65	1.7
	Mounting	rooftop	trolley / inside van	trolley	trolley
T/R Separation [m]		60	35 m / 100 – 3 000	25 / 40	260 – 350
Sampling Interval		λ/4	λ/5	λ/5	λ/10
Analysis Results		DoA	3D DoA	azimuth / elevation APS	PL, diversity gain
Institution		HUT (FIN)	HUT (FIN)	HUT (FIN)	HUT (FIN)
TD References		[Kal98]	[Kal99]	[KaLa99]	[VaSu99]

Table 3.1.4 (continued): Radio channel measurement campaigns reported in COST 259.

		2.154	2.2	5.2	5.2
Frequency [GHz]		2.154	2.2	5.2	5.2
Environment		urban	urban / rural	industry hall	small courtyard
Measured Quantity		ChIR	ChIR	ChIR	ChIR
Bandwidth [MHz]		30	12.5	120	120
Measurement System		6 (direct sequence)	1	13	13
Tx Antenna	Type	discone	n.a.	omni	omni
	Gain [dBi]	2	n.a.	5	5
	Polarisation	vertical	n.a.	vertical	vertical
	Height [m]	1.7	n.a.	1.85	2
	Mounting	stand	mobile	trolley	stand
Rx Antenna	Type	16-ele ziczac patch array	omni, rotating	ULA	ULA
	Gain [dBi]	8 (single element)	n.a.	3 (single element)	3 (single element)
	Polarisation	horizontal / vertical	n.a.	vertical	vertical
	Height [m]	5 25	n.a.	1.7	1.8
	Mounting	trolley	above rooftop	stand	stand
T/R Separation [m]		100 – 600	≤ 800	45 – 70	6 – 30
Sampling Interval		λ/2	n.a.	5 ms	5 ms
Analysis Results		azimuth / elevation DoA, polarisation	DoA statistics	DoA, delay-Doppler	delay spectrum, ADPS
Institution		TUW (A) / HUT (FIN)	CNET (F)	TUI (D)	TUI (D)
TD References		[LHTB99], [KLTH00]	[Paju98]	[THST98], [Lohs98]	[HRSS99]

Table 3.1.4 (continued): Radio channel measurement campaigns reported in COST 259.

		5.2	5.2	5.22	5.2	5.3
Frequency [GHz]		5.2	5.2	5.22	5.2	5.3
Environment		office	office	office / hall / outdoor	office	corridor / office
Measured Quantity		ChIR	transfer function	transfer function	transfer function	ChIR
Bandwidth [MHz]		120	600	160	600	54
Measurement System		7	14	VNA	14	6 (sliding correlator)
Tx Antenna	Type	patch	discone	dipole	discone	discone
	Gain [dBi]	4	0.65	2	0.65	1.3
	Polarisation	vertical / horizontal	vertical	vertical	vertical	vertical
	Height [m]	3	1.3	1.5	1.3	1.8
	Mounting	stand	stand	tripod	stand	stand
Rx Antenna	Type	patch	discone	dipole	discone	discone
	Gain [dBi]	4	0.65	2	0.65	1.3
	Polarisation	vertical / horizontal / circular	vertical	vertical	vertical	vertical
	Height [m]	1.5	1.3	1.5	1.3	1.8
	Mounting	rail	rail	tripod	rail	trolley
T/R Separation [m]		≤ 23	≤ 6	≤ 20	≤ 6	5 – 25
Sampling Interval		$\lambda/10$	$\lambda/8$	14.9 mm	$\lambda/8$	$0.28\,\lambda$
Analysis Results		signal, DS statistics	APDP	PDP, fading statistics, Doppler	APDP	PL / DS statistics
Institution		ENSTA (F), IMST (D)	Uni Kassel (D)	ERA (S)	Uni Kassel (D)	HUT (FIN)
TD References		[Sibi00]	[EnKF99]	[MASA98]	[EnKF99]	[KiVa97]

Table 3.1.4 (continued): Radio channel measurement campaigns reported in COST 259.

Frequency [GHz]	5.3	5.3	5.3	5.3
Environment	office	office / hall	indoor	corridor / library
Measured Quantity	ChIR	ChIR	ChIR	ChIR
Bandwidth [MHz]	54	54	54	54
Measurement System	6 (sliding correlator)	6 (sliding correlator)	6 (sliding correlator)	6 (sliding correlator)
Tx Antenna — Type	discone	discone | patch	discone	discone | discone
Gain [dBi]	1	1 | 6	1	1 | 1
Polarisation	vertical	vertical	vertical	vertical / horizontal | vertical
Height [m]	1.55	1.55	1.55	1.55
Mounting	stand	stand	stand	stand
Rx Antenna — Type	horn	discone	horn	patch | horn
Gain [dBi]	14	1	14	6.7
Polarisation	vertical	vertical	vertical	vertical / horizontal
Height [m]	1.55	1.55	1.55	1.55
Mounting	rotating	trolley	rotating	trolley
T/R Separation [m]	5 – 20	n.a.	5 – 25	5 – 50
Sampling Interval	0.44°	$\lambda/5$	0.44°	$\lambda/4$
Analysis Results	PL, DoA	tapped delay line model	DoA, correlation, PL, diversity	PL, DS, correlation, DoA
Institution	HUT (FIN)	HUT (FIN)	HUT (FIN)	HUT (FIN)
TD References	[KiZV99]	[ZhKV00]	[KiVa99]	[KiVa98a]

Table 3.1.4 (continued): Radio channel measurement campaigns reported in COST 259.

		5.8	5.8	5.8	17.81
Frequency [GHz]		5.8	5.8	5.8	17.81
Environment		urban square	indoor	out- to indoor	corridor / hall / parking lot
Measured Quantity		ChIR	ChIR	ChIR	transfer function
Bandwidth [MHz]		400	400	400	CW / 50
Measurement System		12	12	12	SpA
Tx Antenna	Type	monopole / patch	monopole	horn / monopole	n.a.
	Gain [dBi]	-0.6 / 6	-0.6	17.8 / -0.6	3
	Polarisation	vertical	vertical	vertical	n.a.
	Height [m]	1.2	1.2	13	n.a.
	Mounting	stand	stand	20° downtilt	n.a.
Rx Antenna	Type	3-ele patch	virtual UCRA / 6-sector	virtual UCRA	directional
	Gain [dBi]	9	–	–	3
	Polarisation	vertical	vertical	vertical	n.a.
	Height [m]	3	1.5	1.5	n.a.
	Mounting	turn table	rail	rail	ceiling
T/R Separation [m]		66 – 110	n.a.	23 / 10 – 30	3 – 10
Sampling Interval		71 ms	λ/4	λ/4 / λ/4	5 mm / 20 mm
Analysis Results		PDP, signal power, DS	APS, PDP	ADPS, APS, penetration loss	PL, DS, K-factor
Institution		TR (S), Malmö	TR (S), Malmö	TR (S), Malmö	DUT (NL)
TD References		[KBMH98]	[KaPa99]	[BeKB99]	[NiWP99]

Table 3.1.4 (continued): Radio channel measurement campaigns reported in COST 259.

Frequency [GHz]		24	24	24	24
Environment		office	outdoor, predefined direct + 2 reflections	shielded room	indoor
Measured Quantity		ChIR	ChIR	ChIR	ChIR
Bandwidth [MHz]		500	1 000	1 000	1 000
Measurement System		5	5	5	5
Tx Antenna	Type	monopole	monopole	monopole	monopole
	Gain [dBi]	n.a.	n.a.	n.a.	n.a.
	Polarisation	vertical	vertical	vertical	vertical
	Height [m]	1.5	1.5	1.473	3.25 / 2.45
	Mounting	tripod	tripod	tripod	ceiling
Rx Antenna	Type	virtual 3D-UCRA	virtual 3D-UCRA	virtual 3D-UCRA	virtual 3D-UCRA
	Gain [dBi]	–	–	–	–
	Polarisation	vertical	vertical	vertical	vertical
	Height [m]	1.4	1.4	1.393	1.4
	Mounting	milling positioner	milling positioner	milling positioner	milling positioner
T/R Separation [m]		4	3.5	1.38	5 – 17
Sampling Interval		$\lambda/2$	$\lambda/2$	$\lambda/2$	$\lambda/2$
Analysis Results		PDP, DoA	reference data	DoA, PDP	PDP, azimuth / elevation APS
Institution		ETHZ (CH)	ETHZ (CH)	ETHZ (CH)	ETHZ (CH)
TD References		[TrSi97]	[TsHT98]	[HTTN99], [TBKH99]	[HeTr99]

Table 3.1.4 (continued): Radio channel measurement campaigns reported in COST 259.

	24	28.5 / 29.4		62	62.5
Frequency [GHz]	24	28.5 / 29.4		62	62.5
Environment	office	urban	suburban	street canyon	indoor
Measured Quantity	ChIR	signal power / ChIR	signal power	signal power	signal power
Bandwidth [MHz]	1 000	CW / 200		CW	CW
Measurement System	5	SpA / 12		MBS	n.a.
Tx Antenna Type	monopole	horn		lens	lens
Gain [dBi]	n.a.	15		13.2	13.2
Polarisation	vertical	vertical		vertical	vertical
Height [m]	2.5 / 3.5	85	20	3	2.5
Mounting	ceiling	above rooftop		side walk	ceiling
Rx Antenna Type	virtual 3D-UCRA	horn lens		lens / rod / slot	lens / horn
Gain [dBi]	–	31.5		13.2 / 18.5 / 4.5	13.2 / 16.0
Polarisation	vertical	vertical / horizontal		vertical	circular
Height [m]	1.4	3	3 – 7	2	1.5
Mounting	milling positioner	track / turn table, car roof		car roof	n.a.
T/R Separation [m]	4.3	5 000 / 2 000		0 – 200	≤ 10
Sampling Interval	λ/2	4.77 mm / 2 λ		λ/10	n.a.
Analysis Results	azimuth / elevation DoA statistics	PL, XPD, AS, PDP, signal statistics		space correlation, signal statistics	signal power
Institution	ETHZ (CH)	TR (S), Malmö		IST (P)	IST (P)
TD References	[HeTr00]	[KaEm97], [KaLJ98]		[VaCo97], [VaCo98]	[FeFe99]

Table 3.1.5: Radio channel sounding equipment used in COST 259.

Reference Number (Table 3.1.4)	1	2	3
Institution	CNET (F)	CPK (DK)	CPK (DK)
Frequency [GHz]	0.8 – 2.6	1.739 / 1.89	1.8
Measured Quantity	ChIR	ChIR	ChIR
Measurement Bandwidth [MHz]	12.5 / 25 / 50	44 / 50	0.27 / 4
Measurement Principle	inverse filtering of pn-sequences	correlation of pn-sequences, sliding correlator	correlation of pn-sequences
Measurement Speed	833 ChIR/s	40 ChIR/s	200 ChIR/s
Maximum Excess Delay [µs]	42	6	15
Instantaneous Dynamic Range [dB]	30 – 40	35	36
AGC Range [dB]	60	60	63
Number of Channels	1	2	8 + 1
Transmit Power [dBm]	n.a.	40	40
Modulation	BPSK	BPSK	MSK
Receiver Sensitivity [dBm]	-90	n.a	n.a.
Trigger Modi	time, distance	time	time
Reference Oscillator	Rb (TX), XO (RX)	Rb	Rb
Data Format	n.a.	binary	binary
Processing of Data	offline	offline	offline
TD References	[Paju98]	[EgKO98]	[PeMF98], [PMFF98], [AsBe98]

Table 3.1.5 (continued): Radio channel sounding equipment used in COST 259.

Reference Number (Table 3.1.4)	4	5	6	
Institution	ERA (S)	ETHZ (CH)	HUT (FIN)	
Frequency [GHz]	1.88	24	2.154	5.3
Measured Quantity	ChIR	ChIR	ChIR	
Measurement Bandwidth [MHz]	5	1 000	30	54
Measurement Principle	correlation of pn-sequences	correlation of pn-sequences	sliding correlator	direct sampling
Measurement Speed	9174 ChIR/s (burst)	488 kHz	24 ChIR/s	≤ 10 MChIR/s
Maximum Excess Delay [µs]	109	2.046	19	1 – 200
Instantaneous Dynamic Range [dB]	50	50	30	
AGC Range [dB]	no	no	72	
Number of Channels	1	1	64	32
Transmit Power [dBm]	35	23	40	30
Modulation	BPSK	BPSK	BPSK	
Receiver Sensitivity [dBm]	-92	n.a.	3 dB NF	
Trigger Modi	time	time, distance	time	
Reference Oscillator	n.a.	n.a.	10 MHz Rb	
Data Format	binary	binary	LabView / Matlab	
Processing of Data	offline	offline	offline	
TD References	[AsBe97], [AlBe99a]	[TrSi97], [TsHT98], [HTTN99], [TBKH99], [HeTr99], [HeTr00]	[KiVa97], [HeKa97], [Kall98], [KiVa98a], [Kall99], [KiZV99], [KaLa99], [VaSu99], [KiVa99], [LHTB99], [KLTH00], [ZhKV00]	

Table 3.1.5 (continued): Radio channel sounding equipment used in COST 259.

Reference Number (Table 3.1.4)	7	8	9
Institution	IMST (D)	Telenor (N)	T-Nova (D)
Frequency [GHz]	0.9 – 26 / 59 – 60	1.9 / 2.1 / 5.2 / 59	1.96
Measured Quantity	ChIR	ChIR	ChIR
Measurement Bandwidth [MHz]	≤ 120	≤ 200	≤ 120
Measurement Principle	inverse correlation filtering	frequency chirp	inverse correlation filtering
Measurement Speed	39 ChIR/s, 390 ChIR/s (burst)	728 ChIR/s, 312 500 ChIR/s	n.a.
Maximum Excess Delay [μs]	25.6	164	25.6
Instantaneous Dynamic Range [dB]	35	> 30	35
AGC Range [dB]	75	64	100
Number of Channels	2	2	2
Transmit Power [dBm]	typ. 30	10	40
Modulation	DSB	chirp	DSB
Receiver Sensitivity [dBm]	≥ -100	-70	≥ -100
Trigger Modi	time, distance	time	time, distance
Reference Oscillator	10 MHz OCXO / Rb (GPS)	common XO	10 MHz Rb
Data Format	proprietary	proprietary	proprietary
Processing of Data	offline	offline	offline
TD References	[Sibi00]	[PLNR99]	[Ortg99]

Table 3.1.5 (continued): Radio channel sounding equipment used in COST 259.

Reference Number (Table 3.1.4)	10	11	12
Institution	TR (S), Farsta	TR (S), Farsta	TR (S), Malmö
Frequency [GHz]	0.45	1.8 / 18	28.5 / 5.8 / 1.8
Measured Quantity	ChIR	ChIR	ChIR
Measurement Bandwidth [MHz]	1	≤ 150	400
Measurement Principle	frequency chirp	frequency chirp	correlation of pn-sequences, sliding correlator
Measurement Speed	≤ 6 kChIR/s	≤ 40 kChIR/s	55.6 ChIR/s
Maximum Excess Delay [μs]	163	25	n.a.
Instantaneous Dynamic Range [dB]	25	25	25
AGC Range [dB]	50	50	no
Number of Channels	1	1	1
Transmit Power [dBm]	43	33	31
Modulation	chirp	chirp	BPSK
Receiver Sensitivity [dBm]	-109	n.a	-90
Trigger Modi	time	time	time
Reference Oscillator	10 MHz Rb	10 MHz Rb	GPS
Data Format	proprietary	proprietary	proprietary
Processing of Data	offline	offline	online
TD References	[LOAN99]	[Lars98]	[KaEm97], [KaLJ98], [KBMH98], [KaPa99], [BeKB99]

Table 3.1.5 (continued): Radio channel sounding equipment used in COST 259.

Reference Number (Table 3.1.4)	13	14	15
Institution	TUI (D)	Uni Kassel (D)	Uni Kassel (D)
Frequency [GHz]	5 – 6	1.8 / 5.2	1.8 / 5.2 / 17.2
Measured Quantity	ChIR	transfer function	ChIR
Measurement Bandwidth [MHz]	≤ 120	600	≤ 600
Measurement Principle	inverse correlation filtering	frequency sweep	stepped-frequency correlation of pn-sequences
Measurement Speed	1 kChIR/s, 20 kChIR/s	0.53 ChIR/s	≤ 275 kChIR/s
Maximum Excess Delay [μs]	25.6	0.668	29
Instantaneous Dynamic Range [dB]	35	90	65
AGC Range [dB]	75	n.a.	60
Number of Channels	1	1	1
Transmit Power [dBm]	30	8	20
Modulation	DSB	no	BPSK
Receiver Sensitivity [dBm]	-88	-80	2 dB noise figure
Trigger Modi	time, distance	distance	time
Reference Oscillator	10 MHz Rb	10 MHz internal	10 MHz
Data Format	proprietary	raw ASCII, proprietary	proprietary
Processing of Data	offline	offline	offline
TD References	[THST98], [Lohs98], [HRSS99]	[KaEn97], [LaFr99b], [EnKF99]	[KaWe00]

Table 3.1.5 (continued): Radio channel sounding equipment used in COST 259.

Reference Number (Table 3.1.4)	16
Institution	Uni Lille (F)
Frequency [GHz]	1.8 / 2.1
Measured Quantity	ChIR
Measurement Bandwidth [MHz]	≤ 70
Measurement Principle	correlation
Measurement Speed	3 kChIR/s
Maximum Excess Delay [μs]	10
Instantaneous Dynamic Range [dB]	25
AGC Range [dB]	63
Number of Channels	1
Transmit Power [dBm]	30
Modulation	BPSK
Receiver Sensitivity [dBm]	-75
Trigger Modi	time, distance
Reference Oscillator	10 MHz Rb
Data Format	proprietary
Processing of Data	offline
TD References	[LiDe99], [LiBD00]

3.1.3. Deterministic modelling

Peter Cullen and Dirk Didascalou

Deterministic wave propagation models intend to reproduce the actual physical wave propagation process for a given environment. They are required for proper coverage predictions and are therefore mandatory for radio network planning and the installation of operational mobile radio systems of any kind. Depending on the modelling approach, wideband analyses can be performed including delay, angle of transmission and arrival, as well as polarisation information for each propagation path. These results enable the planning and design of modern wireless digital systems including adaptive antennas. In order to obtain deterministic propagation predictions, suitable formulations of the physical propagation phenomena and effects are applied to deterministically described scenarios. The description of the propagation environment can be performed by 3D-vector databases, topographic or morphographic raster data etc. The acquisition of this data is treated in [COST99]. The approaches for modelling the different propagation phenomena are presented in Section 3.1.3.1. The methods are subdivided into integral equation formulations, differential equation formulations and asymptotic techniques. The application of these techniques and their integration into functional propagation packages is described in Section 3.1.3.2.

3.1.3.1. Modelling of propagation phenomena and effects

Conor Brennan and Peter Cullen

This section reviews the main progress in the area of computational electromagnetics, giving an overview of the main topics encountered in this interesting field and providing a sense of the wide variety of issues involved in tackling electromagnetic problems. The three main formulations that we shall discuss are *integral equation* formulations, *differential equation* formulations and *asymptotic techniques*. These formulations are deterministic in the sense that they postulate the propagation problem using Maxwell's equations and allow direct inclusion of the physical properties of the propagation environment.

3.1.3.1.1. Integral Equation Formulation

Integral equations are especially suitable for the description of wave propagation in the vicinity of inhomogenities (scatterers) that are embedded in some infinite homogeneous medium. The guiding motivation of the integral equation formulation is to replace the physical problem with a physically equivalent problem that produces the same fields, but is more amenable to analysis. Specifically the original problem, that of *known sources radiating in the presence of a scatterer*, is replaced by an equivalent problem involving *unknown sources radiating in infinite homogeneous media*. The unknown sources may be situated on a continuum of points within the scatterer (leading to a volume integral equation for use when the scatterer is inhomogeneous) or on its surface (leading to a surface integral equation for use when the scatterer is homogeneous). The equivalent sources are related to the (unknown) field quantities at these points and produce the same fields in the regions both internal and external to the scatterer as the original configuration. Potential theory provides expressions for the scattered fields everywhere in space, and in particular, on and within the scatterer itself, enabling us to write coupled Electric Field Integral Equations (EFIE) or coupled Magnetic Field Integral Equations (MFIE). These equations can be solved to yield the equivalent sources and to calculate the fields internal and external to the scatterer. When considering scattering from a perfectly conducting body, the integral equations decouple with a corresponding reduction in solution complexity. It is also possible to decouple the equations by invoking the approximate impedance boundary condition. It should be noted that the surface MFIE is strictly valid only for *closed* scatterers and care must be taken to avoid spurious edge effects if one implements it to an open body problem. In addition, when analysing closed perfectly conducting bodies it is advisable to use the Combined Field Integral Equation (CFIE), a linear combination of the EFIE and MFIE, as this eliminates any spurious resonances encountered by the EFIE and MFIE.

The most common numerical procedure for solving integral equations is the method of moments, whereby the unknown currents are expanded as a sum of N known basis functions with unknown coefficients. A numerical procedure is then employed to obtain a matrix equation. Realising that N, the number of basis functions used, increases with the size of the problem we see that the dense nature of the integral equation impedance matrices can lead to storage problems for problems even of quite modest size. Iterative solutions, which do not require the explicit storage of the impedance matrix, quickly become computationally intractable as the computation times scale as N^2. For these reasons, the integral equation formulation was, for many years, mostly used to model scattering from quite small objects. Scattering from larger objects such as terrain were possible only if one imposed the restriction of forward scattering, as in the analysis of scattering by small urban environments situated on a slowly varying underlying terrain profile [KlAn97]. In recent years, however, intense research has prompted many improvements in how these matrices can be compressed by using sophisticated basis functions that model the currents more efficiently. These include the Impedance Matrix Localisation method of Canning [Cann90a], which uses a set of oscillating basis functions that render the impedance matrix sparse when the scatterer is locally smooth. This technique was used to accurately model wave propagation in an indoor environment [deBa97]. Another efficient method, the Natural Basis Set [MoCu95], uses the phase of the incident field to define basis functions defined over large domains. For a certain class of problem (massive smooth structures) these basis functions accurately describe the phase of the unknown surface current. Hence, fewer basis functions are necessary and the associated matrix equation is of a much lower order. The finest wavelets of a complete wavelet basis set can be shown to produce evanescent waves that decay exponentially quickly with distance. The impedance matrix using such a basis will show great sparsity as a result, though the amount of sparsity is largely dependent on the smoothness of the structure [WaCh95]. Other researchers have shown how to dramatically reduce computation times associated with iterative solutions. The Fast Multipole Method (FMM) [Rokh90] and its derivatives [WaCh94], [SoCh95] offer efficient and robust solutions to a wide class of problems. It effectively decomposes the dense impedance matrix associated with standard pulse basis expansions into a product of three sparse matrices. By collecting discretisation points together into large groups the FMM replaces the point to point interactions of standard iterative schemes with a three level process of group-centre to group-centre interactions, coupled with field aggregation and dissemination within groups. Standard FMM implementation reduces the computational complexity of iterative schemes from $O(N^2)$ to $O(N^{3/2})$, while other sophisticated implementations can reduce this complexity further to $O(N^{4/3})$. An excellent introductory discussion of the FMM can be found in [CoRW93]. The Fast Far-Field method (FAFFA) [BrCu98a], a simplification of the FMM, has been further specialised to the Tabulated Interaction Method (TIM) to facilitate the analysis of large-scale UHF propagation problems [BrCu97], [BrCu98b], [BrCu98c], [RoCB99] in realistic computing times. All of these developments have prompted the use of integral equations in the analysis of problems that were previously considered to be beyond their realm, though, like any deterministic method, computational restraints still limit the size of problem that can be analysed.

3.1.3.1.2. Differential Equations

A) Finite Element Method

The finite element method can be considered as a close relative of the volume integral equation. Similarities include the expansion of the unknown field in terms of a basis set and a testing procedure that reduces the problem to matrix equation form. Important differences include the differential equation-based starting point, the need to discretise a larger volume of space, the need to explicitly impose exact or approximate radiation boundary conditions and the sparsity of the resultant matrix. This process differs from the method of moments applied to the integral equation formulation in that the resultant matrix equation involves a matrix that is, apart from a small submatrix, sparse. This is in contrast to the generally full or dense matrices associated with the integral equation formulation. The small dense part of the matrix alluded to is due to the need to incorporate radiation boundary conditions along the domain boundary. This radiation condition ensures that we are mathematically only treating outgoing scattered waves, or scattered waves that propagate away from the scatterer. Exact radiation boundary conditions are available, in the form of eigenvalue expansions for simple (e.g. spherical boundaries) or integral equations for more arbitrary bounding contours, but have the drawback of being *global* conditions. This means that they couple fields along the entire boundary in the form of an integral equation leading to the dense submatrix. However, it has been proposed that these dense sections of the matrix can be efficiently tackled using some of the integral equation acceleration techniques discussed in the previous section, leading to an efficient hybrid method [LuJi96]. Alternatively, it is possible to approximate *local* boundary conditions, which do not couple the fields along the boundary and result in the matrix being totally sparse. An example of this is the Bayliss-Turkel radiation condition [MaBT81], which, while not guaranteeing the scattered field to be outgoing at the boundary, discriminates strongly in favour of outgoing waves. These radiation conditions differ from the Sommerfeld radiation conditions in being applicable in the near field. It can be shown that they work well in absorbing waves that are normally incident upon the boundary, but work less well for grazing incidences [Cann90b]. Hence, the computational domain may have to be enlarged to ensure that $\delta\Gamma$ is far enough away to ensure that all scattered waves impinge normally. Enlarging the computational domain in this way of course increases the computational burden.

B) Finite Difference Method

A finite difference solution to an electromagnetic problem discretises the computational domain on a lattice and calculates the field values at each lattice point in accordance with the governing equations. In this it is similar to the finite element method. Further similarities include the representation of the field interactions by means of a sparse matrix, the ability to model arbitrarily heterogeneous propagation environments and the need to carefully treat fields at the edge of the computational domain. It differs from the finite element method in that it uses Taylor's theorem to calculate linear approximations to the partial derivatives occurring in the governing equations. These linear approximations relate field values at each discretised lattice point to neighbouring ones, resulting in a sparse matrix equation. Applying this theory to a propagation problem leads to a system of lattice points, the field values at each point being locally linked to those at neighbouring lattice points. Time domain finite difference applications discretise both time and space and, as they track the time-varying fields, they lend themselves well to graphical visualisation. Time domain modelling can be either explicit, where the update at each time step is given in terms of previous ones with no interaction between unknowns within the same time-step, or implicit which does allow this interaction, but can involve more computation. The implicit approach has the advantage of not being subject to Courant instability. This instability can manifest itself in explicit schemes unless certain bounds are placed on lattice spacing. Once again, given that only a finite-sized computational domain can be discretised, the problem of the treatment of fields at the boundary of the computational domain arises. Much effort has been applied to the development of suitable absorbing boundary conditions to achieve this aim. A crude implementation introduces a buffer zone of absorbing lattice points around the computational domain in which the fields are smoothly attenuated to zero. This buffer zone can be considered as a sort of numerical anechoic chamber. In this fashion the spurious reflections can be damped down somewhat. A more rigorous approach is to introduce perfectly matched layers [Bere94] that introduce material into the buffer zone, which gives no reflections and smoothly attenuates the fields to zero. Unfortunately, finding a suitable material to achieve this for arbitrary incident waves is not an easy task. Another method worth noting is the introduction of one way wave equations at the boundaries which support propagation in a preferred direction (out of the computational domain) [Lind75], [EnMa77], [Mur81]. Again, the proper attenuation of arbitrary non-normal waves is a potential problem.

A central feature of the wave equation is that it is hyperbolic in nature, a mathematical feature that essentially means that every lattice point interacts with all its local neighbours. A more attractive type of equation is a parabolic equation, where each lattice point interacts only with some of its neighbours. Under certain circumstances the wave equation can be approximated by an equation that is parabolic. It can be shown that central approximation is reasonable when the fields are deemed to be propagating primarily in a specific direction, as happens when energy is propagating over slowly undulating rural terrain for instance. The advantage of adopting this approximation is that the discretised parabolic equation leads to a lower triangular operator, which lends itself to solution by a process of marching forward over the computational domain. This is in contrast to the discretised Helmholtz equation, which, including both forward and backward propagating waves, leads to a more general operator that cannot be inverted so readily. Accurate assumptions about the phase behaviour of the field allow these range steps to be quite sparsely spaced allowing for a very efficient solution. Depending on what form of finite difference approximation introduced the main computational task at each step is the inversion of a tridiagonal matrix. The Fourier Split step solution method was used to analyse terrain scattering in [Jana97], where computations are carried out on a novel numerically generated mesh. This analysis was repeated in the case where buildings are added to the underlying terrain and the use of a wide-angle propagator is considered [Jana98]. A related formulation is that of Physical Optics (PO), where propagation is modelled as diffraction past a series of perfectly absorbing screens. The PO approach can be shown to be closely related to the parabolic equation, in that it too assumes forward propagation confined to a narrow angular region. A combination of such formulations (as well as some RT) was used to model propagation in an urban environment in [GoCo98], where the absorbing screens simulate very tall buildings. Another study [IHAP99] examined the relative success of four different PO based models in predicting propagation in a city, modifying some of the techniques where appropriate. The generic problem of propagation past a series of absorbing screens has been tackled using other formulations, and we shall return to it in the next section.

3.1.3.1.3. Asymptotic Methods

Geometrical Optics (GO) is an approximate high frequency method for determining wave propagation, which postulates the existence of direct, reflected and refracted rays. These rays follow paths which make the optical distance from source to observation point an extreme. The field at a point is the superposition of the fields associated with each ray passing through that point, giving an asymptotic approximation to the field at that point. By asymptotic, we mean that the fields postulated are the leading terms in a power series expansion in inverse powers of the wave number. Thus, as the wavelength λ approaches zero this high frequency solution becomes increasingly accurate, and we expect Geometric Theory of Diffraction (GTD) to be a good model. However, at lower frequencies its usage is more suspect. We can write the GTD solution as the sum of a geometrical optics term, which exists only in certain regions of space and a diffracted field that exists everywhere. The GTD diffracted field is invalid along the incident and shadow boundaries, and gives unreliable answers in the *transition regions* around these lines, a shortcoming overcome by the Uniform Theory of Diffraction (UTD), which introduces transition functions to remove the singularities associated with the shadow boundaries. The UTD solution for the diffracted field is thus smoothly continuous over the shadow boundaries. A very slight additional computational expense is incurred in that calculation of the transition function necessitates calculating a Fresnel integral. However, accurate series approximations exist for such integrals and numerical quadrature is not needed.

Utilising UTD in a scattering problem involves modelling the scatterer boundary as a connected series of canonical shapes for which UTD solutions exist. Rays can be direct, reflected or diffracted, or multiple combinations of these, in which case reflection or diffraction points are treated as line sources illuminating subsequent points of reflection or diffraction. Obviously, an upper limit on the number of multiple combinations must be imposed. Implementation of the UTD thus requires some sort of RT engine, which can efficiently find all such ray paths to any particular point and simple computation of the associated fields. Such is the lack of computational overheads associated with ray-based methods that they have found applications in areas such as radio propagation in urban micro-cells, where the more computationally intensive full-wave solutions offered by integral and differential equation formulations are simply too slow or memory-intensive. Compromises involving hybrid schemes, where UTD is used in certain areas alongside full wave solutions in other more complicated regions, have met with mixed success and may yet offer a framework to exploit the virtues of the complementary approaches.

2D UTD is applied to micro-cells in one of two ways: by considering obstacles in either the horizontal or vertical plane, depending on whether propagation is deemed to be around building or over rooftop. A study of this led to a quasi-3D deterministic model of [LiWL97]. Wideband propagation was studied using RT techniques in [HeKa97]. Full 3D RT suffers from a high computational overhead in determining the ray paths, an overhead that can be reduced significantly by intelligent pre-processing of the database [HoWL99]. This pre-processing takes the form of precomputing and storing all mutual visibility conditions between the interacting facets, and can speed up computation significantly. Another fast 3D RT model is discussed in [OBKC99], [OBKC98], which suggests processing many RX points in parallel, thereby reducing the RT burden. A 3D ray trace is used in [LaFr99a] to predict the influences of a movement of the RX on the spatial properties of an indoor radio channel. Binary space partitioning algorithms and image theory is used in [TVDL99a] and [TVDL99b] to perform an efficient 3D ray trace in urban and indoor environments.

Problems arise when diffraction points lie in the transition region of previous diffraction points. When this occurs the ray is no longer a so-called ray field and conventional UTD fails. [LlCa97] examines the attenuation function pertaining to propagation past buildings. Another recent innovation tackling this problem is the incorporation of higher order diffracted fields. For instance the inclusion of slope diffraction can improve UTD's capability to describe propagation over a series of knife-edges, as shown by Andersen [Ande94]. Slope diffraction adds to the standard UTD diffracted field, a contribution due to the slope of the incident field. The diffraction of this slope field is described by a diffraction coefficient, which is based on the angular derivative of the diffraction coefficient. Slope diffraction was used in micro-cell propagation analysis by [RVCG98]. Holm [Holm96] heuristically generalised Furutsu's expression for double knife diffraction to generate expressions for higher order diffraction coefficients. His expression agreed with the slope diffraction expression and he presented results that included up to 32 higher order terms. This interesting double knife-edge problem was also examined in [Berg99], using an integral equation approach with a forward scattering approximation. This gave results similar to those of UTD when the incident field is not parallel to the lines between the edges. When the incidence is parallel to this line the result is computationally quite unattractive, but can be used as a 'benchmark' result to gauge the accuracy of other approximate approaches, such as slope diffraction.

3.1.3.1.4. Future Directions in Computational EM

The previous sections have reviewed some of the progress achieved in computational electromagnetics in recent years. Computational advances have facilitated a strong trend towards less abstraction, which is of more realistic modelling incorporating a wider range of physical effects. This trend is likely to continue, as is the growing sophistication of computational techniques, with even more fruitful interplay between analysis and numerical simulations. As well as analysing larger and larger problems (using 3D formulations rather than 2D, for example), we can expect the level of detail that can be modelled using future computational techniques to increase. This is indicated by the growing interest in the inclusion of scattering in more detailed environments, such as the inclusion of lamp-posts and panel scattering in micro-cell applications [RWLG97], or scattering from humans in indoor environments [LaFr99b]. The interesting question of how accurate the propagation environment description must be (for the case of indoor propagation) was addressed in [DCFR97]. Beyond a certain level of refinement, statistical descriptions are necessary. One such example is the analysis of random scattering from a tree in [ToLa99] using the approach of Foldy and Lax. Fast fading statistics and local mean power levels were measured and predicted (using RT) in [LTDV99], [TLDV99], [LTVD00] for indoor wireless propagation. A more 'brute force' approach to the problem of randomness is the use of Monte Carlo simulations, where, once a probabilistic description of the problem at hand has been chosen, the statistics are calculated by performing ensemble averaging over many realisations. Each realisation represents one particular member of the space of all problem configurations, and can be tackled numerically with any one of the computational schemes that we have discussed. Ensemble averaging allows us to sample from this set of problem configurations and compute relevant statistics. More formal results in the area of scattering from randomly rough surfaces include Kirchhoff theory, where any point on a rough surface is treated as if it belonged to an infinite plane parallel to the local surface tangent. Perturbation theory, valid for small height variations from a mean surface height, expands the relevant fields in a power series in terms of the local height fluctuations. The number of terms retained in the power series differentiates between first and higher order perturbation theories. However, in practice, second and higher order theories have limited usage. These techniques can be considered as single scattering theories and encounter difficulties when applied to situations where multiple scattering dominates. Many results remain broadly formal in nature and are restrictive in what statistics they offer (for example many of the more general theories offer predictions for the mean field only). Much work remains to be done to incorporate these results into more general scenarios.

3.1.3.2. Application of deterministic modelling

Dirk Didascalou

In the previous subsection, the different approaches to model electromagnetic wave propagation with the corresponding approximations have been presented. This section is devoted to the application of the diverse modelling techniques yielding complete prediction packages for various scenarios and/or environments. Unfortunately, a unique 'all-in-one' simulation tool, able to cope with any scenario at any frequency, is infeasible due to the enormous complexity. Instead, the different propagation environments, e.g. rural, urban, indoor or tunnel areas, are treated separately. One of the difficulties is the choice of the appropriate modelling technique, which mainly depends on the desired accuracy and the available computing-power or simulation-time restrictions. Whereas in the previous COST Action 231 [COST99], the focus was on the general feasibility aspect of deterministic modelling and the achievable accuracy, a lot of the work during COST Action 259 concentrates on the efficiency of deterministic propagation modelling. Several novel approaches have been proposed to speed-up simulation times considerably, retaining the precision of the prediction or even enhancing it. Furthermore, several benchmarks and comparisons of different approaches have been performed in order to ease the selection of the appropriate modelling technique.

There is an enormous – and still increasing – capacity demand of the established second generation systems. Furthermore, they will have to operate in coexistence with the upcoming (broadband) future generation systems. The resulting trend to smaller cell sizes (i.e. micro- or even pico-cells) leads to the necessity of sufficiently precise prediction tools in order to efficiently plan and install operational systems, and to combat interference. The propagation characteristics in such small cells differ significantly from the ones in larger areas [COST99], [Kürn99]. Additionally, much higher accuracy is needed, not only in propagation modelling, but also in the description of the propagation scenario (i.e. the databases). Therefore, the impact of the underlying databases, together with the utilised modelling techniques has to be investigated.

The remainder of this sub-section is organised as follows. First, methodologies to evaluate the performance of propagation modelling are presented. This is followed by the description of different propagation packages/approaches, which are gathered/classified according to their areas of application. The performance and the applicability of the various models are evaluated in the corresponding subsections.

3.1.3.2.1. Methods of Comparison and Evaluation

Traditionally, the evaluation of a prediction model is carried out by comparing predicted results with measurements and quantifying the errors by the statistical parameters mean μ_e, standard deviation σ_e, and/or the RMS of the prediction error [COST99], [Stee92]. The prediction error e is defined by

$$e = L_{pred} - L_{meas},\tag{3.1.27}$$

where L_{pred} and L_{meas} denote the predicted and the measured path loss, respectively. Generally, the values of L_{pred} and L_{meas} are averaged to remove the effects of the so-called fast fading, which is due to multipath from the local scattering environment. The necessary spatial averaging depends on the environment and is usually performed over tens of wavelengths [Lee82], [Stee92]. In urban micro-cellular or indoor environments, however, the 'deterministic' signal fluctuations could be very sudden as in the case of a mobile turning out of the line-of-sight around a corner. Ideally, these abrupt deterministic fluctuations should be predicted. Although rather local, rapid fluctuations may lead to large prediction errors. Furthermore, the limitations in the accuracy of the geometrical databases and of the measurement conditions (e.g. traffic) additionally affect the quality of a prediction. Thus, the conventional μ_e and σ_e characterisation may not give a representative view of the accuracy of a prediction model.

As an example, Figure 3.1.9 [WaLL97] shows a prediction (thin line) performed by a 3D RT tool compared to measurements in an urban environment (thick grey line). Also shown is the result from a simple power law relation (thick line) fitted to the measurements. Compared to the measurements, both the RT prediction and the power law lead to the same value for the standard deviation of the error, i.e. to about 11 dB. However, it is obvious from Figure 3.1.9 that the RT result fits much better than the power law curve.

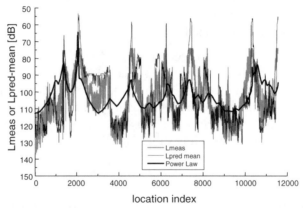

Figure 3.1.9: Measurements (Lmeas), 3D RT results (Lpred) and results using a power law curve fit (Power Law) for the case of a European city. $\sigma_e \approx 11$ dB for both prediction results.

In [WaLL97], a figure of merit F is proposed as the percentage of locations where coverage and non-coverage have been correctly predicted, i.e.

$$F = (N_{cc} + N_{cnc}) / N_{tot}, \qquad\qquad (3.1.28)$$

where N_{cc}, N_{cnc} and N_{tot} denote the number of locations with correctly predicted coverage, correctly predicted non-coverage and the total number of locations. Obviously, the actual value of F depends on the adopted threshold, since a location is considered as being either covered or non-covered by comparing the received signal level with the chosen threshold. Results obtained in [WaLL97] and [BaGi98] indicate that the figure of merit provides a good measure of the prediction accuracy of different models in the case of large standard deviations. Whenever the standard deviation σ_e is sufficiently small (< 7 dB), the accuracy of a model is well described by σ_e, whereas F should be used with care.

Additional criteria to compare different propagation models are, for example, the cumulative distribution function of the received power level, or techno-economical considerations (i.e. cost-functions) [BaGi98].

3.1.3.2.2. Propagation Models for Rural Terrain

Propagation models for rural terrain are extensively treated in [COST99]. However, some advances concerning the split-step Parabolic Equation Method (PEM) have been reported in [Jana97], [JaAn00], leading to a reduced computational complexity and allowing for irregular terrain topography.

3.1.3.2.3. Propagation Models for Urban environments

Especially in densely populated urban areas, highly sophisticated radio network planning methods are needed to efficiently explore the use of the scarce spectrum.   They require more accurate prediction models and detailed digital databases, which describe the propagation environment properly. Additionally, the different environments (e.g. macro- and micro-cells), cannot be treated separately any more, since results from all cell types have to be considered simultaneously for interference calculations and handover (HO) planning. However, for the complete planning of a large city, modelling of all propagation effects together with high-resolution data is not appropriate in a practical implementation due to computing-time constraints. Therefore, several approaches are undertaken in order to reduce the computational complexity.   As an example, the most relevant propagation mechanisms/paths can be determined for a given scenario retaining the prediction accuracy. Intelligent pre-processing of the databases and RT acceleration techniques are additional options.

A) Ray-Tracing Based Propagation Models

A convenient way to handle ray-optical based (including GTD/UTD) wave propagation modelling in urban areas is to sub-divide the propagation paths into different classes, which can be treated separately, Figure 3.1.10 [GoCo98], [RVFC98], [Kürn99]. A vertical plane 2D model accounts for over-rooftop contributions, a lateral 2D model is used for the propagation inside streets, whereas a full-3D (or multipath) model is able to determine all

MPCs in three-dimensional space. Obviously, the latter model is most accurate with the highest computational complexity. Furthermore, various derivatives of these different types of models are available [COST99], [GeWi98]. The influence of vegetation is generally included by additional path loss contributions for each ray interception [GoCo98], [ToBL98], [Kürn99], and/or additional diffraction contributions over the vegetation [Kürn99].

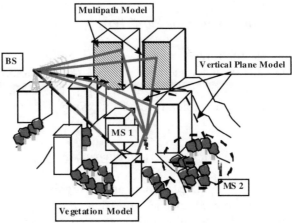

Figure 3.1.10:   Propagation paths in urban areas.

For BS locations above rooftop level (often referred to as macro-cells) it was found that the prediction accuracy increases significantly by considering multipath, i.e. 3D propagation, and vegetation effects in the vicinity of the TX up to distances of 500 to 1 000 m from the BS [Kürn99]. At larger distances, over-rooftop contributions dominate the propagation behaviour and thus a vertical plane model seems appropriate for sufficiently accurate prediction results [RVFC98], [Kürn99]. Consequently, the simulation time of large areas, relevant for interference calculations, decreases drastically by neglecting 3D multipath effects at larger distances. For BS locations at or below rooftop level (referred to as micro-cells), the same conclusions can be drawn. One exception is BS locations significantly below rooftop level, where a lateral propagation model might suffice close to the TX up to distances of 200 to 400 m [FCGB98]. Another factor relevant for the prediction accuracy is the quality and resolution of the underlying database, which describes the propagation environment. Although for typical macro-cellular applications the prediction accuracy in urban areas is reasonable even with low-resolution data, detailed knowledge of building heights and structures is absolutely necessary for BS antennas mounted at, or only a few metres above, rooftop level in dense urban areas [KüFa97]. The application of such prediction packages [FCGB98], [GoCo98], [RVFC98], [Kürn99] leads to mean values $\mu_e \leq 3$ dB and standard deviations $\sigma_e \leq 8$ dB between measurements and predictions.

A different approach to reducing computational complexity by retaining the accuracy of full-3D models is an intelligent pre-processing of the building database [HeKa97], [HoWL99]. Prior to any propagation calculation, the walls of all buildings are discretised and visibility relations amongst all elements are established, which are independent of the position of the BS. The results of the pre-processing are stored in a tree structure. Once the pre-processing of the building database is accomplished, the time consuming task of path finding reduces to a search in a tree structure, which results in a reduction of the computational complexity by a factor of 1000 [HoWL99].

If no pre-processing of the database is desired, the efficiency of three-dimensional RT can be enhanced using computational geometry and/or ray acceleration techniques [HeKa97], [OBKC98], [OBKC99].

B) Neural Network Based Propagation Models

An alternative to RT are neuronal networks [FrCa97], [CaFr98], [FrCa98a], [FrCa98b]. Although completely different in their approach, neuronal networks – especially Multilayer Perceptrons (MLP) – lead to a reasonable precision in the prediction, comparable to the COST 231 propagation model [FrCa97]. Their great advantage is the reduction in simulation time. The inconveniences are a lack of insight in the physical propagation behaviour and the mandatory training based on measurement results.

C) Semi-Empirical Propagation Models

For many applications, especially with large cell sizes or in the absence of appropriate databases, semi-empirical models deliver a reasonable accuracy [IHAP99], [Ortg99], [ChOC00]. In contrast to purely empirical models (e.g. log-distance, Okumura-Hata [GeWi98], COST231-Hata [COST99], etc.) semi-empirical models (e.g. Walfish-type [COST99], Xia and Bertoni [XiBe92], Saunders and Bonar [SaBo94], etc.) or simple diffraction models (e.g. Vogler [Vogl82], Epstein-Peterson [COST99], [GeWi98], etc.) can be adopted to a variety of scenarios, without the implicit demand of prior calibration. Often semi-empirical approaches are used for the vertical plane model in RT based prediction packages [COST99], [GoCo98], or as fast models to evaluate the influence of the resolution of databases or other 'external' parameters [KüFa97]. However, one has to ensure that the respective models are applied to scenarios for which they were adapted.

Otherwise this would lead to unacceptable deviations of the prediction results from the real situation [IHAP99], [Ortg99], [ChOC00].

3.1.3.2.4. Indoor Propagation Models

Indoor wireless communications are rapidly expanding and gaining importance. Amongst the several applications are cordless telephony (DECT), portable computing and wideband wireless local area networks (WLAN). Simple field strength predictions are insufficient, because information on the directional channel properties can be very important for planning purposes [LaFr99a]. Moreover, movements of TXs, RXs and/or persons may result in a time-variance of the transmission channel. Means to predict wave propagation with adequate precision, able to deliver angle-of-arrival information and to generate time-series, are given by 3D RT [COST99]. In fact, the completely man-made indoor environment can usually be represented by a set of plane walls, and is thus suitable for RT field predictions. Due to the very localised propagation environment, however, the RT prediction results are very sensitive to environment modelling. For example, the introduction of a rough description of furniture in the model helps in reducing both the mean error μ_e and the standard deviation of the error σ_e [DCFR97]. Also the influence of the human body on the radio field cannot be neglected. In [LaFr99b] modelling of the human body by a homogeneous dielectric cylinder with finite extent led to satisfying results compared to measurements. In general, RT based indoor prediction packages [DMSV97a], [DMSV97b], [TVDL99a], [LTDV99] permit us to estimate accurately enough both the mean level of the received signal and its variations (fast fading characteristics) around the mean [LTDV99], [Zwic00].

Due to the limited extend of the reflecting surfaces in indoor environments and the (generally) close distances between the surfaces and RX and/or TX, the approximation of plane wave reflections is not necessarily valid. In order to increase the accuracy of indoor RT, an analytical approximation of near-field physical optics for surface scattering (i.e. 'diffuse reflections') can be used instead of the conventional GO reflection calculation [DVLP97], [DVLV98], [DeVV98], [DVLD98]. This approximation is valid for distances r up to

$$r \approx D/2 \, (D/\lambda)^{1/3}, \tag{3.1.29}$$

where D denotes the extend of the reflecting aperture and λ is the free space wavelength.

In order to speed up simulation time, a neural network based dominant path method has been proposed [WoLa99], [WoWL99]. However, in contrast to MLP for urban areas, this approach still allows a physical interpretation of the propagation process. The basic idea is to group rays together, which traverse the same walls, rooms, etc. The relations between adjacent rooms are determined and a tree structure of the rooms can be established. The root of the tree represents the room in which the TX is located. The so-called dominant paths are in the uppermost layers of the tree, since each layer represents a penetration of a wall, and thus the lower the layer, the higher the experienced attenuation. The neural networks are actually used to determine the predicted field strength based on the dominant paths.

3.1.3.2.5. *Tunnels and Other Special Environments*

A) *Tunnels*

Realistic tunnel geometries are generally of rectangular cross-section or arched shape, i.e. of elliptical cross-section with a raised floor and eventually an additional ceiling. Furthermore, they are mostly curved. Due to the curved boundaries, wave propagation in tunnels is usually modelled by different empirical models [COST99]. Since the application of conventional RT approaches is restricted to planar boundaries, these techniques can only be applied to most tunnel scenarios under simplifying assumptions [COST99], [LiDe99], [LTDV99]. In contrast, a novel ray-optical method is presented in [DDZW99], [Dida00], which is not restricted to planar surfaces. It allows a sufficiently accurate RT based coherent calculation of the electromagnetic field in tunnels of arbitrary shape. The modelling is based on GO. Contrary to classical RT, where the one ray representing a locally plane wave front is searched, the new method requires multiple representatives of each physical electromagnetic wave at a time. The contribution of each ray to the total field at the RX is determined by the so-called Ray Density Normalisation (RDN). This technique has the further advantage of overcoming one of the major disadvantages of geometrical optics, the failure at caustics. The approach has been verified theoretically with canonical examples (waveguides), by various measurements at 120 GHz in scaled tunnel models and in real subway tunnels at mobile communications frequencies [Dida00], [DiMW00].

B) *Construction Sites*

A construction site is a special environment where several propagation related issues have to be considered. In particular, an important problem concerns the coverage of buildings under construction and therefore changing continuously. The deployment of wireless systems (e.g. for supply and production supervision purposes) in such scenarios cannot be carried out by the usual cable-based BS links. Obviously no initial measurements can be performed to determine the optimum BS antenna locations, since initially the building is non existent. Means to solve these problems are proposed in [RuSA99] by a specific DECT system architecture together with a semi-empirical propagation model, which can easily be incorporated in a standard CAD tool.

3.2. Directional Channel Modelling

Martin Steinbauer and Andreas F. Molisch

Models for the mobile radio channel are vital for the study of smart antenna systems – both for the design of algorithms, system-testing purposes, and for network planning.   These models have to reproduce the typical characteristics observed in measurement data from a number of different representative environments.   Bearing in mind the importance of the COST207 wideband channel models [COST89] for the GSM standardisation, we can expect that directional channel models for future mobile radio systems like UMTS or HIPERLAN will have a similar importance. Due to the directional component, and the larger bandwidth required for future systems, the new channel model strives to fulfil the following requirements:

- Details of the Directional Channel Impulse Response (DCIR), i.e. statistics of the MPCs should be reproduced correctly (e.g. to study the behaviour of RAKE RXs). → Accuracy.
- It should be as realistic and physical as possible, but simple enough to allow theoretical studies and not to place too high requirements on computing power. → Simplicity.
- For non-directional or narrowband purposes, it should be as good or better than current channel models. It should contain the COST207 models as special cases to allow easy comparison with earlier system simulations. → Consistency.
- The directional properties both at the BS and the mobile station should be reproduced. Short- and long-term fading, channel variations due to non-stationary scenarios and dynamic evolution of paths should be reproduced. The main application areas are in the 1, 2, and 5 GHz band and for macro-, micro- and pico-cells (outdoor, penetration and indoor). → Completeness.

There might be a contradiction between simplicity and accuracy. Thus, we have to carefully trade off these requirements.

Within the COST259 project, approaches for directional parametric stochastic models have been presented in [MoLK98] and [HeBF97]. The first reference characterises a propagation scenario by the locations of the scatterers and the position of the BS as well as the MS. In the second reference, a propagation scenario is described by the parameters of the impinging waves, i.e. their delay, incidence direction, carrier phase, and complex amplitude. A transformation between the scattering geometry and the resulting delay-angle distribution can be computed in a straightforward manner [MoLK98]. Based on this work, the first general proposal for a channel model has been elaborated within the framework of the COST259 subgroup on directional channel modelling (SWG 2.1) [SACH99]. The model provides a more general and flexible framework that also covers large-scale variations of the channel. The parameter selection relies on previous models as well as recent measurement results. It uses the findings of Hata-Okumura and Walfisch-Ikegami, as well as COST231 for modelling the path loss in different environments; COST207 for global power delay profiles in macro-cells; TSUNAMI-II in respect to the azimuth-delay-power spectrum shape of clusters in macro-cells [MPFF98], and severely influenced by the CODIT radio environment selection in micro- and pico-cells [PéJi94]. The approach of distinction between global/local and location-dependent instead of time-dependent channel functions stems from the Magic WAND project [HeBF97], the clustering approach from the METAMORP project [WMMK99].

In Section 3.2.1 the modelling concept will be presented. After these theoretical-mathematical basics, measurement results for the necessary model parameters are presented, separately for outdoor (Section 3.2.2) and indoor (Section 3.2.3). Finally, in Section 3.2.4, all model parameters are summarised, together with guidelines on their correct usage for system-oriented implementations. Due to the large variety of system simulations, no model on the system-level is given here. For the design of system-level models based on link-level models, see Chapter 2 (Radio System Aspects).

3.2.1. Modelling concept

Andreas F. Molisch and Ralf Heddergott

3.2.1.1. Radio channel simulation methods

Three solutions have been proposed to simulate the radio channel, i.e. reproduce its IR: stored channel IRs (ChIRs) [Goll94], ray-optical methods, and stochastic models. Stored ChIRs incorporate all details of the radio channel. However, it turned out to be very difficult to find a small set of reference scenarios that included all of propagation effects encountered in a certain environment. Ray-optical methods are a very powerful tool to perform a coverage prediction in a particular propagation scenario. The main drawback is the need for a precise description of the location, which makes it difficult to translate the result to other situations.

Stochastic radio channel models describe the IR as realisations of a stochastic process. The WSSUS model [Bell63] characterises the channel by means of correlation functions or power spectral densities. The radio channel is represented by a tapped delay-line, where each tap represents a fading MPC following Rayleigh or Ricean statistics [Cox72]. This is also used in the COST207 model [COST89]. Recently, Bello's terminology has been extended to describe directional channels [Katt99]. In [Mart96] the tapped delay-line model is extended to also include also DoAs.

A different approach is given by *parametric* stochastic models [TCJF72]. The received signal consists of a superposition of several waves. Thus, it is simple and realistic to characterise the ChIR by means of the parameters of the incident waves. Moreover, these parameters remain constant for movements of the MS over a distance of a few tens of a wavelength. The WSSUS property is not required. However, it is assumed that the parameters of different waves are statistically independent of each other.

3.2.1.2. COST 259 modelling framework

Radio propagation depends on extremely different topographical and electrical features of the surroundings. To account for this variety, a 3-level structure according to Figure 3.2.1 has been defined for the COST259 directional channel model (COST259-DCM). The latter provides a framework from which channel models can be deduced. At the top level, a first distinction has been made by the cell type. For each cell type, a number of *Radio Environments* (REs) have been identified, where all RE names begin with the word 'general'. This has been chosen to stress that a RE stands for a whole class $\mathfrak{R}$ of propagation constellations that exhibit similar or typical features that can be related to the surroundings in which a communication system operates. The topographical features of a RE are given by a number of *external parameters*, such as the frequency band, the average height of BS and MS, their average distance, average building heights and separations, etc. Furthermore, in micro- and pico-cells it has been defined whether the propagation has a LoS or NLoS property.

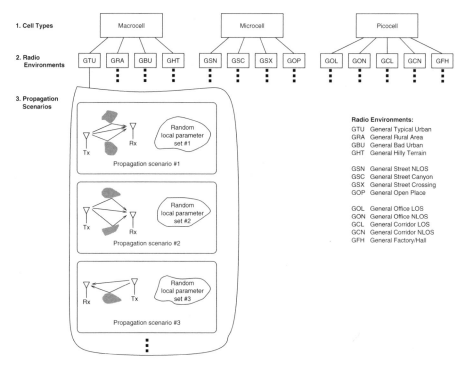

Figure 3.2.1: Structure of COST 259 directional channel model.

The propagation conditions encountered in each RE are characterised statistically by a set of PDFs and/or statistical moments. Since the members of this set characterise propagation conditions of the entire RE, they are referred to as *Global Parameters* (GPs). They serve as key channel parameters that provide the necessary information for the basic system design decisions on modulation technique, burst length, coding scheme, etc. The GPs have to be extracted from extensive measurement data. Recently, RT simulations have also been used for global parameter extraction [ZDDW99].

Usually, the performance of a communication system is evaluated by means of extensive Monte-Carlo simulations. For the latter, realistic realisations of the channel response have to be generated. To this end, the third level of DCM consists of *propagation scenarios*, which are defined as random realisations of incidence constellations (see Figure3.2.1). The latter are specified by random *Local Parameters* (LPs). A possible set of LPs may be given by the parameters of the waves impinging at the location of the RX antenna, i.e. their number, complex amplitude, delay, and incidence direction [HeBF97], or equivalently, by describing the location of the BS, MS, and the scattering objects interacting with the electromagnetic field

[FuMB98] (following established notation in the literature, the word 'scattering' is not only used for diffuse scattering, but even for those processes that are strictly speaking 'specular reflections'). The statistical properties of the LPs are given by a set of global parameters defined in the second level of COST259-DCM.

It is reasonable to assume that for movements of the MS over a sufficiently small *local area*, say *A*, not larger than some tens of a wavelength, the LPs determining the propagation scenario remain approximately constant. As a consequence, the spatial variations of the IR are modelled as changes of the phase of the impinging waves. Since, with a finite system bandwidth, not all multipath arrivals can be resolved, the phase variations the impinging waves cause rapid fluctuations of the resulting amplitudes and phases of the IR. These fluctuations are called *small-scale fluctuations*.

When the MS moves over large areas, the distance and the direction of the scatterers may vary with respect to its instantaneous location. Moreover, some propagation paths may be partly obstructed or disappear while new ones may arise. The variations due to MS location changes over wider areas are called *large-scale variations*. In the COST259-DCM, they are modelled by means of a drift of the LPs, and an appearing/disappearing process of MPCs controlled by transition functions, respectively.

Moreover, transitions between REs are defined by a 'random walk' process, see the end of Section 3.2.1.

3.2.1.3. Directional channel impulse response

In a radio link, a part of the energy radiated by the TX antenna reaches the RX station through different paths. Along these paths, interactions may occur between the electromagnetic field and various objects. Possible interactions are specular reflection on large plane surfaces, diffuse scattering from surfaces exhibiting small irregularities or from objects of small size, transmission through dense material, shadowing by obstacles, etc. The attributes 'large' and 'small' are to be understood with respect to the wavelength.

At the position of the RX antenna, attenuated and phase-shifted replicas of the transmitted signal are received from different directions with distinct time delays. It is assumed that the RX antenna is located in the far-field region of the TX antenna and all scatterers. Thus, all impinging waves are plane transverse-electromagnetic. The dispersion of the propagation channel in the temporal and angular domain is described by the DCIR.

$$h(\vec{r},\tau,\Omega) = \sum_{\ell=1}^{L(\vec{r})} h_\ell(\vec{r},\tau,\Omega).$$

(3.2.1)

Here, $\vec{r}$ denotes the location of the RX antenna, and τ is the delay variable. The direction of arrival is characterised by Ω. The latter may be denoted by its azimuth angle ϕ and its elevation angle ϑ, where ϑ is positive above the horizontal plane if not stated otherwise. The DCIR is decomposed into a finite number L of components, each of which corresponding to an impinging wave. The components may embody a specular and/or a diffuse part. Note that in (3.2.1) directions of departure and polarisation have been omitted.

3.2.1.4. Channel impulse response

In the previous section the channel was described as being location-dependent. A system theoretical characterisation as given in [Bell63] is *time-dependent*. Moreover, the channel includes the TX and RX antennas. A transition from location-dependency to time-dependency can be accomplished by assuming a time-dependent trajectory $\vec{r} = \vec{r}(t)$, where the scatterers are assumed to be non-moving. This yields the *time-variant* DCIR $h(t,\tau,\Omega)$. The non-angle resolved time-variant IR is obtained by integrating $h(t,\tau,\Omega)$ weighted by the complex antenna pattern $G(\Omega)$ of the RX antenna over directions:

$$h(t,\tau) := \int_\Omega G(\Omega)\, h(t,\tau,\Omega)\, d\Omega.$$

(3.2.2)

Strictly speaking, this function is not the response of the channel at a certain time instant, but the response at the RX antenna output at time t if a Dirac impulse was transmitted at time $t-\tau$. Bello [Bell63] has called $h(t,\tau)$ the input *delay spread function*. However, the name 'time-variant IR' has become rather popular.

This DCIR can be further decomposed into

$$h(t,\tau,\Omega) := \int_{\Omega_T} G_T(\Omega_T)\, h(t,\tau,\Omega,\Omega_T)\, d\Omega_T \,,$$

(3.2.3)

where Ω_T is the angle at the TX antenna [Stei98], [SHSS00]. The $h(t,\tau,\Omega,\Omega_T)$ is called the double-directional IR. Measurement results of a similar campaign are presented in [ZHMR00].

A further refinement of the model can be achieved by including the polarisation of the waves. The IR then becomes a matrix, e.g. describing the relation between horizontal and vertical field components [ZwDF98], [ZFDW00]:

$$\begin{pmatrix} h_{VV}\left(t,\tau,\Omega,\Omega_T\right) & h_{VH}\left(t,\tau,\Omega,\Omega_T\right) \\ h_{HV}\left(t,\tau,\Omega,\Omega_T\right) & h_{HH}\left(t,\tau,\Omega,\Omega_T\right) \end{pmatrix}. \tag{3.2.4}$$

Note that the model parameterisation in Section 3.2.4 is not truly double directional and assumes a vertical BS antenna polarisation. To simplify notation, we will use the scalar IR h_{VV} in the remainder of the section if not stated otherwise.

3.2.1.5. Clustering

Measurement results show that in some radio environments the MPCs are not always uniformly spread in the τ, Ω space but typically arrive in clusters. In outdoor environments, clustering appears because some groups of scatterers give comparatively high receive amplitudes, either due to their closeness to the RX, or their large scattering cross-section. In indoor NLoS scenarios, a clustering pattern arises due to openings such as doorways or due to angular spreading at wall transmissions. Moreover, back wall reflections of clustered MPCs leads to additional clusters. Thus, clustering can be motivated by the physical propagation process in the sense that components of a cluster experience the same large-scale behaviour, while the large-scale behaviour for components of distinct clusters is independent.

Formally, the indices of the MPCs $h_l(t,\tau,\Omega)$, $l = 1, ..., L$ can be grouped into $M \leq L$ *disjoint* classes (or clusters) $C_1,...,C_M$, where each class has $N_m \geq 1$ elements, and

$$\sum_{m=1}^{M} N_m = L. \tag{3.2.5}$$

With this notation, (3.2.1) can be re-written as

$$h(\vec{r},\tau,\Omega) = \sum_{m=1}^{M} \sum_{n \in C_m} h_n(\vec{r},\tau,\Omega). \tag{3.2.6}$$

In the case of clustering, a statistical description of the propagation conditions by means of global parameters must include a description of the spreading of the waves *within* a cluster as well as the spreading of the clusters themselves.

3.2.1.6. Definition of channel models

The structure of the COST 259-DCM, consisting of the cell type, the RE, and the random propagation scenarios, defines a basic framework from which models for the behaviour of the MPCs given in (3.2.1) can be deduced. This requires the definition of particular sets of local parameters and their statistics by means of global parameters. The latter definitions should be based on the propagation conditions encountered in a given radio environment.

The parameters of the DCIR can be given by means of a delay-angle distribution [FlBH96], [HeBF97] or a scatterer geometry [FuMB98], [MPFF98], [MoLK98]. In the sequel, both approaches and a transformation between the delay-angle distribution and the scatterer geometry are briefly sketched. The particular settings of local and global parameters for COST 259-DCM are given in Section 3.2.4.

3.2.1.7. Definition of local and global parameters based on the delay-angle distribution

For a specular wave, the l-th contribution to the DCIR in (3.2.1) is given as

$$h_l(\vec{r},\tau,\Omega)=\alpha_l\,\delta(\tau-\tau_l)\delta(\Omega-\Omega_l),\tag{3.2.7}$$

where α denotes a complex amplitude.

The local parameters are given by the parameters controlling the DCIR, i.e.

$$\{L(\vec{r}),\alpha_1,\tau_1,\Omega_1,\ldots,\alpha_L,\tau_L,\Omega_L\}\tag{3.2.8}$$

According to the definition of the LPs, all parameters in (3.2.5) are random. It is usually assumed that the parameters of different MPCs are statistically independent from each other. However, to account for clustering, the parameters of impinging waves belonging to distinct clusters may depend on different cluster mean values.

Within the local area A, it is assumed that the variation of the propagation delay and the attenuation of the impinging waves due to the modification of the propagation path lengths can be neglected. Moreover, it is assumed that plane waves are impinging within A, i.e. the variations of the incidence angles can be neglected. Thus, the spatial variations of the DCIR are modelled as changes of the phases of the components $h_l(\vec{r},\tau,\Omega)$.

The power contribution of the l-th wave is given as

$$P_l(\tau,\Omega) = E_{\vec{r} \in A}\left\{\left|h_l(\vec{r},\tau,\Omega)\right|^2\right\} = P_l\,S(\tau-\tau_l,\Omega-\Omega_l), \qquad (3.2.9)$$

where $E_{\vec{r} \in A}\{\cdot\}$ denotes the expectation over all locations within A. The parameter P_l denotes the mean power of the l-th component. A possible spreading of a wave's power in delay and direction due to diffuse scattering is described by the scattering function $S(\tau,\Omega)$. In Section 3.2.4 this function is modelled to be equal for all clusters and normalised to unit power.

The *local* Power Delay-Direction Profile (PDDP) reads

$$P_A(\tau,\Omega) := \sum_{l=1}^{L} P_l(\tau,\Omega) = \sum_{l=1}^{L} E_{\vec{r} \in A}\left\{\left|h_l(\vec{r},\tau,\Omega)\right|^2\right\}. \qquad (3.2.10)$$

It is obvious that the underlying stochastic process is only US in the sense of Bello [Bell63] if the equality

$$\sum_{l=1}^{L} E_{\vec{r} \in A}\left\{\left|h_l(\vec{r},\tau,\Omega)\right|^2\right\} = E_{\vec{r} \in A}\left\{\left|h(\vec{r},\tau,\Omega)\right|^2\right\} \qquad (3.2.11)$$

holds. In this case, the local average of $h_l(\vec{r},\tau,\Omega)$, yields uncorrelated components. In [PéJi94], [IwKa93] it has been stated that this is not always appropriate, at least for indoor channels. Note, however, that the realisations of the DCIR rely on the local parameters (3.2.8) but not on (3.2.10). Thus, the non-validity of the US property does not lead to a more complex model.

A *global* characterisation of the RE is given by the *global* PDDP

$$P_\Re(\tau,\Omega) := E_{A \in \Re}\left\{\frac{1}{P_A}P_A(\tau,\Omega)\right\}, \qquad (3.2.12)$$

where the expectation is taken over the scenarios identified with their local area A within the radio environment $\Re$. The local PDDPs are normalised with their power P_A.

The generation of DCIR realisations requires additional global information, such as the PDF of the number of impinging waves $f(L)$, the PDF of the multipath arrivals $f(\tau_l, \Omega_l)$ and the expectation of the power of the l-th wave given their delay and incidence direction $E\{|\alpha_l|^2|\tau_l,\Omega_l\}$ [IwKa93], [PeMF99b]. The latter PDFs and power profiles can be obtained from extensive measurement campaigns.

3.2.1.8. Definition of local and global parameters based on the scatterer distribution

In the Geometry-based Stochastic Channel Model (GSCM), Figure 3.2.2, the probability density function of the geometrical location of the scatterers (diffuse scatterers or specular reflectors) is prescribed [FuMB98], [BlJu98]. For each channel realisation, the scatterer locations are taken at random from this PDF and then the multipath delays and DoAs are computed by a simple RT approach, assuming that each of the scatterers carries one MPC to the BS. Of course, the PDF has to be selected in such a way that the resulting Power-Delay Profiles (PDPs) and Azimuth-Power Spectra (APSs) at the BS agree reasonably with the measured values.

An extension of the basic model places scatterers not only around the BS, but uses also distant group of scatterers (far scatterer clusters). This corresponds to groups of high-rise buildings or mountains that can work as efficient scatterers, especially if they have LoS both to the BS and the MS.

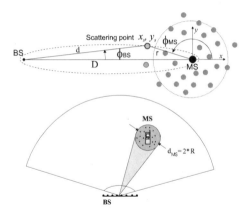

Figure 3.2.2: Principle of the GSCM.

An alternative implementation, called NSCS (Nonuniform Scattering Cross Section) uses a PDF of the scatterers that is uniform throughout the whole relevant cell area, but has the cross-sections of the scatterers different (weighted), and may also change with time. This not only corresponds better to physical reality, but is also more efficient computationally [MKLH99].

The GSCM principle allows a very efficient implementation, especially since small-scale and large-scale effects can be easily separated. Small-scale fading results from the interference of the waves from the scatterers, whose phases change when the MS moves over a distance smaller than one wavelength. The amplitudes of the partial waves stay constant over a range over some ten wavelengths. The large-scale effects are included by changing the scatterer amplitudes, and by the appearance and disappearance of clusters. Also the large-scale effects are usually strongly related to geometry, and can thus be implemented by inspection of the appropriate configurations.

3.2.1.9. Transforms between implementation methods

The Delay-Angle-Distribution approach contains two philosophies that can be transformed into each other by mathematical transformations [Egge98], [MoLK98]. Using the notation of Figure 3.2.2, we have

$$\phi(x_s, y_s) = \text{atan}\left(\frac{y_s}{D + x_s}\right) \tag{3.2.13}$$

$$\tau(x_s, y_s) = \frac{\sqrt{x_s^2 + y_s^2} + \sqrt{(D + x_s)^2 + y_s^2}}{c} \tag{3.2.14}$$

$$pdf_2(\tau, \phi_{BS}) = pdf_1\left(x_s(\tau, \phi_{BS}), y_s(\tau, \phi_{BS})\right) \cdot J(\tau, \phi_{BS}), \tag{3.2.15}$$

where $x_s(\tau, \phi_{BS})$ and $y_s(\tau, \phi_{BS})$ are given below, and the Jacobian $J(\tau, \phi_{BS})$ is

$$
\begin{aligned}
J(\tau, \phi_{BS}) &= \begin{vmatrix} \dfrac{d}{d\tau} x_s(\tau, \phi_{BS}) & \dfrac{d}{d\tau} y_s(\tau, \phi_{BS}) \\ \dfrac{d}{d\phi_{BS}} x_s(\tau, \phi_{BS}) & \dfrac{d}{d\phi_{BS}} y_s(\tau, \phi_{BS}) \end{vmatrix} \\
&= \frac{(\tau c - D) \cdot (\tau c + D) \cdot \left(-2\tau c D \cdot \cos(\phi_{BS}) + D^2 + (\tau c)^2\right)}{4 \cdot \left(\tau c - \cos(\phi_{BS}) \cdot D\right)^3}.
\end{aligned} \tag{3.2.16}
$$

We can also employ the same kind of transformations in the opposite direction [MoLK98]. For example, the inverse transformation can be used to provide the scatterer geometry from a measured joint delay-angle distribution. This allows a straightforward adaptation of the GSCM to correspond to specific measured environments.

After some manipulations, the relation between the random variables can be written as

$$x(\tau, \phi_{BS}) = \frac{\left(D^2 - (\tau c)^2\right)}{2D - 2\tau c\sqrt{1 + \tan^2(\phi_{BS})}} - D \tag{3.2.17}$$

$$y(\tau, \phi_{BS}) = \frac{\left(D^2 - (\tau c)^2\right) \cdot \tan(\phi_{BS})}{2D - 2\tau c \sqrt{1 + \tan^2(\phi_{BS})}} .$$

(3.2.18)

The PDF transformation employing the Jacobian can be now written

$$pdf_1(x_s, y_s) = pdf_2\left(\tau(x_s, y_s), \phi_{BS}(x_s, y_s)\right) \cdot J(x_s, y_s) ,$$

(3.2.19)

where $\phi_{BS}(x_s, y_s)$ and $\tau(x_s, y_s)$ are shown in (3.2.13) and (3.2.14), and the Jacobian $J(x_s, y_s)$ is

$$J(x_s, y_s) = \begin{vmatrix} \dfrac{d}{dx_s} \phi_{BS}(x_s, y_s) & \dfrac{d}{dx_s} \tau(x_s, y_s) \\[2mm] \dfrac{d}{dy_s} \phi_{BS}(x_s, y_s) & \dfrac{d}{dy_s} \tau(x_s, y_s) \end{vmatrix} .$$

(3.2.20)

The GSCM approach can also be combined with the Delay-Angle-Distribution approach for the implementation: the GSCM is used to compute the small-scale averaged Azimuth Delay Power Spectra (ADPS) (i.e. averaged over the Rayleigh fading). The realisations of the IRs are then computed by the Delay-Angle-Distribution method. This combines the computational efficiency of the former approach with the easy physical interpretation of GSCM [MKLH99].

The COST259 model is thus implementation-independent. Some model parameters (Section 3.2.4) will be given in a way that suggest implementation in either of the two described methods, but the above transformation formulas are always applicable.

3.2.1.10. Transitions between radio environments

The REs are defined by their common propagation properties, but basically no statement is made on how long a MS is situated in one RE. In a macro-cell, this duration is typically very long. For micro- and pico-cells, transitions from one RE to another (e.g. from a street canyon to a street crossing) can easily appear during a period of interest for channel tracking applications. Transitions between REs in micro-cells (or also in macro- and pico-cells) are thus defined by appropriate transition functions. We suggest the use of a transition function

$$\frac{1}{2} - \frac{1}{\pi} \arctan\left[\sqrt{2} \frac{2y}{\sqrt{\lambda x}}\right] .$$

(3.2.21)

Here, y and x are the distances of the antenna from the transition point, where x is the distance along the direction of the wave propagation, and y the direction normal to it. These distances can be determined from the map of the virtual environments and the prescribed route of the MS. Simultaneously, all contributions from the second RE are increased by this function. Of course, more complicated transition functions can also be specified.

Additional to the transition function, the probability of transitions and the duration between them also have to be specified. The approach taken in COST 259 is to specify a 'typical' map of a micro-cellular city environment, and a typical office building, and define a route and speed that the mobile takes in that environment. This gives in a deterministic way the transitions between the environments. Different realisations can be obtained by specifying alternative routes of the mobile.

A more general approach is the definition of the probability of transition as a Markov process. This assumes that the probability of changing from one RE to another depends only on what RE you are currently in, but not on the long-term history. While this is still not completely exact, it seems like an approximation that is reasonable for most applications. The distance between the RE transitions (e.g. the distance between street crossings) could be given by its probability density function. However, this approach requires a lot of input from city planners, so that it was decided not to use it for the COST 259 model.

3.2.2. Outdoor measurement results

Henrik Asplund, Klaus I. Pedersen and Preben E. Mogensen

The directional properties of the outdoor mobile radio channel have received a comparably large interest within COST 259. Advanced direction-sensitive channel sounders have been built, using different measurement techniques and set-ups. Channel sounders switched over several antenna elements have been used by several institutions [HeKa97], [Kall98], [THST98], [Lars98], [Kall99], [HRSS99], [KaLa99], [PLNR99], [NLAL99], to obtain pseudo simultaneous measurements of parallel channels. Measurements based on synthetic aperture techniques are addressed in [Paju98] and [KBMH98]. In [LHTB99], [LKTH00] the two techniques are combined to further increase the number of effective elements. True parallel channel sounding on multiple antenna elements was used in [PeMF98], [PMFF98], [MPFF99].

The measurements have been in three areas, macro- and micro-cellular directional measurements at the BS, and directional measurements at the

MS. In this section some of the results from these measurement campaigns are presented.

3.2.2.1. Power azimuth spectrum in typical urban macro-cells

Two examples of the estimated APS from measurements in Aarhus, Denmark, and Stockholm, Sweden, are illustrated in Figure 3.2.3 [PeMF00]. The azimuth $0°$ corresponds to the azimuth towards the MS. In both cases the incident power is highly concentrated around $0°$, even though the measurements are obtained in a NLoS situation. This indicates that a significant fraction of the received power propagates from the MS to the BS via rooftop diffractions. It is further observed that the Laplacian function matches the estimated APS for signal levels higher than -12 dB [PeMF97]. Due to the limited dynamic range of the measurement system, it is not possible to determine whether the APS continues to decrease for large azimuths in accordance with the tails of the Laplacian function. Additional APS estimates from measurements made in Stockholm and Aarhus also support the thesis that the APS is accurately modelled by a Laplacian function.

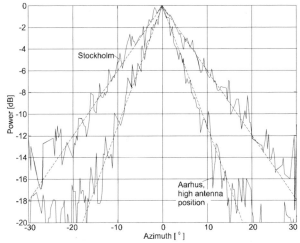

Figure 3.2.3:   Two examples of measured power azimuth spectra obtained in Aarhus, Denmark, and Stockholm, Sweden.

However, notice that the results in Figure 3.2.3 represent typical cases of measured APS from typical urban environments. Rare cases due occur where the APS looks differently [PeMF97]. Moreover, in bad urban environments the APS is frequently found to consist of two Laplacian functions, i.e. a two-cluster model is more appropriate for this type of environment [PeMF00].

Azimuth spread versus BS antenna height

The empirical CDF of the estimated azimuth spread (AS) is reported in Figure 3.2.4 for different antenna heights in typical urban environments [PeMF98], [PeMF00]. Here, the AS is defined as the root second moment of the APS. It can be observed that the AS changes significantly, but the shape of the APS still follows a Laplacian function. It is observed from the measurements in Aarhus that the AS increases significantly when the antenna height is reduced. The 50 % quartile of the AS equals 5° and 10° for the high and low antenna position, respectively. Similarly, the 90 % quartile of the AS equals 14° and 23°, respectively. The average received signal power drops at the same time by 8 dB. This loss is in part caused by an increased degree of obstruction between the MS and the BS, which also explains the larger AS. The higher AS observed when the antenna height is low results in a larger decorrelation between the signals at the output of the antenna array compared to the high antenna position.

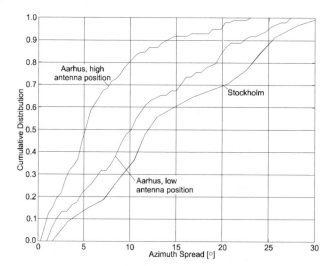

Figure 3.2.4: Empirical CDF of estimated AS in TU environments. High and low antenna height correspond to rooftop-level of the surrounding buildings and 12 m above rooftop.

The results of measured AS presented within COST 259 have been compiled in Table 3.2.1 in terms of medians and 10 % and 90 % percentiles. Values vary over a considerable range, however this may partly be a result of the variety of antennas and measurement techniques being used. The lowest AS is found in rural environments, while in suburban and urban cells the median AS is mostly in the range 5° to 20°.

Table 3.2.1: Azimuth spread in degrees.

Environment	Reference	10 %	Median	90 %
Urban small cell	[Paju98]	7	17	35
Urban micro-cell LoS	[Paju98]	3	7.5	15
Urban micro-cell NLoS	[Paju98]	7	19	40
Rural	[Paju98]	1	2.5	5
Urban low BS	[PeMF98]	2	10	22
Urban med BS	[PeMF98]	2	7.5	20
Urban high BS	[PeMF98]	1	5	14
Dense urban	[Lars98]	4	9	17
Suburban	[Lars98]	1	7.5	14
Suburban	[PLNR99]	6	16	27
Irregular urban, low BS	[PLNR99]	5	15	24
Irregular urban, high BS	[PLNR99]	8	15	20
Regular urban, low BS	[PLNR99]	21	28	40
Regular urban, high BS	[PLNR99]	6	16	25
Suburban micro-cell	[LSJB99]	–	18.3[1]	–
Univ. campus, micro-cell	[LSJB99]	–	8.7[1]	–
Crossroads, micro-cell	[LSJB99]	–	12.9[1]	–
Urban 4 m BS[2]	[LiKo99]	25	70	140
Urban 8 m BS[2]	[LiKo99]	10	60	130
Urban 11 m BS[2]	[LiKo99]	5	30	100
Urban 13m BS[2]	[LiKo99]	5	20	110
Urban	[NLAL99]	4	8	17
Suburban	[NLAL99]	2	5	7

[1]Mean value.

[2]Azimuth spreads include system response (array pattern).

Azimuth spread as a function of distance

The AS is reported vs. the MS-BS separation for three different situations in
Figure 3.2.5 [PeMF98].

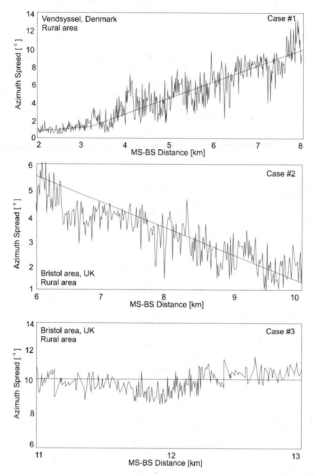

Figure 3.2.5:   AS vs. the distance between the MS and BS for three
                 different measurement scenarios in rural environments.

Notice that these results are obtained from measurements recorded in rural
environments. For case #1 the AS is increasing, in case #2 the AS is
decreasing, and in case #3 the AS is constant as a function of distance. For
all three cases the APS has been observed to be Laplacian shaped, although
it is not illustrated here. In order to explain this behaviour, let us compare
the results to that of the wellknown single scatter model, where it is assumed
that the received signal at the BS originates from scatterers within a circle of

radius r centred around the MS, R meters from the BS. The beam width is thus decreasing with R assuming that r is constant, and so is the AS. Consequently, the AS will also decrease as the BS-MS separation is increased. More complex geometrical models also predict a decreasing behaviour of the AS as a function of distance. However, this model only matches the trend observed for case #2, while cases #1 and #3 contradict the simple theoretical model.

However, models that predict the behaviour observed for cases #1 and #3 do exist [Ande98e]. For case #1, the MS is moving towards a small city, which acts as a cluster of scatterers. As the MS closes in on the city, the result of back scattering from the city and back to the BS becomes more dominant and consequently the AS increases. A model that reproduces this behaviour was presented in [Ande98e]. For case #3, the direct path is obstructed by a small village, which results in the relatively high AS of $10°$. This behaviour is also predicted by the model proposed in [Ande98e]. For urban environments, no correlation between AS and the MS-BS separation has been found.

Azimuth spread in two orthogonal polarisations

Pairs of estimated ASs from Aarhus are reported in Figure 3.2.6 [PeMF99b] for polarisations of $±45°$. It is observed that these pairs are symmetrically distributed around the diagonal, and that the variability of these points increases with the magnitude of the AS. The same conclusion can be drawn from the results presented in [NLAL99].

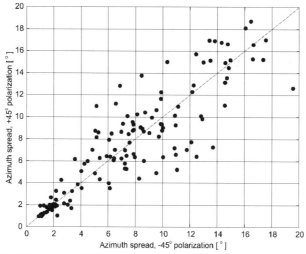

Figure 3.2.6: Measured AS for two linear polarisations of $±45°$. These results are obtained in a typical urban environment in NLoS.

Azimuth spread in the uplink and downlink band

Figure 3.2.7 shows a scatter plot with estimated pairs of ASs for the up- and downlink channel [PeMF99a]. Here the duplex distance equals 68 MHz. These results are obtained from measurements in typical urban environments. For comparison, the linear regression line and the expected one-to-one mapping are also plotted. It is observed that the solid and dashed lines are close together, which indicates that the AS in the up- and downlinks can be assumed to be identical, although there are some data points which deviate from this behaviour. The correlation coefficient is computed to 0.83, which supports the thesis that the AS is identical for the up- and downlink channels. From an array processing point of view, these results indicate that estimates of the APS obtained in the uplink can be utilised to create an appropriate beam pattern in the downlink channel.

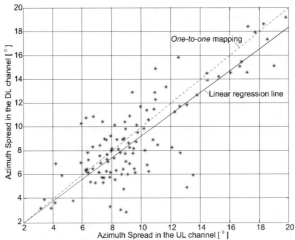

Figure 3.2.7: Scatter plot with the estimated AS for the up- and downlink channels.

Correlation between azimuth spread and delay spread

A scatter plot of the estimated DS and ASs is shown in Figure 3.2.8 for $N = 187$ measurement locations along routes in both Stockholm and Aarhus [PeMF00]. The correlation coefficient between the estimated DSs and ASs is computed to be 0.72. The linear regression line is found to be DS = (0.058 AS + 0.12) [μs], where AS is expressed in degrees. The root-mean-square residual error is computed to be 0.18 μs, which indicates that the linear regression yields a reasonable approximation. These results furthermore indicate that the potential space and frequency diversity gains are highly correlated.

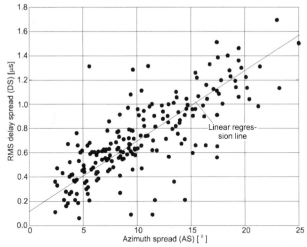

Figure 3.2.8: Scatter plot of estimated AS and DS obtained along measurement routes in Stockholm and Aarhus.

3.2.2.2. Power azimuth spectrum in micro-cells

In LoS situations the direct path carries a major part of the signal energy, resulting in very small AS [Paju98]. A different situation arises in NLoS situations, where the dominating propagation mechanism is not the same as in macro-cells. Due to the low height of the BS in micro-cells, waves propagating over rooftops are strongly attenuated compared to propagation around street corners. The result is that the majority of energy does not come from the general direction of the mobile, but rather from the street orientation at the BS. A typical example of this is shown in Figures 3.2.9 and 3.2.10 [LHTB99], [LKTH00]. Here we see how the APS contains three distinct clusters, each corresponding very well in azimuth to the direction of a street. The absence of components with elevation angles higher than approximately 0 degrees is evidence of the stronger attenuation experienced by waves propagating over rooftops.

Nevertheless, high elevation components do occur even in micro-cells, but they are typically associated with reflections or scattering from prominent high-rise buildings [LHTB99], [LKTH00].

Since several propagation paths, all with roughly similar power, may be present the AS tends to be higher than in LoS. In [Paju98] the AS in NLoS urban micro-cells is more than twice as large as in LoS, Table 3.2.1. [LSJB99] also reports AS values in the same range.

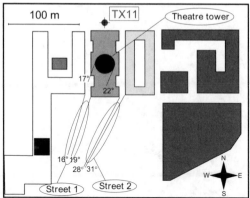

Figure 3.2.9: Map of urban micro-cellular environment in Helsinki, corresponding to the measured power-azimuth-elevation spectrum in Figure 3.2.10. Azimuth directions to the street ends are indicated to facilitate comparison with the measured power-azimuth-elevation spectrum (© 2000 ESA).

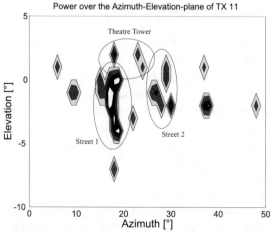

Figure 3.2.10: Power-azimuth-elevation spectrum measured in a micro-cell in Helsinki (© 2000 ESA).

3.2.2.3. Joint power azimuth-delay spectrum

The joint ADPS, $P(\phi,\tau)$ has been measured for both macro- and micro-cell configurations. For typical urban macro-cells with one cluster it is found that $P(\phi,\tau)$ can be decomposed, so $P(\phi,\tau) = P(\phi) P(\tau)$, where $P(\phi)$ is the APS and $P(\tau)$ is the power delay spectrum [PeMF00]. However, for macro-cell bad urban environments (several clusters), this simple decomposition is typically not valid.

The ADPS in micro-cells has been studied in [LHTB99], [LKTH00]. Figure 3.2.11 shows the ADPS corresponding to the situation in Figure 3.2.9. Again we notice the strong correlation to the geometry of the micro-cell. It is also evident that the APS will be different at different delays. This leads us to conclude that the relation $P(\phi,\tau) = P(\phi) \, P(\tau)$ is in general *not* true, however for each of the clusters the relation may still hold approximately. This is a quite important result for channel modelling.

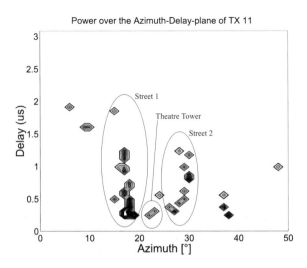

Figure 3.2.11: ADPS measured in a micro-cell in Helsinki (© 2000 ESA).

3.2.2.4. Directional results at the mobile station

Contrary to the BS, the mobile is usually located in a strong scattering environment. The directional spectra at the mobile therefore contain components from many directions, leading to rapid variations in the signal as the mobile moves just a short distance. Using an assumption of homogenous distribution of uncorrelated waves in azimuth yields the well known classical Doppler spectrum for the signal variations. In Figures 3.2.12 and 3.2.13 APS measured for indoor and outdoor mobiles in an urban micro-cell are shown [KaLa99]. Both spectra are non-homogenous, although in about half of the spectra the power level exceeds –10 dB. The two major directions in the indoor spectra correspond to the orientation of the corridor in which the mobile was located.

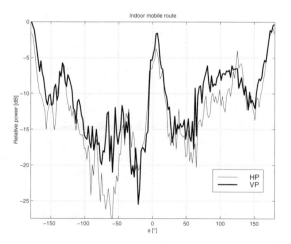

Figure 3.2.12: Measured average APS at MS for an indoor route in an urban micro-cell.

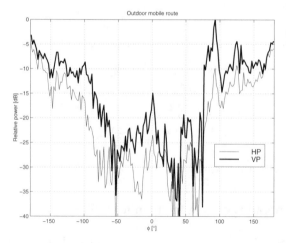

Figure 3.2.13: Measured average APS at MS for an outdoor route in an urban micro-cell.

Elevation spectra at the mobile are much more confined, since all but a few of the scatterers are located at angles close to the horizontal plane. Figures 3.2.14 and 3.2.15 exemplify this with results from [KaLa99]. Almost all signal power arrives in the interval [+20°, -20°]. The results provide some justification for Taga's [Taga90] assumption of Gaussian distribution of power in the elevation angle.

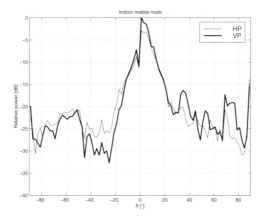

Figure 3.2.14: Measured average power-elevation spectra at MS for an indoor route in an urban micro-cell (© 2000 ESA).

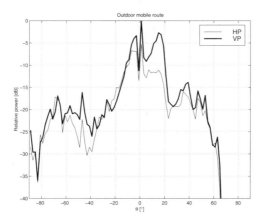

Figure 3.2.15: Measured average power-elevation spectra at MS for an outdoor route in an urban micro-cell (© 2000 ESA).

3.2.3. Indoor measurements results

Thomas Zwick and Peter Karlsson

In this sub-section we present results from indoor measurements in the frequency range from 900 MHz to 24 GHz. The results can be used for a large set of wireless and cellular systems, such as DECT, IMT-2000, WLANs and future mobile broadband systems. It is obvious that these systems will more or less be operating in indoor environments, where usage and demand for flexible broadband access are growing.

There has been a clear focus on measurements in office environments and in the frequency bands above UHF. A considerable amount of results are from the 5 GHz frequency band, where the emerging HIPERLAN/1 and HIPERLAN/2 systems, standardised by ETSI, have designated allocations. The global interest in 5 GHz propagation results is further emphasised by the allocation of spectrum for MMAC in Japan and IEEE 802.11a in the USA.

The results are mainly given as path loss values for coverage and interference analysis and as wideband characteristics for modem evaluation and optimisation. As a consequence, the data is useful both for system-level and link-level evaluations and simulations. Some general propagation characteristics for indoor environments have been identified and verified in the measurement campaigns reported below.

3.2.3.1. Path loss results

Multifrequency measurements at 2, 5, 17 and 60 GHz have been taken on several floors of an office building [Nobl99]. The power decay with LoS inside a room is less than free space. The decay exponent n ranges from 1.3 to 1.7 for all frequencies with TX-RX separations ranging from 1 to 10 m. A similar propagation effect, due to reflections from walls, floor and ceiling, was also found in a LoS corridor up to 30 m separation at 5.3 GHz [KiVa97]. It is interesting to note that this waveguide effect is common in other studies, e.g. [KaEM92], and is found irrespectively of carrier frequency.

The path loss in OLoS and NLoS environments is of course strongly dependent on furniture, the wall material and the building geometry. For light wall structures like plasterboards the average wall attenuation for each wall, i.e. in excess of the free space path loss L_{FS}, is in the range from 2.5 to 4 dB [Nobl99], [KiZV99]. The results are in agreement with the linear behaviour of the total wall attenuation as a function of the number of penetrated walls according to the Motley-Keenan model [KeMo90].

In [Zwic00], path loss measurements have been performed in a multi-floored office building. Figure 3.2.16 shows the measured excess path loss of 6 TX-RX-configurations at 900 MHz, 1.8, 2.5 and 5.2 GHz. The excess path loss (EPL) L_{Ex} is given by:

$$L_{Ex} = L - L_{FS} = L - 20 \log_{10}(4\pi d/\lambda) \text{ [dB] ,} \tag{3.2.22}$$

where L is the total loss in dB and d is the distance between TX and RX. The free space loss and the wavelength are denoted by L_{FS} and λ respectively. It can be seen that in many cases, the excess loss is between 20 and 30 dB, while the dependency on the frequency is rather low.

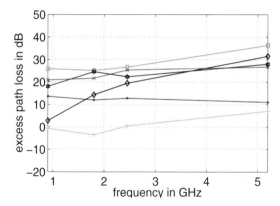

Figure 3.2.16: Excess path loss in dB of 6 different TX-RX-configurations measured at 900 MHz, 1.8, 2.5 and 5.2 GHz.

Other path loss measurement campaigns are reported in [WoLa99], [LTDV99].

3.2.3.2. Wideband results

In the following, directional wideband measurement results at a centre frequency of 24 GHz are shown in two different scenarios. The measurements are described in detail in [HTTN99], [HeTr99], [HeTr00] (see also Section 3.1.2). At both, TX and RX wideband λ/4 monopole antennas are used. The TX and RX antennas have been mounted at 3 and 2.6 m above floor level, respectively. Spatial information at the RX is obtained by moving the RX antenna along a virtual 3-D cross array. The path properties (loss, delay, azimuth and elevation) can then be extracted by the use of the SAGE algorithm [FeHe94].

In Figure 3.2.17 the map of scenario A is shown. The TX and RX positions are given, as well as the orientation of the virtual array at the RX. As can be seen in Figure 3.2.18 in scenario A, most of the power propagates through the doorway. These waves are less attenuated than the direct component. Waves at larger delays are reflected at a wall of the room or at objects in the corridor, but also propagating through the doorway. A review of all measured scenarios reveals that the direct component is often not visible in such OLoS cases [HeTr00].

Figures 3.2.19 and 3.2.20 show a map of scenario B and the corresponding results respectively. Most of the energy propagates through the wall with an angular spreading. Reflections from other walls can also be seen clearly.

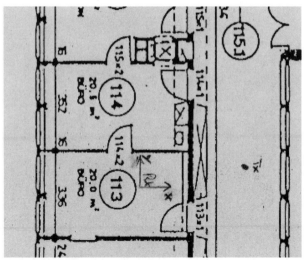

Figure 3.2.17: Scenario A.

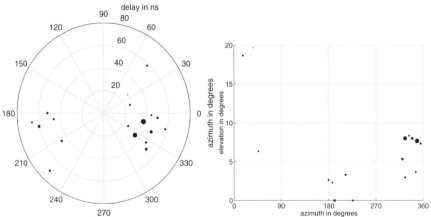

Figure 3.2.18: Measured delay-azimuth and azimuth-elevation profile in scenario A. The thickness of the dots indicates the field strength of the path.

In the building shown in Figure 3.2.21, directional wideband measurements have been performed at 5.8 GHz [BeKa98], [KaPa99], [BeKB99], [KBTB99] (see also Section 3.1.2). At TX and RX omnidirectional monopole antennas are mounted. Using a virtual array at RX together with the MUSIC algorithm the angular information is extracted. In Figures 3.2.22 and 3.2.23 the measured PDP and power-azimuth-profile at the RX are given for seven neighbouring TX locations (see Figure 3.2.21). A slight changing of the propagation channel can be observed during the movement of the TX. Some propagation paths appear and/or disappear. This effect has to be modelled by a time-variant stochastic channel model.

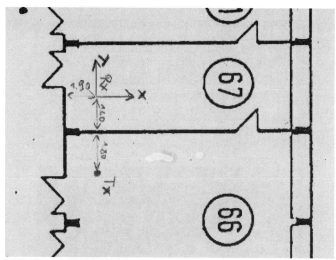

Figure 3.2.19: Scenario B.

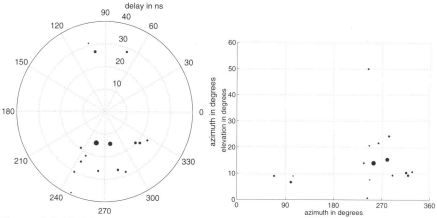

Figure 3.2.20: Measured delay-azimuth and azimuth-elevation profile in scenario B. The thickness of the dots indicates the field strength of the path.

The directional indoor measurements show that the indoor scenario behaves differently from urban environments, because of their closed structure. Multipath propagation including multiple reflected or scattered paths play an important role. Therefore, waves are impinging at any RX from all directions.

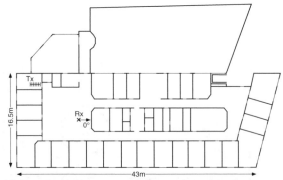

Figure 3.2.21: Office building.

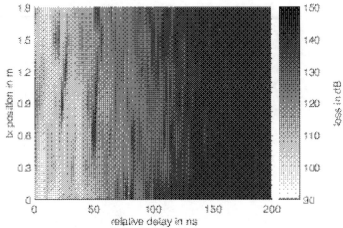

Figure 3.2.22: Measured PDP.

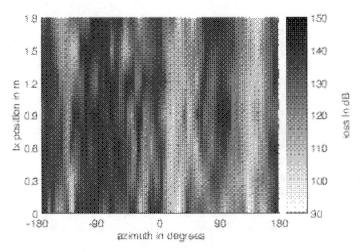

Figure 3.2.23: Measured power-azimuth-profile.

In [KaPa99], [BeKB99] spatial wideband measurement results are presented (see also Section 3.1.2). Figure 3.2.24 shows the distribution of the RMS DS at 5.8 GHz measured in four different scenario types. In Figure 3.2.25 the distribution of the RMS DS is given for the OLoS scenario at the two frequencies 1.8 and 5.8 GHz. The average DS is dependent on the environment rather than on the carrier frequency. Power variations are rapid due to the fading of the dominant component and reflections from walls, floor and ceiling. The instantaneous DS corresponds to power variations, i.e. increases when the power of the dominant component is weak relative to the multiple reflections. Comparable results about the statistics of the RMS DS in different indoor environments can be found in [KiZV99], [KiVa99].

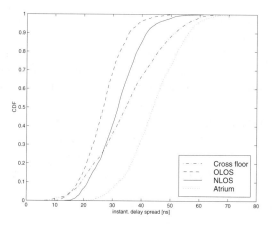

Figure 3.2.24: Example of RMS DS distributions at 5.8GHz in four different scenarios in an office building.

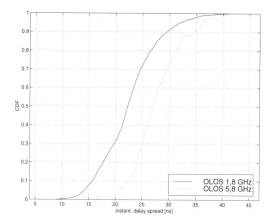

Figure 3.2.25: Example of RMS DS distributions at 1.8 and 5.8GHz in an office building.

The basic time-variant propagation aspects due to human movement in the indoor radio channel were investigated in [LaFr99b]. The dominant signal of the IRs changed significantly when the direct path between TX and RX was obstructed by a person.

For WLAN applications at 5 GHz, the ETSI BRAN group proposed a simple model to describe the PDP that is based on an exponential power decay [MASA98]. In [EnKF99], wideband channel sounding measurements have been performed to investigate the validity of the model. Thus, the assumption of exponential power decay has been shown to provide a good description in general.

In [Sibi00] wideband channel sounding measurements using antennas with sectored radiation patterns and different polarisations are described. It could be observed that the selection between the antenna sectors can lead to an essential reduction of DS. This coincides with the wide angular spread of the propagating waves in indoor environments mentioned above. In most of the cases, the sector of minimum DS is not identical with that of maximum power. In [KiVa98a], [KiVa98b], [KiZV99] and [KiZV99], the ChIR has been measured with a rotating horn antenna (30° half power beam width). It could be seen that in the direction of maximum power a small DS occurs.

3.2.4. Parameter settings

Martin Steinbauer

3.2.4.1. Radio environment characterisation

As explained in Section 3.2.1, we have identified a set of radio environments for each cell type. In order to unambiguously define large-scale parameters that are realistic for European morphology, a bottom-up approach has been chosen. The bottom layer represents the morphology of the actual environment where BS and MS reside. This can be an ensemble of scatterers as in macro-cells or a more regular structure, as is appropriate for micro- and pico-cells. For micro- and pico-cells, the following two-dimensional *Virtual Cell Deployment Areas* (VCDAs) have been defined, Figures 3.2.26–28: the BS position for the respective LoS REs is marked by a big dot. For the NloS, REs the BS of the corresponding LoS RE is the active one, except for the indoor case (GSN: use BS of GSL, GCN: use BS of GCL, GON: use BS of GCL).

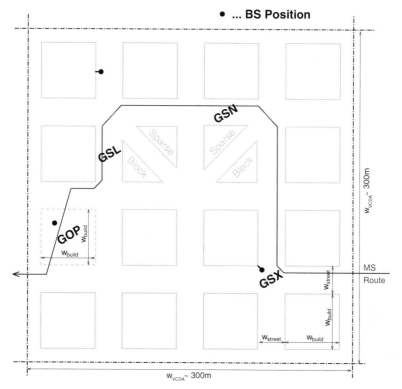

Figure 3.2.26: VCDA for micro-cells.

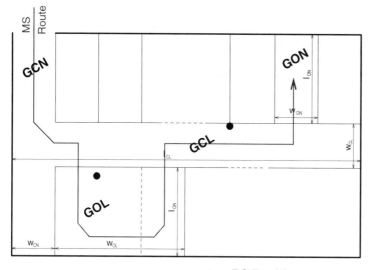

Figure 3.2.27: VCDA for indoor environments (1).

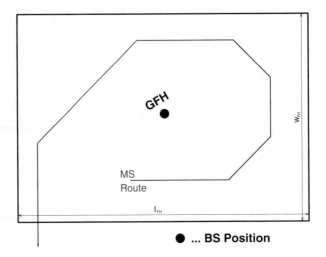

Figure 3.2.28: VCDA for indoor environments (2).

By defining such a virtual map or floor plan, the transitions between radio environments that are induced by large-scale movements of the MS (transition lengths) can be derived depending on a chosen mobile route through this environment. The VCDAs are further defined through the parameters in Tables 3.2.2–3.2.4.

The parameters in Table 3.2.2 for macro-cells are chosen to correspond to the validity range of the proposed path loss models. The 45° street orientation has been chosen to reflect the worst-case situation.

The counts given in Table 3.2.3 for micro-cells are absolute numbers related to about 1 km² cell deployment area. These parameters should be evaluated only over those urban regions that are likely to contain micro-cells. An open place or square has the size of a building block; it is intended as BS position. The chosen building width corresponds to those for the GBU macro-cell RE. The rooftop height corresponds to the GTU macro-cellular RE.

Small office rooms as position of the MS are expected to be covered indirectly by a BS mounted on the corridor. Therefore, this situation suffers from NLoS propagation →GON. Small offices are separated by thin walls (LW), however, the corridor is shielded by a thick wall (HW). A BS on a corridor might supply several small office rooms due to waveguiding. The corridor length of 30 m allows exactly 10 small offices at each side of the corridor to be covered. If the MS is moving, the GON RE is followed or preceded by the GCL RE. The NLoS-corridor as MS position (GCN) is included for completeness. It is separated from the LoS by a 90° edge giving rise to many diffracted components.

Table 3.2.2: External parameters for cell type MACRO-OUTDOOR.

Parameter	Symbol	Radio Environment				Valid	Source/
		GTU	GBU	GRA	GHT	range	Comment
BS height [m]	h_{BS}	30	50	50	50	GRA, GHT: 30 – 200 GTU, GBU: 4 – 50	COST231 WI/Hata
MS height [m]	h_{MS}	1.5	1.5	1.5	1.5	GRA, GHT: 1 – 10 GTU, GBU: 1 – 3	COST231 WI/Hata
Roof top height [m]	h_{roof}	15	30	NA	NA	8 – 60 (2 floors+roof – 20 floors)	COST231 WI
Building separation [m]	w_{sep}	30	50	NA	NA	20 – 50	COST231 WI
Cut-off distance [m]	d_{co}	500	500	5 000	5 000		[AsBe99b]
Visib. area radius for LoS occurrence [m]	$r_{c,\,LoS}$	30	30	100	100		[AsBe99a]
Visib. area radius for cluster shadowing [m]	$r_{c,,\,SF}$	100	100	300	300		[AsBe99a]
Street orientation [°]	Φ	45	45	NA	NA		COST231 WI
Typical BS-to-MS distance [m]	d_{MS}	500	500	5 000	5 000		COST231 WI/Hata
Carrier frequency [MHz]	f_c	900 / 1 800					COST231 WI/Hata

Table 3.2.3: External parameters specifying the virtual micro-cellular deployment area (MICRO-VCDA).

Parameter	Symbol	Value
Number of empty lanes	N_{el}	0
Number of open places	N_{op}	1
Number of 'sparse' blocks	N_{sbl}	2
Cell size	w_{cell}	$4 \times w \sim 300$ m
Building width	w_{build}	50 m ($\leftrightarrow w_{sep}$ in GBU)
Street width	w_{street}	20 m
Street grid	w	70 m ($=w_{build} + w_{street}$)
Square dimension		$(w_{build} + 2\, w_{street})^2$
Empty lane		$w_{build} + 2\, w_{street}$
Rooftop height		15 m (urban macro-cells GTU)
Angle of street corner		$45°, 90°$ (typical: $90°$)
BS position		Figure 3.2.26
BS height		3 – 10 (typical: 5 m)
MS height		1.5 m
Typical BS – MS distance		1.5 – 500 m
Carrier frequency		1 / 2 / 5 GHz (typical: 2 GHz)

A conference room is equivalent to a large office. Both names can be used alternatively for GOL. For a conference room the BS antenna placement in the centre of the room below the ceiling is a promising alternative. A large office might be divided into two smaller offices with a light wall (dashed in the figure). This, however, is expected not to change the delay-angular characteristics dramatically [HeTr99]. The deterministic LoS component might however be attenuated causing the Rice-factor of the first cluster to be lower in practice. All offices and corridors have the same height h_0.

The factory or hall (GFH) RE is different from above in height and the propagation conditions. In GFH the LoS path might be often obstructed by the MS moving around/behind machines. The dynamics in this RE depend on the minimum distance of the MS to obstructions while moving. Changes of the propagation scenario meaning updates of the scatterer ensemble seem to become frequent in this RE as opposed to the other indoor-REs.

Table 3.2.4: External parameters specifying the virtual pico-cellular deployment area (PICO-VCDA).

Parameter	Symbol	Value
Small office room length	l_{ON}	6 m
Small office room length	w_{ON}	3 m
Small office wall type		LW
Corridor width	w_{CL}	3 m
Corridor length	l_{CL}	30 m
Corridor NLoS length	l_{CN}	15 m
Corridor NLoS width	w_{CN}	3 m
Corridor wall type		HW
Conference room length	l_{OL}	6 m
Conference room width	w_{OL}	9 m
Conference room wall type		HW
Office height	h_O	3 m
Factory / hall length	l_{FH}	90 m
Factory / hall width	w_{FH}	30 m
Factory / hall height	h_{FH}	10 m
Angle of corners		90°
Thick (heavy) wall	HW	Concrete
Thin (light) wall	LW	Plasterboard
BS position on corridor		on the wall
BS position on conference room		under the ceiling, centred
BS height in offices		$h_{BS} = h_O - 30$ cm
BS height in factory / hall		$h_{BS} = h_{FH} - 2$ m
MS height (typical)		1.5 m
MS height (nomadic mobility)		1 m
Carrier frequency		2 / 5 / 24 GHz

Many dimensions in Table 3.2.4 are interrelated. For example, the length of the NLoS corridor follows from the length of the small and large office and the LoS-corridor width. The basic distance grid is 3 m, which seems to be a reasonable simplification even in a vertical direction. For the GFH cell-type, the effective cell-size depends on the actual BS placement.

The typical MS position is that of a human using a telephone, i.e. 1.5 m. For nomadic mobility the MS is not a phone but a notebook with antenna(s). Communication in this situation takes place while the MS is fixed, and located on a table, i.e. at 1 m in height.

A summary of the values of the global parameters for the various scenarios is presented in Tables 3.2.5, 3.2.6 and 3.2.7.

3.2.4.2. Path loss

In macro-cells we have to distinguish different stages in the model/simulation. At first, the average path loss for the actual BS-MS distance has to be computed from the path loss models given here (taken from the open literature).

Macro-cells: For the macro-cell radio environments, narrowband path loss models are given in [COST99]. For the desired 2 GHz frequency band and keeping in mind that the BS antenna is always above the average rooftop height in macro-cells, we have chosen the *COST231-Hata Model* for GRA, GHT and the *COST231-Walfisch-Ikegami Model* for GTU and GBU. Both methods give the mean path loss (large-scale averaged) and are described in [COST99, Chapter4].

The ratio of the path loss of additional clusters to the path loss of the first cluster is given by [MSAB00]

$$L_i = L_1 + L_{add} \, [\text{dB}]$$
$$L_{add} = U(0,20) + (\tau_i - \tau_0)/\mu s \, [\text{dB}].$$

(3.2.23)

where $U(a, b)$ signifies a uniform distribution between a and b.

Micro-cells: For micro-cells, the overall path loss is given by Feuerstein *et al.* [FBRS94] as a one- or two-slope log-distance law, depending on whether LoS exists or not.

Table 3.2.5: Global parameters for cell type MACRO-OUTDOOR.

Parameter	Symbol	Radio environment				Source/
		GTU	GBU	GRA	GHT	Comment
Minimum number of clusters	$N_{cl,\,min}$	1	1	1	1	(3.2.28)
Mean number of additional clusters	m	0.17	1.18	0.06	1	(3.2.28)
Probability of LoS	p_{LoS}	(3.2.29)				[AsBe99b]
Path loss (NB)	L_1	(4.4.6) [COST99]		(4.4.3) [COST99]		[COST99]
Additional cluster path loss	L_{add}	(3.2.35)				[Aspl00b]
Cluster power / shadow fading distribution	P_{cl}	$N\,(0,\,\sigma_{sf})$ [dB]				
Cluster shadow fading standard deviation	σ_{sf}	9 dB	9 dB	6 dB	6 dB	
Cluster shadow fading decorrelation distance	d_{sf}	5.5 – 11 m	5.5 – 11 m	500 m	500 m	[Sore98], [MET00]
Rice factor distribution (NB)	K_0	$N\,(\mu_K,\,\sigma_K)$ [dB]				(3.2.30) [Aspl00b]
Rice factor mean (NB)	μ_K	(26 – EPL) / 6 [dB]				(3.2.30) [Aspl00b]
Rice factor spread (NB)	σ_K	6 dB				(3.2.30) [Aspl00b]
Cluster position	$\vec{x}_{ci,\,i}$	$U\,(a,\,b)$				
Maximum cluster distance	d_{max}	3 km	3 km	5 km	5km	
Cluster spread mean delay	$M_{S\,\tau,\,i}$	0.4 µs	0.4 µs	0.1 µs	0.1 µs	(3.2.35)
Cluster spread delay spread	$S_{S\,\tau,\,i}$	2 µs				(3.2.35)
Cluster spread mean azimuth	$M_{S\,\varphi,\,i}$	10°	10°	15°	5°	(3.2.35)
Cluster spread azimuth spread	$S_{S\,\varphi,\,i}$	2°				(3.2.35)
Elevation	$S_{\delta,\,i}$	0°				
DoA spectrum at MS	$\varphi_{MS,\,1}$	$U\,(-\pi,\,\pi) + \delta(<(-\vec{x}_{MS}))$				
DoA spectrum at MS	$\varphi_{MS,\,I>1}$	$<(\vec{x}_{cl,i} - \vec{x}_{MS}) +$ $L(-S_{\varphi,max},\,S_{\varphi,max})$				
XPD mean	M_{XPD}	6 dB	6 dB	12 dB	12 dB	
XPD spread	S_{XPD}	6 dB	6 dB	3 dB	3 dB	

Table 3.2.6: Global parameters for cell type MICRO-OUTDOOR.

| Parameter | Symbol | Radio environment | | | | Source/ |
		GSL	GSX	GSN	GOP	Comment
Minimum number of clusters	$N_{cl,\,min}$	2	4	2	1	
Mean number of additional clusters	m	0	0	1	4	
Probability of LoS	p_{LoS}	1	1	0	1	
Path loss (NB)	L_1	(3.2.24)				[FBRS94], [Rapp96]
Propagation coefficients	$n/n_1/n_2$	-/2.2/3.3	-/2.2/3.3	2.6/-/-	-/2.2/3.3	[FBRS94]
Additional cluster path loss	L_{add}	(3.2.25)				
Cluster power / shadow fading distribution	P_{cl}	$N\,(0,\,9)$ [dB]				
Cluster shadow fading decorrelation distance	d_{sf}	5 m	5 m	5 m	5 m	
Rice factor distribution (NB)	K_0	$N\,(\mu_K,\,\sigma_K)$ [dB]				[Diet00]
Rice factor mean (NB)	μ_K	(3.2.33)				[FDSS94]
Rice factor spread (NB)	σ_K	(3.2.33)				[FDSS94]
Fixed cluster positions: delay / azimuth / elevation	$\tau/\varphi/\upsilon$	from geometry of ray paths in VCDAs				
Inter-cluster delay distribution		$U\,(0,\,1\;\mu s)$				
Inter-cluster azimuth distribution		$U\,(-\pi,\,\pi)$				
Inter-cluster elevation distribution		$U\,(<(\vec{x}_{MS}),\,0)$				
Intra-cluster delay distribution		$\exp(-\tau/S_\tau)$				
Intra-cluster azimuth distribution		$L\,(S_\varphi)$				
Intra-cluster elevation distribution		1-sided exp.				
Intra-cluster delay spread	$S_{\tau,\,i}$	120 ns				[ArTP94], [FBRS94]
Intra-cluster azimuth spread	$S_{\varphi,\,i}$	$3-5^\circ$				[Paju98], [Diet00]
Intra-cluster elevation spread	$S_{\upsilon,\,i}$	$1-2^\circ$				[LHT+99]
DoA spectrum at MS		$\varphi \sim U\,(-\pi,\,\pi),\; \upsilon \sim U\,(0,\,-<(-\vec{x}_{MS}))$				
XPD mean	M_{XPD}	9 dB				[LHT+99], [Toel00]
XPD spread	S_{XPD}	3 dB				[LHT+99], [Toel00]

Table 3.2.7: Global parameters for cell type PICO-INDOOR.

Parameter	Symbol	Radio environment					Source/
		GOL	GON	GCL	GCN	GFH	Comment
Minimum number of clusters	$N_{cl,\,min}$	1	0	1	0	1	(fixed cluster
Mean number of additional clusters	m	16	16	16	16	TBD	
Probability of LoS	p_{LoS}	1	0	1	0	0.5 r_c=5m	[AsBe99b]
Path loss (NB)	n	2	3	1.5	2	2.2	[SaVa87], [Rapp96], [Rapp89]
Additional loss	L_{add}	compute from inter-cluster distrib.					
Shadow fading distrib.	P_{cl}	$N\,(0,\,7)$ [dB]					[Rapp96]
Cluster shadow fading decorrelation distance	d_{sf}	according to room dimensions					[ZFDW00]
Rice factor distrib. (NB)	K_0	$N\,(\mu_K,\,\sigma_K)$ [dB]					
Rice factor mean (NB)	μ_K	1.98≈2 / 3 dB	0	2 / 3 dB	0	by EPL	[HeTr99], [Aspl00a]
Rice factor spread (NB)	σ_K	1.98≈2 / 3 dB	0	2 / 3 dB	0	by EPL	[HeTr99], [Aspl00a]
Fixed cluster position		$\vec{x}_{MS}$ (in QLoS direction)					
Inter-cluster delay distribution		$\exp(-\tau/S_\tau)$					
Inter-cluster azim. distr.		see reference					[ZFDW00]
Inter-cluster elev. distr.		see reference					[ZFDW00]
Inter-cluster delay spread		40 ns	40 ns	120 ns	120 ns	360 ns	
Intra-cluster delay distribution		$\exp(-\tau/S_\tau)$					
Intra-cluster azim. distr.		$U(-\pi,\pi)$	$L(S_\varphi)$	$U(-\pi,\pi)$	$L(S_\varphi)$	$U(-\pi,\pi)$	[HeTr00]
Intra-cluster elev. distr.		1-sided exp.	$L(S_\varphi)$	1-sided exp.	$L(S_\varphi)$	$L(S_\varphi)$	
Intra-cluster delay spread		5 ns					[SaVa87]
Intra-cluster azim. spr.		–	20°	–	6°	–	[HeTr00]
Intra-cluster elev. spr.		6°					[HeTr00]
DoA spectrum at MS		same as for BS (relative to $<(-\vec{x}_{MS}))$					
XPD mean	M_{XPD}	6 dB					[KnPO00], [Olss99b]
XPD spread	S_{XPD}	6 dB					

$$L(d) = \begin{cases} 10\,n_1\log_{10}(d/\text{m}) + L_0(\text{1m}) & , \text{ for } \text{1m} < d < d_f \\ 10\,n_2\log_{10}(d/d_f) + 10\,n_1\log_{10}(d_f/\text{m}) + L_0(\text{1m}), & \text{ for } d > d_f \end{cases}, \text{ LOS}$$

$$L(d) = 10\,n\log_{10}(d/d_0) + L_0(\text{1m}), \quad \text{OLOS} \tag{3.2.24}$$

$$L_0(\text{1m}) = 20\log_{10}(4\pi(\text{1m})/\lambda)$$

n, n_1 and n_2 are empirical regression coefficients given in the table of global parameters for 2 GHz. The attenuation of the latter clusters is given by

$$L_i = L_1 + L_{add} \,[\text{dB}] \tag{3.2.25}$$

$$L_{add} = U(0,10) + 10(\tau_i - \tau_0)/\mu s\,[\text{dB}].$$

Pico-cells: In pico-cells, the general model [Rapp96]

$$L = L(\text{1m}) + 10\,n\log_{10}(d/\text{1m}) + L_{floor}\,[\text{dB}] \tag{3.2.26}$$

is proposed for the overall (narrowband) path loss (including all clusters). For L_{floor}, values ranging from 13 to 24 dB [Rapp96], [PéJi94] are reported.

The average path loss per cluster is taken from a one-sided exponential distribution, whose decay time constant is in the range 40–360 ns, according to the environment (see Table 3.2.7).

All further parameters are cluster-specific. Therefore, clusters have to be fixed in the next step.

3.2.4.3. Cluster generation

Before generating clusters, the number of clusters must be known. This is done differently for macro-cells on one hand, and micro- and pico-cells on the other hand.

Macro-cells: Clusters are to be distributed uniformly in space, but only within a maximum distance from the BS, which is set equal to the cell size. Clusters are generated independently (in future, a conditional cluster occurrence PDF could also be defined to avoid clusters appearing in groups), the minimum number of clusters is unity, corresponding to the cluster around the MS.

The position of the clusters in space can be mapped to the delay-azimuth plane. Assuming single-scattering for clusters, the cluster delay and azimuth are evaluated as $\left(\vec{x}_{BS} = \vec{O} \rightarrow d = |\vec{x}_{MS}| \right)$

$$\tau_i = \left(\left| \vec{x}_{BS} - \vec{x}_{cl,i} \right| + \left| \vec{x}_{cl,i} - \vec{x}_{MS} \right| \right) / c,$$

$$\varphi_{cl,i} = \arg\left(\vec{x}_{cl,i} \right) - \arg\left(\vec{x}_{MS} \right),$$
$$\vartheta_{cl,i} = 0.$$
(3.2.27)

The cluster around the MS moves as the MS moves; the other clusters stay fixed in space. For a motivation of this approach, see also Section 3.2.1.

For the appearance and disappearance of clusters, we use the concept of 'visibility regions' [AsBe99b]. This requires the generation of (circular) regions with a radius given in Table 3.2.5. If the MS is in such a region, it 'sees' the cluster; otherwise it does not. For each cluster, a separate set of visibility regions must be generated. The maximum number of clusters must be chosen in such a way that (averaged over all MS positions in the cell) the mean number of 'active' clusters equals the cluster number given in Table 3.2.5.

An alternative and simpler, but not completely equivalent, way is to choose the number of clusters from a Poisson distribution (see also micro-cells).

Micro-cells: In micro-cells, we have 1 (GOP), 2 (GSL, GSN), or 4 (GSX) clusters with fixed mean DoAs. The mean DoA is in the direction of the MS-BS connection for the LoS clusters, and along the street grid for the other clusters. Additional clusters with random mean DoAs are also possible, where the number of additional clusters is Poisson distributed. Taking into account a minimum number of clusters $N_{cl,min}$, the number of clusters then results as

$$N_{cl} = N_{cl,min} + P(m)$$
(3.2.28)

where m gives the parameter of the Poisson distribution $P(m)$, i.e. the mean number of additional clusters, whose parameters are given in Table 3.2.6. The number of additional clusters stays constant within a RE, but is assumed to be statistically independent from one RE to the next and directions of the waveguiding clusters do not change while we remain within one RE.

For determination of the delays of the clusters, we first have to determine which of the clusters results in the smallest runtime; this can be computed from the map of the virtual cell environment. All other clusters exhibit an excess delay that is uniformly distributed between 0 and 1 µs.

Pico-cells: For pico-cells, we use a slightly different modelling approach based on a birth/death process with inheriting properties. The probability for appearance and disappearance is taken from a Poisson process.

Implementation aspects, especially with respect to step width and the inclusion of moving scatterers, are given in [Zwic99].

3.2.4.4. Large-scale fading

Macro-cells: Knowing the cluster positions, cluster-specific realisations of the shadow-fading process can be drawn. The resulting value (in dB) is added to the path loss. For macro-cells, we use an exponential function for the Autocorrelation Function (ACF). The correlation length is given in Table 3.2.5.

The correlation of the large-scale fading of the different clusters is neglected. The variance of the shadowing is assumed to be identical for all clusters. This variance must be chosen in such a way that the shadowing variance of the narrowband signal attains the values given in Table 3.2.5. This normalisation can be achieved either by Monte-Carlo simulations, or by an appropriate analytical procedure [MSAB00].

Micro-cells: For the micro-cells, we essentially use the same model for the large-scale fading as for the macro-cell; the correlation length is the same as for the GTU and GBU environment.

Pico-cells: Again, the pico-cells use a somewhat different approach. Shadow fading is assumed to stem only from the appearance and disappearance of paths, without changes in the mean amplitude of each path. Since the number of pico-cell paths is much higher than the number of micro-cell clusters, this is a good assumption. The birth rate and the lifetime of each cluster are uniquely related to the variance and correlation length of the shadowing, as given in Tab. 3.2.7. For more details, see [Zwic99].

3.2.4.5. Small-scale fading

The small scale fading (to be observed in distance, time or frequency) is the result of incoherent superposition of MPCs at different delays and angles of incidence. This superposition depends on the system-specific bandwidth and antenna aperture. For flexibility, not the fading but the MPCs as the underlying source of fading are modelled. In what follows, the necessary parameters for a complete characterisation are derived.

Macro-cells: In case there is LoS or Quasi-LoS (QLoS) between the MS and the BS, the power of the deterministic QLoS-component is derived from the Ricean K-factor (Narrowband K-Factor). Otherwise, if there is NLoS, the K-factor is much lower. The probability (not the PDF!) for LoS is given in [AsBe99b]

$$P_{LOS}(d) = \begin{cases} \dfrac{h_{BS} - h_B}{h_{BS}} \cdot \dfrac{d_{co} - d}{d_{co}}, & d < d_{co} \text{ and } h_{BS} > h_B \\ 0 & d \geq d_{co} \text{ or } h_{BS} \leq h_B \end{cases} \qquad (3.2.29)$$

The Rice factor is modelled as log-normal distributed random variable [Aspl00a], whose mean depends on the excess path loss EPL,

$$K_0(EPL) = N\left(\frac{26 - EPL\,[\text{dB}]}{6}, 6 \right)[\text{dB}]. \qquad (3.2.30)$$

The excess path loss is the deviation (in dB) between the narrowband free space path loss and the narrowband path loss.

$$EPL(d) = L_1 - 20\log_{10}\left(\frac{4\pi d}{\lambda} \right) \ [\text{dB}]. \qquad (3.2.31)$$

Assuming an exponential distribution of delays in the cluster with a decay-constant of σ_τ, the fading for any bandwidth B can be predicted as [Aspl00a]

$$K(B) = \frac{K_0}{1 - \exp\left(-\dfrac{1}{B\sigma_\tau} \right)} \geq K_0. \qquad (3.2.32)$$

Micro-cells: For the (narrowband) Ricean K-factor in micro-cells, a simple distance-dependent model for LoS and distance independent model for NLoS can be derived, based on a measurement report in [FDSS94]

$$\begin{aligned} K(d) &= N(13 - 0.03\,d\,/\,\text{m}, 13 - 0.03\,d\,/\,\text{m})[\text{dB}]\,\text{for LOS} \\ K &= N(-3, +3)[\text{dB}]\,\text{for NLOS} \end{aligned} \qquad (3.2.33)$$

where $N(a, b)$ is a normal distribution with mean a and standard deviation b.

Pico-cells: For the LoS-office and corridor, the probability for LoS is unity, while for the NLoS office and corridor, it is zero. For the factory hall, the LoS component is treated statistically by means of the visibility region concept, with an average probability for visibility of 0.5, and a visibility region diameter of 5 m.

3.2.4.6. Number of MPCs

The number of resolved MPCs depends on the system bandwidth and antenna aperture. From 500 up to 3 GHz carrier frequency, up to 20 MPCs have been observed by Turin in macro-cells in the San Francisco Bay area

[TCJF72]. The delay resolution for this evaluation was 100 ns. In indoor environments a minimum number of 30 separate paths (over all clusters) have been identified in the delay-azimuth plane in [KiVa99].

For a simulation, it is often necessary that the number of MPCs per cluster or per delay bin is sufficiently large so that each cluster/bin still exhibits small-scale fading. For the reconstruction of this fading, a minimum of seven properly selected components has been shown to be sufficient [Pätz99]. For a whole cluster the number of components should be larger to still achieve a certain (scenario-specific) amount of fading in delay-azimuth bins resolved by a system. [MPFF99] shows that even in a delay-azimuth bin of 250 ns / $5°$ as low as 0 dB Ricean K-factor is possible at the BS.

3.2.4.7. Delay/azimuth/elevation dispersion

The clusters are characterised by their position, spread, power and shape. In all cell types, clusters have exponential decay in delay τ, Laplacian shape in azimuth φ and in elevation ϑ, so that the normalised angular power spectrum is

$$P_s\left(\tau,\varphi,\vartheta\right)=\frac{1}{1-\exp(-\sqrt{2}\pi/S_\varphi)}\frac{1}{1-\exp(-\sqrt{2}\pi/S_\vartheta)}\frac{1}{\sqrt{2}S_\tau S_\varphi S_\varphi}\cdot$$
$$\cdot\exp\left(-\tau/S_\tau\right)\cdot\exp\left(-\sqrt{2}\left|\varphi/S_\varphi\right|\right)\cdot\exp\left(-\sqrt{2}\left|\vartheta/S_\vartheta\right|\right),\quad(3.2.34)$$

except for pico-cells where the elevation power profile and PDF are one-sided exponential. For clusters with larger delay, a different definition of the shape of the decay in delay and azimuth is still under discussion. A definite version will be given in [MAHS00].

Macro-cells: In macro-cells, the DS, azimuth spread, and shadowing of each cluster are correlated. The model is based on [GEYC97], with some modifications [AsBe99a]. The cluster spreads are derived from

$$P_i[dB]=S_{sf}\cdot X-L_i\ \left(\text{from stochastic process}\left(SP\right)X\right)$$
$$S_{\varphi,i}[deg]=M_{S_{\varphi,i}}\cdot10^{\wedge}[S_{S_\varphi}\cdot Y/10]\left(\text{from } SPY\right)\qquad(3.2.35)$$
$$S_{\tau,i}[s]=M_{S_{\tau,i}}\cdot d^{e_g}\cdot10^{\wedge}[S_{S_\tau}\cdot Z/10]\left(\text{from } SPZ\right)$$

where e_g is a parameter of the model [GEYC97], here set equal to 0.5. X, Y, Z are normal random variables with zero mean, unit variance, and correlation coefficients

$$\rho_{XY} = -0.75$$
$$\rho_{YZ} = +0.5 \qquad\qquad (3.2.36)$$
$$\rho_{XZ} = -0.75$$

Values for the means and variances are given in Table 3.2.5. Note that d is $\tau_i c$ according to (3.2.27). A possible modification, especially for far clusters, is still under discussion [MAHS00].

Micro-cells: For micro-cells, the intra-cluster delay and azimuth profile is the same as for the macro-cell, but the elevation: in case of LoS: is a 1-sided exponential in elevation down from LoS-direction (truncated if needed). The spreads are all independent, the actual values are given in Table 3.2.6.

Pico-cells: The shape of the clusters is again exponential in delay, Palladian in azimuth, but is now single-exponential in elevation, where the reference axis is the connection line between BS and MS. The shape as seen from the MS is identical when taking the appropriate reference axis for the elevation. The values for the spreads are given in Table 3.2.7.

3.2.4.8. Polarisation

We studied only cases where the primary polarisation is vertical. Therefore, co-polarisation in this context means Vertical-Vertical, and cross polarisation Vertical-Horizontal. We assume horizontal and·vertical component of the received field to have uncorrelated small-scale fading because of different propagation paths. Moreover, the XPD,

$$XPD = P_{vv} / P_{vh} \qquad\qquad (3.2.37)$$

then varies rapidly on a wavelength scale. This implicitly requires drawing new realisations of this fading process on a small-scale basis. The proposed distribution of the XPD is

$$XPD \sim \mathrm{N}\left[M_{XPD}, S_{XPD}\right] \quad [\mathrm{dB}], \qquad\qquad (3.2.38)$$

where M_{XPD} is the mean, and S_{XPD} is the standard deviation of the normal distribution, both in dB. For a simulation, the XPD-value of each MPC in each cluster has to be chosen independently from this distribution, and afterwards it remains constant over the whole lifecycle of the MPC.

3.2.4.9. Spreading at the mobile sation

For the spreading at the MS, we can assume the following: (i) the delay spreading is identical to that at the BS; (ii) the azimuthal power spectrum is uniform over the whole angular range for the local cluster in macro-cells, and for all clusters in a micro-cell; (iii) the azimuthal power spectrum is Laplacian with a spread S_{max} (as given in Table 3.2.5) for the additional clusters in macro-cells; (iv) the elevation power spectrum is a delta function in macro-cells, uniformly distributed between 0 and the LoS direction in micro-cells, and one-sided exponential in pico-cells; (v) the azimuthal power spectrum is identical to that at the BS in pico-cells.

3.3. Smart Antennas

Martin Weckerle

3.3.1. Introduction

'Smart antennas' is of course an absurdity [Ande98c]. An antenna itself can never be smart or intelligent. Antennas are called smart if at least two or more signals to or from different antenna elements are combined or commonly processed in a more or less sophisticated manner in order to exploit the directional inhomogeneity of the mobile radio channel. The combination of antenna elements with a signal processing unit is finally called a smart or intelligent antenna, whereas the intelligence lies in the signal processing [Ande98c]. The terminology 'smart antenna' for a combination of antenna elements with a signal processing unit is based on the classical approach in which the signal processing consists of combining the signals received at the different antenna elements to one signal, or in distributing one transmit signal to the different antenna elements. Therefore, the antenna elements, together with their 'intelligence', can directly replace the conventional single antenna at the TXs or RXs. This implies the advantage that the smart antenna can be designed and evaluated independently of the TX or RX structure for which it is considered. In this classical approach, the spatial signal processing, which is performed by the signal processing unit of the smart antenna, and the temporal signal processing are performed separately, see Figure 3.3.1.a). In the literature [Ande98c], such antennas are called one-port antennas. In the case of a one-port antenna, e.g. a symbol sequence is transmitted sequentially, and the antenna weights are used to optimise some property like a specific shape of the radiation pattern, or an optimum combination of the uncorrelated signals at different antenna elements, but the information transmitted or received is the same for all elements. In Figure 3.3.1.b) a multi-port antenna is shown, where several symbols are treated at the same time, combined or delayed. Therefore, the information symbols fed to the individual elements are different. The combination could be a base band coding, where the information is spread over the antennas, essentially leading to a number of quasi-independent channels [Ande98c]. Also, if the channels are not independent, space-time signal processing in tems of the spatial and temporal correlation properties of the signals and interference can be performed, in order to optimise the multi-antenna RXs in the space and the time domains. For the multi-antenna TXs, a pre-distortion approach can be applied if channel information is available.

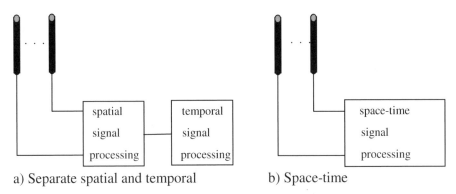

a) Separate spatial and temporal   b) Space-time

Figure 3.3.1: Two versions of signal processing (© 1999 IEEE).

Depending on the scenario, different smart antenna techniques follow different methods of exploiting the directional inhomogeneity of the radio channel:

- Diversity techniques:
  - Increasing the mean SNR
  - Reducing the standard deviation of the channel fading
- Adaptive array processing and beamforming:
  - Interference reduction
  - Reduction of time dispersion and time variance [PFBH97]
  - Optimisation of the antenna pattern in order to maximise the SNR

The choice of antenna type and the smart antenna concept depends on the scattering environment. The Rayleigh type fading of the mobile radio channel could be uncorrelated at different antenna elements if the antenna elements are spaced at large distances, or it could be correlated if the distances of the antenna elements are in the order of one half of the carrier wavelength. Therefore, the use of different smart antenna techniques like diversity techniques or beamforming depends strongly on the scenario and environment. In particular, the angular spreads play an important role. To be able to evaluate the different smart antenna concepts in terms of their performance gain dependent on the scenario, the following classification of different angular spreads at the TXs and RXs is considered [Ande98c]:

- *Small angular spread at both arrays*
  In the receive case, all the signals will be correlated, and the incoming signals apparently come from a point source. This is true for LoS propagation or scattering from a dominant scatterer. In a single user scenario there will be the benefit of full array gain, if the mean direction of incidence or departure, respectively, is known. Powerful techniques (ESPRIT, MUSIC) are available in this situation to determine the DoA in the receive case. The same direction can than be used for the transmit case. In the multi-user case, the radiation pattern of the array may be adjusted to maximise the Signal over Interference plus Noise Ratio (SINR), assuming that the angular distribution of the interferer power is known.

- *Narrow spread at base, large spread at mobile*
  This is a typical macro-cell situation with a fairly high BS antenna and the MS at the clutter. Signals at the antenna elements at the BS are correlated, while at the MS they are uncorrelated. Seen from the base, the situation is the same as above. In the uplink case, transmit diversity is possible. In the downlink case, depending on the number of elements, huge diversity gains may be achieved at the MS in the receiving case.

- *Large spreads at both arrays*
  This is the most advantageous situation from a signal processing and coding point of view, since it gives us many new dimensions to explore. In practice, it will occur in micro-cell and indoor cases, where there is scattering from all around the BS antenna and the MS. Receive diversity is now obvious and possible at both ends, and virtual nulling of interferers in receiving for both uplink and downlink may be performed. For the transmit case, possible solutions depend on the availability of channel information at the TX.

The design of smart antennas requires a DCM. Several DCMs have been presented in Section 3.2.1, and a COST 259-DCM is proposed in Section 3.2.4. To represent the channel models in Section 3.2.1, one model should be mentioned, which comprises a unified consideration of both flat fading and frequency-selective fading [FuMB98]. It considers the aforementioned spatial distribution of the scatterers in macro-, micro- and pico-cells, even for moving scatterers. Further impact on BS antenna system performance of spatial channel characteristics can be found in [PeMF98].

Investigations of antenna diversity and polarisation diversity schemes are presented in Sections 3.3.2 and 3.3.3. Though antenna diversity gains are also achievable with antenna arrays, there is a separate Section 3.3.4 about antenna arrays, where the focus is on the exploitation of dependent or correlated channels at the different antenna elements, and on space-time signal processing.

3.3.2. Space and antenna diversity

In the following, transmit and receive diversity, with two antenna arrays illuminating a scattering environment, is considered [Ande98a], [Ande98b], [Ande98d], [Ande99a], [Ande99b]. In the receive case Maximal Ratio Combining (MRC) is applied. Assuming the channel is reciprocal, which will be the case for stationary or slowly moving terminals in a TDD system, the antenna weights obtained in the receive case are used to transmit from the transmit array. Figure 3.3.2 shows the mean antenna gain for two K_a element arrays with correlated ($\rho = 1$) and uncorrelated ($\rho = 0$) channels at the antenna elements in a reciprocal channel, where the channels are known at the TX. Due to MRC, the total average gain will be the product of the two antenna gains K_a^2 for correlated signals (small angular spread), while the diversity gain achieved by uncorrelated signals (large angular spread) is smaller, see Figure 3.3.2 [Ande98d]. It is interesting to study the basic capacity from an information theoretic point of view [Ande98c]. Following [Ande98c], the capacity C depending on SNR and the number K_a of antenna elements for the highly correlated case is given by

$$C = \log_2\left(1 + K_a^2 \cdot SNR\right),\tag{3.3.1}$$

where SNR is the signal-to-noise ratio for one antenna at both ends. For a random set of n deterministic independent parallel channels, the capacity equals

$$C = n \cdot \log_2\left(1 + SNR\right).\tag{3.3.2}$$

This indicates that a random channel with uncorrelated signals at the different antenna ports leads to large gains in capacity, since the number of antennas now appears outside the log function. For a system with two K_a element arrays at the RX and TX, the mean capacity for a reciprocal (known) and a non-reciprocal (unknown) channel and for uncorrelated ($\rho = 0$) and correlated ($\rho = 1$) antenna signals is shown in Figure 3.3.3.

For a large number of transmit antennas C approaches

$$C = K_a \cdot \log_2\left(1 + SNR\right).\tag{3.3.3}$$

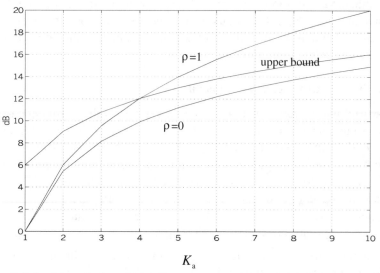

Figure 3.3.2: Mean gain of two linear arrays transmitting through a random medium giving ride to uncorrelated antenna signals.

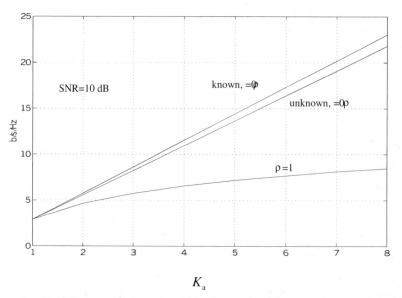

Figure 3.3.3: Mean capacity of a (K_a , K_a) array for a reciprocal and non-reciprocal channel for uncorrelated and correlated antenna signal.

The capacity grows linearly with the number K_a of receiving antennas. It can only be concluded that by having many uncorrelated antennas, the theoretical spectral efficiency is very large. An interesting reciprocity relationship has been found in [Tela95]. If $C(R, T, P_T)$ is the capacity of a channel with R RXs, T TXs and the total transmit power P_T, then

$$C(a,b,P_b) = C(b,a,P_a).$$

(3.3.4)

Applying several antennas in an array configuration at one or both ends in a communications link may give huge benefits in gains or spectral efficiency. The benefits depend on the correlation between the elements or the angular spread as seen from the arrays. For high correlations the patterns give high gain and suppression of interference, but no diversity. For low correlations the seemingly bad, fading channel may be turned into a number of independent channels, leading to a higher spectral efficiency. Transmit diversity for one-port antennas is not as efficient as receive diversity, when the channel is not known at the TX, but when known large gains and diversity orders may be achieved equal to the product of the number of antenna elements on the receive and transmit side.

Assuming independent Rayleigh fading channels between each TX and RX antenna, remarkable capacity gains in Multiple-In, Multiple-Out (MIMO) systems close to the Shannon capacity can be obtained by increasing the number of antenna elements [Tela95]. This implicitly assumes a high degree of scattering of signals between RX and TX arrays, and in particular, that there is a substantial angular spread in the signals at both sites. However, in a practical situation where a limited number of significant scatterers is present, the capacity of MIMO systems cannot be increased indefinitely by increasing the number of TX/RX antennas [Burr00a]. A limit is reached which is given by the capacity of n_s uncoupled Rayleigh fading channels. This means that we cannot always provide uncorrelated fading at the different antenna elements by increasing the element spacing of the antennas. However, if the number of antenna elements is much less than the number of scatterers, the Rayleigh channel capacity is a good approximation [Burr00a]. If the scatterers lie in a cluster around the RX antenna, capacity is improved by increasing the transmit element spacing [Burr00a].

For deriving the impact on capacity of providing BS diversity in a power controlled CDMA system, some initial work is presented in [KeOB99]. The goal is to determine the expected increase in the number of BSs necessary for a particular level of BS diversity. The BSs are randomly distributed and a triple spatial diversity is proposed. The statistical analysis aims to calculate the average expected power that will be received at all the BSs. By making any particular one of these BSs control the MSs transmit power, the total received power for a certain level of diversity can be calculated. The total received power can then be related to the power level for standard CDMA systems [KeOB99].

For the applicability of space diversity to a given antenna arrangement in a certain environment, the correlation between signals, received by two antennas in a BS, is a good measure. Measurement results performed at 60 GHz in a canyon-like street are presented in [GaCo98]. The distance between antennas (27λ horizontal, 25λ vertical) has been shown to be enough to decorrelate the signals and to make space diversity work. Diversity is supposed to work on a frame-by-frame basis since a digital transmission is used. For diversity to be applied, a low value for the correlation coefficient ρ over a short, rather than long, path is required. In [GaCo98] it is shown that a 1 m path distance is the right length to determine an upper bound of ρ between two signals. The frame-by-frame analysis has shown that diversity is applicable in a canyon-like street, since ρ is always below 0.5. The average investigated gains applying the combining techniques Selection Combining (SC), Maximal Ratio Combining (MRC) and Equal Gain Combining (EGC) range from 0–5 dB, where MRC performs best.

Further investigations of space and frequency diversity techniques at 60 GHz for indoor WLANs are presented in [DRFF97]. The selected environments are an ideal empty room and a real room with doors, windows and columns. The computation of received fading envelopes is performed by means of a 3D RT tool. The electromagnetic field components are adequately processed to obtain single branch and combined signal envelope. Due to the high frequency, results for the diversity gain can be obtained even adopting small space or frequency separation with both techniques.

For antenna diversity systems at handsets, two prototypes with dual-antenna systems are evaluated for diversity combining, using sampled RF antenna signals [BrNM99a], [BrNM99b]. Both simulation and measurements are performed. Prototype 1 consists of a quarter wave monopole and a shorted patch antenna on a Printed Circuit Board (PCB). Prototype 2 consists of a monopole and a planar meander antenna on a similar PCB. The distance between the antenna feeding points was about 20 mm for both prototypes. The SINR improvements from using the diversity combining schemes SC, EGC and Interference Rejection Combining (IRC) are presented in Table 3.3.1 [BrNM99b], which shows both the simulated and measured values, and the cases with one and two interferers. The values shown are the SINR improvements compared to the antenna having the best SINR, which is equivalent to the diversity gain.

Table 3.3.1: SINR improvements in dB at the 1 % outage level for
prototypes 1 and 2 (P1, P2), for one (I) and two (II) interferers.

		Simulation			Measurement		
		SC	EGC	IRC	SC	EGC	IRC
P 1	I	11	13	23	7	10	24
	II	10	12	15	8	10	16
P2	I	10	13	22	9	12	26
	II	10	12	14	9	12	16

3.3.3. Polarisation diversity

If two or more antennas at the same position receive or transmit
electromagnetic waves with different polarisations, and if the channels of the
different diversity branches are uncorrelated, we could benefit from
polarisation diversity. Especially in environments with scatterers around the
TX and RX, like in pico-cell and indoor environments, polarisation diversity
is expected to be a promising diversity scheme. Measurements are required
to evaluate the benefits of polarisation diversity. A measurement technique
which allows simultaneous sounding for polarisation matrix investigations in
pico-cell environments at 1 890 MHz is presented in [NeEg99]. The
measurement technique is based on two transmitting signals (separated by an
artificial Doppler shift of 20 Hz), and a dual branch correlation sounder,
allowing the simultaneous sounding of four individual diversity branches.
The investigations of the polarisation matrix are conducted in terms of
correlation coefficients and XPD (defined as the average cross-coupled
power relative to the average co-polarised power). The measurements show
that in the LoS case, the co- and cross-polarised power levels and thus the
XPDs are a function of the overlapping footprints of both RX and TX
antenna patterns in the azimuth and elevation plane, respectively. In the
NLoS outdoor cases, the co-polarised vertical power components are larger
than the corresponding horizontal ones, caused by the higher number of
reflections in the elevation plane. The cross-polarised power levels are
found to be similar for both transmitted polarisations. It can also be
observed that power cross-coupling is higher in the indoor case than in the
outdoor pico-cell situation. Due to the small height and width of the indoor
environment, more scattering effects takes place, since the first Fresnel zone
is not clear of obstacles, and higher power cross-coupling is observed.

Achievable gains for BS polarisation diversity are determined by measurements in [EgKO98]. The aim of the investigations is to disclose the factors determining BS polarisation diversity gain. Measurements were performed with a dual branch correlation sounder, with a 44 MHz bandwidth at 1739 MHz in urban and suburban micro-cell environment. The dual polarised BS receive antenna had a horizontal/vertical 3 dB beam width of about 60°/5° for both polarisations, and a bore sight XPD better than 20 dB. At the MS the TX handset with three different antennas (dipole, helix and patch, see [KoEO98]) were used with three different elevation of the handset (fixed 0°, fixed 60° and sinusoidal modulation between 0° and 60°). The effect of body proximity on the polarisation purity of the handset was tested by doing measurements with a phantom head, see [EgKO98], [KoEO98]. The following conclusions on BS polarisation diversity gain can be drawn from the handset experiments [EgKO98]:

- Building penetration encountered by indoor coverage has a minor influence on the XPD perceived at the BS. Thus, the diversity potential remains fairly unchanged when applying power control to compensate for the penetration loss (absolute power levels).
- Handset orientation and antenna type (polarisation purity and radiation pattern) are dominant factors determining the diversity gain.
- Though power cross-coupling is higher in the indoor case than in the outdoor situation, [NeEg99], the environment cross-polarisation and body proximity effects have less importance for the diversity gain.
- Random handset orientation and typical handset antennas yield XPDs less than 5 dB both indoors and outdoors.
- Simulated signal-to-noise diversity gains at the 10 % cumulative level are approximately 4 and 5 dB, respectively, for SC and MRC.
- No change in diversity gain due to penetration for indoor coverage is found when comparing it to the outdoor case.
- The radio propagation behaviour of typical handset usage provide a polarisation diversity potential of the same order as traditional space diversity, see Section 3.3.2, thus providing a much more compact antenna system for the task.

In [GDVP98] polarisation diversity schemes, antenna diversity schemes and the combination of both for an indoor scenario are compared concerning their diversity gain when applying a power-combining scheme. Polarisation diversity and antenna diversity are performed by combining the output signals of a dual polarised antenna and two vertical polarised antennas, respectively. The distance between the two vertical polarised antennas is chosen such that the correlation of the two received signals is smaller than 0.5 (in this case one wavelength). The diversity gain is determined by comparing the CDF of the combined signal and the CDF of one received signal. The improvements of the diversity schemes with respect to one antenna in 99 % of the cases are 5.7 dB for the two vertical polarised antennas, 3 dB for the polarisation diversity scheme and 4.2 dB for the combination of both schemes.

3.3.4. Antenna arrays

3.3.4.1. Calibration

Systems that exploit DoA information suffer from mutual coupling and amplitude or phase errors of the antenna array elements. Therefore, calibration schemes are necessary. One possible approach is calibration via the wave fronts that are impinging on an antenna array [PeNo97a]. The solution offered for DoA estimating systems is based on wave fronts that are impinging on the antenna array from surrounding BSs. The number of these wave fronts has to be minimised for the calibration procedure to have reasonable calculation expense, and is, in principle, different for up- and downlinks. If at least one direction is known, the errors due to mutual coupling or different amplifications of the antenna signals can be eliminated. All wave fronts that arrive (uplink) or are sent (downlink) at the same time will be called a 'scenario', Figure 3.3.4 [PeNo97a]. The index s belongs to scenario s, where S denotes the total number of scenarios.

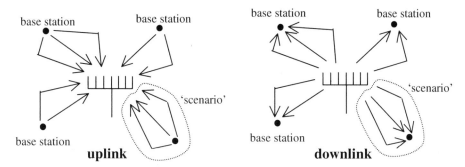

Figure 3.3.4: Uplink and downlink calibration (© 1998 IEEE).

In the uplink, the modulated signals **S** of W wave fronts impinging on a uniform linear array are mapped by the steering matrix **A** onto the matrix

$$\mathbf{X} = \mathbf{K} \cdot \mathbf{A} \cdot \mathbf{S} + \mathbf{N}, \tag{3.3.5}$$

where **N** represents the thermal noise and **K** the mutual coupling and amplification errors. The aim is to find a correction matrix $\mathbf{K}_{corr}$ that eliminates the calibration error. In the uplink, this can be achieved by the following insight: if the number of antenna elements is K_a, according to spatial smoothing for DoA estimation, it is possible to select K_a- $K_{a,sub,s}$ +1 subarrays of $K_{a,sub,s}$ antennas for each scenario $s = 1, ..., S$, until $W_s \geq K_{a,sub,s}$. With the sum of all autocorrelation matrices of the subarrays and all scenarios $\overline{\mathbf{M}}$ the minimisation of the power received by the antenna array leads to an eigenvalue problem

$$\overline{\mathbf{M}} \cdot \mathbf{k}_{corr} - \lambda_{min} \mathbf{I}^{K_a^2 \times K_a^2} \cdot \mathbf{k}_{corr} = \mathbf{0}, \tag{3.3.6}$$

where $\mathbf{k}_{corr}$ is the eigenvector of the corresponding smallest eigenvalue λ_{min} . Since $\mathbf{k}_{corr}$ is obtained by diagonal wise vectorisation of $\mathbf{K}_{corr}$ an improved update of the correction matrix $\mathbf{K}_{corr}^{(l)}$ in the l-th step of the estimation can be determined in the $l + 1$-th step by

$$\mathbf{K}_{corr}^{(l+1)} = \mathbf{K}_{corr} \cdot \mathbf{K}_{corr}^{(l)} . \tag{3.3.7}$$

In the downlink case the power received by the surrounding BSs has to be minimised. Uplink and downlink share the iterative procedure. However, there are three major differences. The minimisation cannot apply eigenvalue decomposition, but has to run around by a numerical optimisation method (e.g. Conjugate Gradient (CG)), that minimises an objective function defined by the connection between eigenvalue decomposition and Rayleigh quotient. For each numerical evaluation of the gradient, a measurement has to be taken. Each subarray is loaded with identical weight vectors and sends uncorrelated wave fronts.

Decomposition into subarrays, path delay differences that decorrelate the wave fronts and a proper exploitation of the mutual coupling structure significantly reduce the required number of BSs.

The discussion that followed the presentation of [PeNo97a] raised some doubt about the practicability of this approach, because it was felt that mutual coupling of the antenna elements and recurrent electronic balancing of the different RX-TX-trains can be accounted for in simpler ways.

3.3.4.2. DoA estimation

One promising approach to exploiting the directional inhomogeneity of the mobile radio channel with antenna arrays consists of estimating the DoA of impinging information carrying wave fronts, and in benefiting from the knowledge of the DoA in the signal processing. Due to the importance of precise and reliable estimation, especially when considering SDMA systems, techniques are proposed in [PeNo97b] to improve the DoA estimation with respect to accuracy and robustness when compared to a simple TLS solution. Subspace techniques like MUSIC or ESPRIT [RoKa89], [HaNo95], which are based on eigenvalue decomposition and on the separation between a signal subspace and a noise subspace spanned by the eigenvectors, show a significant DoA estimation improvement, if the signal power is much larger than the noise power [PFBH97]. With increasing noise power, the influence of disturbances increases the closer the smallest eigenvalue of the signal subspace is and the larger the eigenvalue of the noise subspace is. The results are high standard deviations in the direction estimation or even the loss of the directions [PeNo97b]. The hard decision subspace methods show an obvious sensitivity to errors in the assumed number of wave fronts and low stability in the case of high noise scenarios. The use of a directional optimisation criterion exhibits better results in this respect, which minimises the estimation error [PeNo97b]. A second optimisation method is based on minimising the error by soft weighting. This algorithm combines both advantages of the previous methods. The detection of direction is enhanced due to eigenvalue decomposition, but instead of selecting some eigenvectors the whole space is taken into account.

The afore-mentioned results are obtained by assuming fixed punctual sources of the signals. However, adaptive array techniques are very sensitive to changes in the velocity direction of the mobiles [SeLS99]. Algorithms that also estimate the angular spread, besides the DoA, are assumed to be better adapted to real environments.

Ambiguity problems also appear in DoA estimation. In order to overcome these problems, and to make algorithms like MUSIC consistent in presence of angularly close sources, polarisation information is also estimated and additionally considered [SeLS99].

Another promising DoA estimation scheme is given by the space-alternating generalised expectation maximisation (SAGE) algorithm [DJFH97]. Besides the DoA $\phi_{k,l}$, the SAGE algorithm allows the estimation of all relevant channel parameters like the delay $\tau_{k,l}$ and the attenuation $A_{k,l}$ of the l-th path, $l = 1, ..., L_k$, of the k-th user, $k = 1, ..., K$. For each user k the channel parameters are combined in a vector

$$\theta_k = \left[\tau_{k,1}, \phi_{k,1}, A_{k,1}, ..., \tau_{k,L_k}, \phi_{k,L_k}, A_{k,L_k} \right], \quad k = 1...K , \qquad (3.3.8)$$

which finally forms the vector

$$\theta = [\theta_1, ..., \theta_K] . \qquad (3.3.9)$$

The channel parameter estimation is based on the symbol estimates **B** for a given time period, shorter than the coherence time of the channels of all K users. Together with the received signal **y** for a given observation time θ is estimated by maximising the log-likelihood-function with respect to θ and **B**:

$$\hat{\theta} = \arg \min_\theta (\theta, \mathbf{B}, \mathbf{y}).$$ (3.3.10)

Since the estimation quality of θ depends on the BER, in [DJFH97] the SAGE algorithm is combined with a Multistage Detector (MS) with multiple antenna input in order to obtain a sufficient bit error performance.

3.3.4.3. Beamforming

3.3.4.3.1. Adaptive beamforming in the uplink

In the following single transmit and multiple receive antennas in the uplink are considered. In [PFBB99], [PaFB97] multi-antenna arrays, in conjunction with a Joint Detection (JD) CDMA system, are investigated by link level simulations in the uplink. If K_a antenna elements are considered at the BS, the conventional joint channel estimator (Steiner estimator) of a JD-CDMA system estimates the ChIRs valid for the K users that are simultaneously active and separated by the CDMA codes, at each of the K_a antenna elements, independently by using the *a priori* information of the user-specific transmitted midamble training sequences. By applying a combined DoA and joint channel estimation scheme (see also Section 3.3.4.2), the channel estimation and, consequently, the data detection can be improved by a beamforming technique. The DoA of the *l*-th path, $l = 1, ..., L_k$, of the *k*-th user, $k = 1, ..., K$, is estimated by applying the Unitary ESPRIT [HaNo95]. Together with the knowledge about the array geometry and the DoAs $\theta_{k,l}$ of the *l*-th path of the *k*-th user, the user- and path-specific array steering vector $\mathbf{a}(\theta_{k,l})$ can be obtained [PFBB99], [PaFB97]. It is assumed that there are K_i directional intercell interfering signals present, which are pair wise temporally uncorrelated. Each of the K_i interfering signals is assigned to one single DoA, i.e. $L_{k_i} = 1, k_i = 1...K_i$.

With $\sigma_{k_i}^2$ is the interference power of the k_i-th interferer, we obtain the spatial covariance matrix of the interference by

$$\mathbf{R}_\mathrm{I} = \sum_{k_i=1}^{K_i} \sigma_{k_i}^2 \cdot \mathbf{a}\left(\theta_{k_i}\right) \cdot \mathbf{a}^{*\mathrm{T}}\left(\theta_{k_i}\right).$$ (3.3.11)

$\mathbf{R}_I$ contains only the correlation properties of the intercell interference, because intracell multiple access interference is eliminated by joint detection. Therefore, $\mathbf{R}_I$ is the same for all K users, which is, not true for SDMA systems, for example, where the intracell interference must be included in $\mathbf{R}_I$. Under consideration of the spatial covariance matrix of the interference $\mathbf{R}_I$ the user- and path-specific weighting vector $\mathbf{w}_{k,l}$ is given by

$$\mathbf{w}_{k,l} = \left(\mathbf{R}_I^{-1}\right)^* \cdot \mathbf{a}^* \left(\theta_{k,l}\right). \tag{3.3.12}$$

The beamformer of (3.3.12) maximises the ratio between the instantaneous power of the signal of user k having DoA $\theta_{k,l}$ and the instantaneous power of the sum of the noise signals that constitute the intercell interference. Finally MRC of all L_k paths is performed for each user k, and ISI and MAI is cancelled based on joint channel estimation and joint data detection, respectively. In the simulations, single DoAs in rural area for each of the K equal to 8 user signals are assumed. The E_b/N_0 improvement due to the application of the enhanced RX is approximately 6 dB at a coded BER of 10^{-3} for a Uniform Linear Array (ULA) with K_a equal to eight antenna elements compared to the ULA that employs the conventional scheme. The simulation results also indicate that the use of two-dimensional array configurations at the BS RX can improve the system performance considerably, simultaneously offering the potential of whole range coverage, i.e. both in azimuth and in elevation. Further simulations with the same RX structure show that the performance of systems with smart antennas can be severely degraded when the user signals impinge on the BS RX from single DoAs that are not spatially separated well [PSWB98]. It is assumed that all DoAs of the K equal to eight user signals are equidistant and, without loss of generality, they impinge on the BS within an azimuthal sector of width ϕ_s. A cross array of eight-by-eight array elements is used at the BS RX and the system performance enhancement by beamforming varies from approximately 4 dB ($\phi_s = 0°$) to 6 dB ($\phi_s = 360°$) [PSWB98]. There is almost no performance difference for $\phi_s > 90°$. This is due to the fact that with $\phi_s = 90°$, two successive DoAs differ by $12.857°$, which is very close to the 3 dB beam width of the main lobe of a cross array antenna diagram with K_a equal to 16 antenna elements (approximately $12°$) [MoMi80].

A simple algorithm for spatial filtering in WCDMA systems, without the need of explicit estimation of the DoAs, is introduced in [Morg99]. By correlation of the received signal based on the transmitted pilot sequence after descrambling and despreading with the known pilot sequence at each antenna element a vector $\mathbf{r}$ is obtained. Instead of resolving the DoAs of each user k, the vector $\mathbf{r}$ could be seen as a steering vector for the superposition of all DoAs of the k-th user. By using the received chips before despreading a pre-correlation matrix $\mathbf{R}_x$ is calculated. Now, in a similar way as in (3.3.12) the spatial weights of the interference suppression filter are obtained by

$$\mathbf{w} = \mathbf{R}_x^{-1} \cdot \mathbf{r}. \tag{3.3.13}$$

Simulations are performed considering four antenna elements with an inter-element spacing d of half the carrier wavelength λ and 7λ and up to 20 users are simultaneously active. The results indicate that, for a relatively low number of users, spatial diversity seems to be better than beam steering. However, for the expected large number of users of UMTS systems beam steering by the use of antenna arrays with sufficient narrow beam widths may improve the system performance.

3.3.4.3.2. Adaptive beamforming in the downlink

For systems with TDD, almost the same channel conditions exist for up- and downlinks, if the transmission frame is short compared to the coherence time of the channel [Czyl00], [Utsc99]. Therefore, beamforming is optimised instantaneously, corresponding to the temporal fluctuations of the channel and is identical for up- and downlinks. In the case of systems with FDD the fast fading process in up- and downlinks are different. However, if the frequency separation is not too large, the up- and downlink waves exhibit the same directional dependence [Czyl00]. Therefore, downlink beamforming could be carried out in an average sense based on the uplink channel measurements. From the uplink channel measurements, the DoAs and the average power transfer factor of the desired and undesired downlink paths can be obtained. With the array steering vector $\mathbf{a}(\theta, f)$ depending on the azimuth angle θ and the carrier frequency f, the complex amplitudes $A_{k,l}$ and the DoA $\theta_{k,l}$ of the l-th propagation path of the k-th user signal the spatial covariance matrix of the k-th user at the frequency f can be written as [Hugl99], [HuLB99]

$$\mathbf{R}_{S,k}(f) = \sum_{l=1}^{L_k} A_{k,l}^2 \cdot \mathbf{a}(\theta_{k,l}, f) \cdot \mathbf{a}^{*T}(\theta_{k,l}, f), \tag{3.3.14}$$

where L_{kl} is the total number of signal path of the k-th user. In (3.3.11) a spatial covariance matrix $\mathbf{R}_I$ is calculated by considering only the intercell interference correlation properties, which is sufficient in the uplink, if joint detection is applied. In this special case, for all K desired users we have the same interference. For downlink beamforming all other users, besides the k-th user must be treated as interference, as in SDMA systems, which leads to a user specific spatial interference covariance matrix [Hugl99], [HuLB99]

$$\mathbf{R}_{\mathrm{I},k}\left(f\right) = \sum_{i \neq k} \mathbf{R}_{\mathrm{S},i}\left(f\right) + \sigma_{\mathrm{N}}^{2}\mathbf{I}, \tag{3.3.15}$$

where σ_{N}^{2} denotes the thermal noise variance and $\mathbf{I}$ the $K_{\mathrm{a}} \times K_{\mathrm{a}}$ identity matrix, respectively. If estimated DoAs are utilised for determining the covariance matrices $\mathbf{R}_{\mathrm{S},k}$ and $\mathbf{R}_{\mathrm{I},k}$, respectively, this approach restricts the application to systems where the number of dominant angles of arrivals is smaller than the number of used antenna elements [Hugl99]. Furthermore, the DoA estimation accuracy strongly depends on the physical propagation environment.   Therefore, approaches shown in [Hugl99], [Utsc99], [WePa99a], [AnMF99] determines the matrices $\mathbf{R}_{\mathrm{S},k}$ and $\mathbf{R}_{\mathrm{I},k}$ from the received uplink signals at the different antenna elements by utilising a known signal sequence transmitted for channel estimation purposes or the already detected data.

However, the goal of maximising the received average power at the desired mobile while maintaining the sum of the total interference power at all undesired mobiles leads to a generalised eigenvalue problem [Czyl00],

$$\mathbf{R}_{\mathrm{S},k}\mathbf{w} = \lambda \cdot \mathbf{R}_{\mathrm{I},k}\mathbf{w}, \tag{3.3.16}$$

where the optimum weight vector $\mathbf{w}_{k}$ is the eigenvector associated with the largest eigenvalue λ. In the calculation of the matrices $\mathbf{R}_{\mathrm{S},k}$ and $\mathbf{R}_{\mathrm{I},k}$ the angular distribution of the individual paths could be correctly taken into account, or the paths are described only by discrete waves without any angular spread. In [Czyl00] it is shown that there is only a small difference in the resulting path diagrams. Especially for the strong paths, the difference can be neglected if the RMS angular spread is smaller than $1°$. The azimuth power distribution is assumed to be Laplacian with a standard deviation $\sigma_{\theta} < 1°$. The results of the system level simulations show hardly any difference in the CDF of the SIR for optimised beamforming where all paths are included in $\mathbf{R}_{\mathrm{S},k}$, where only the strongest path is included in $\mathbf{R}_{\mathrm{S},k}$ and where only the strongest desired path is included in $\mathbf{R}_{\mathrm{S},k}$ and all other desired paths are considered as interference and included in $\mathbf{R}_{\mathrm{I},k}$.

In a SDMA mobile radio system, the beamforming vectors of all users could be jointly calculated such that all terminals receive their required signals with a given SINR [Utsc99]. This approach may suffer from the frequency gap between uplink and downlink in FDD systems.   Therefore, a combination of a least squares approximation of beam patterns and a power minimisation beamforming based on spatial covariances is proposed in [Utsc99] to overcome the difficulties of FDD. Besides the less complex beamforming concept proposed in [Czyl00], this approach opens the possibility of an instantaneously optimised downlink beamforming. With a least squares approximation of the array manifold at the uplink frequency by means of beamforming at the downlink frequency and a power minimisation approach, which requires the transposition of the spatial covariance matrix obtained in the uplink to a spatial covariance matrix in the downlink, the downlink beamforming vector can be instantaneously optimised. The results in [Utsc99] show that the effects of the frequency gap $f_{up}/f_{down} = 0.915$ between uplink and downlink frequency in FDD-WCDMA radio systems must not be neglected. The approximated power minimiser ensures the desired SINR of each user without increasing the transmitted power at the BS. $K = 2$ users and $K_a = 7$ antenna elements of a uniform linear antenna array and three dominant paths with the DoAs 60°, 40°, 0° and 50°, 20°, 10° are considered in the simulations. The SINR for user 1 and 2 with approximation of the beam pattern at downlink frequency f_{down} by means of the Least Square (LS) approach and the SINR for both users by reusing the uplink beamforming vector in the downlink (REUSE) is shown in Table 3.3.2 [Utsc99]. As a reference (REF) the SINR of both users for a beamforming under the condition $f_{up} = f_{down}$ is also included in Table 3.3.2.

Table 3.3.2: SINR for the downlink beamforming at f_{down}.

SINR	REF	LS	REUSE
User 1	10.00 dB	9.67 dB	6.83 dB
User 2	10.00 dB	9.64 dB	9.68 dB

Another possibility of transforming the frequency dependent spatial covariance matrix $\mathbf{R}_{S,k}(f_{up})$, obtained in the uplink, to its corresponding spatial covariance matrix $\mathbf{R}_{S,k}(f_{down})$ in the downlink is shown in [HuLB99], [Hugl99]. Since fading is uncorrelated between up- and downlinks, due to the frequency duplex distance, only mean values of the channel parameters are considered for beamforming. With $\mathbf{R}_{S,k}(f)$ from (3.3.14) and the array steering vector $\mathbf{a}(\theta, f)$ the transformation to the downlink frequency f_{down} is performed by first estimating the APS

$$P_k(\theta) = \frac{1}{\mathbf{a}^T(\theta, f_{up}) \cdot \mathbf{R}_{S,k}^{-1}(f_{up}) \cdot \mathbf{a}(\theta, f_{up})} \qquad (3.3.17)$$

of each user k by applying the Minimum Variance Distortion less Response filter (also called *Capon's beamformer*) [Capo69]. The magnitudes $|A_{kl}|$ and the angles of arrival θ_{kl} of the signal paths are invariant to carrier frequency shifts. Therefore, the spatial covariance matrix $\mathbf{R}_{S,k}(f_{down})$ at the downlink frequency f_{down} can be determined under consideration of the estimated APS from [Hugl99], [HuLB99]:

$$\mathbf{R}_{S,k}(f_{\text{down}}) = \int_{\theta} P_k(\theta) \cdot \mathbf{a}(\theta, f_{\text{down}}) \cdot \mathbf{a}^{\mathrm{T}}(\theta, f_{\text{down}}). \qquad (3.3.18)$$

The spatial covariance matrix of the interference $\mathbf{R}_{I,k}(f_{\text{up}})$ can be transformed to $\mathbf{R}_{I,k}(f_{\text{down}})$ in the same way as it has been done for $\mathbf{R}_{S,k}(f_{\text{up}})$. Having now $\mathbf{R}_{S,k}(f_{\text{down}})$ and $\mathbf{R}_{I,k}(f_{\text{down}})$ for the downlink frequency f_{down}, one can find the optimum antenna weights $\mathbf{w}_k$ for the transmit signal towards the k-th user by solving (3.3.16).

Beamforming investigations of the downlink spectrum efficiency of a TD-CDMA mobile radio system with array transmit antennas have been made in [ScPa98a]. Investigations concerning the spectrum efficiency η require both system level and link level simulations. With the transmission gains and the DoAs of a certain BS to all MSs, beamforming for each transmitted signal is performed within the link level simulations. The antenna pattern is adjusted in such a way that the power transmitted to the MS that is served by this signal is maximised vs. the sum of the power received by all MSs in the other cells. As JD is the preferred detection scheme in TD-CDMA, the signals transmitted to the MSs of the same cell do not influence or disturb each other. After the beamforming is done for each transmitted signal, a power control follows. The power control algorithms considered are the C/I balancing algorithm and the constant received power algorithm (C-control) [ScPa98a]. Since the scenario, the power control algorithm, the beamforming strategy, and the re-use factor have influenced the system level simulations, only the statistics of C/I have to be considered in the link level simulations. The QoS criterion is fulfilled if the outage probability is less than 5 % at a maximum coded BER of 10^{-3}. In these cases, the spectrum efficiency η is evaluated with the system bandwidth B, the re-use factor r, the number of users which are simultaneously active in one cell K, the bit rate per user R, the frame duration T_{fr}, the burst duration T_{bu}, according to [ScPa98a]:

$$\eta = \frac{K \cdot R \cdot T_{\text{fr}}}{r \cdot B \cdot T_{\text{bu}}}. \qquad (3.3.19)$$

Table 3.3.3 [ScPa98a] shows the best combination of r and K for a macro rural environment. At each MS only one single antenna is considered.

Because of the diversity of the channels in the uplink, the uplink always offers better results than the downlink. In all simulations the power control scheme $C/I = const$ gives better results than the power control scheme $C = const$ [ScPa98a].

Table 3.3.3:   Spectrum efficiency in the uplink η_{up} and the downlink η_{down} of a TD-CDMA system with one and four BS antennas; $v = 30$ km/h; C/I balancing; $B = 1.6$ MHz; $T_{fr} = 6$ ms; $T_{bu} = 0.5$ ms; $R = 84$ kbit/s.

Number of BS antennas	r	K	$\eta_{down} / \dfrac{bit}{s \cdot Hz}$	$\eta_{up} / \dfrac{bit}{s \cdot Hz}$
1	3	2	0.0350	0.045
4	4	8	0.0525	0.120

3.3.4.3.3. Switched-beam antennas

The impact on capacity when using switched-beam antennas with 6, 12 and 24 beams, respectively, in a GSM 1800 system compared to usual tri-sectorised antennas is shown in [BaSa97]. A switched-beam antenna has its narrow beams pre-defined. It selects the best-received beam that is used for both reception and transmission, Figure 3.3.5 [BaSa97]. Adaptive antennas are more 'intelligent' and, therefore, more complex. They modify dynamically the antenna pattern to steer its main lobe towards the mobile and to form side lobes and nulls with the aim to minimise the interfering signals. The switched-beam can be seen as a multi-sectorised site where the resource management is done at the site level, while it is done at the sector level in the case of the usual multi-sectorised antenna. Taking into account GSM system constraint and mobility features, the switched-beam antenna can improve the capacity by a factor of 2.29 compared to the conventional solution [BaSa97]. The simulations show that the number of beams, NB, (6, 12, 24) has no impact on the overall capacity.

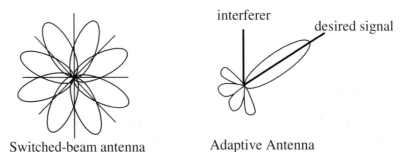

Figure 3.3.5:   Principle of switched-beam antenna and adaptive antenna.

To compare the performance of switched-beam approaches and optimum combining in the case of uplink reception in GSM, simulation results under consideration of measured ChIRs are shown in [LeMo98]. *NB* equal to 8 and 22 fixed beams, respectively, can be steered with a uniform linear array with K_a equal to 8 antenna elements to cover a sector of 120°. The measured ChIRs have been obtained from the TSUNAMI II stand-alone testbed [MFDO96]. For the beam steering algorithm the *C/I* performance dependent on K_a, *NB*, the antenna height and the AS is shown in Table 3.3.4 [LeMo98]. The beam steering algorithm gives a *C/I* performance gain of nearly a factor K_a for low values of the azimuth spread (AS < 5°). Increasing the number of beam directions *NB* from 8 to 22 does not change the performance significantly, since the spatial coverage is almost uniform.

Table 3.3.4:   The absolute *C/I* and the *C/I* relative to single element at 1 % outage, for the different beam steering configurations and base antenna heights (AS: measured azimuth spread) (© 1998 IEEE).

Configuration	BS height					
	32m (AS=5°)		26m (AS=7.5°)		20m (AS=10°)	
	Abs. C/I [dB]	Rel. C/I [dB]	Abs. C/I [dB]	Rel. C/I [dB]	Abs. C/I [dB]	Rel. C/I [dB]
Single element	-7.2	0	-9.2	0	-9.5	0
$K_a = 8$, NB = 22	0.9	8.0	-2.5	6.7	-4.0	5.5
$K_a = 8$, NB = 8	-0.3	6.9	-2.8	6.3	-4.5	5.0

Simulation results presented in [LeMo98] show that optimum combining by adjusting the antenna weights in a MMSE sense performs better than beam steering for a *C/I* outage threshold of less than 10 %. Optimum combining provides some diversity effects over beam steering, but the mean gain is lower. For a synchronised network, a significant performance improvement by optimum combining can be obtained by modifying the GSM training sequences with the purpose of achieving better cross-correlation properties.

For beam orientated SDMA systems, a grid of beams is required, which is the base for both initial search and prolonged link user-tracking [Egge98a]. In [Egge98a] it is described how virtual BS Doppler domain beams can be used in beam orientated SDMA. The advantage is straightforward usage of FFTs instead of DFTs on the element signals, providing equally Doppler spaced beams with a lower numerical complexity [Egge98a]. Considering a ULA with a spacing d equal to $\lambda/2$ between the K_a antenna elements, the spatial sampling frequency is

$$f_s = \frac{1}{d}.$$
(3.3.20)

Like in [LeMo98], a 120° sector with its centre being perpendicular to the ULA centre is considered (i.e. from 30° to 150°). In the TSUNAMI II stand alone testbed [MFDO96] a grid of NB equal to 21 beams within the 120° sector has been implemented via DFTs. The numerical complexity of DFTs is $K_a \cdot NB$. However, applying FFT instead of DFT it is required to pad the array with K_a elements with zeros (elements) up to the number of required beams NB. The numerical complexity of the FFTs is in this case $NB \cdot \log(NB)$. In the FFT approach NB beams are spaced equidistantly from $-f_s/2$ to $+f_s/2$, but only the N_{eff} beams with $\pm f_{d,\max} \cdot \sqrt{4/3}$, where the maximum Doppler shift $f_{d,\max}$ is $1/\lambda$ and $f_{d,\max} \cdot \cos(30°)$ is $\sqrt{4/3}$ and $f_{d,\max} \cdot \cos(150°)$ is $-\sqrt{4/3}$, are used. To compare the Doppler beam FFT approach with the direct angular DFT approach, only the N_{eff} beams are considered for the calculation of the complexity of the DFT approach. If the maximum Doppler shift $f_{d,\max}$ is normalised to 1, the number of Doppler beams within $\pm\sqrt{4/3}$ is

$$N_{\text{eff}} = NB - 2 \cdot \text{TRUNC}\left(\frac{f_s/2 - \sqrt{3/4}}{\Delta f_d}\right) - 1$$
$$= NB - 2 \cdot \text{TRUNC}\left(\frac{1/2 - d \cdot \sqrt{3/4}}{NB}\right) - 1$$
(3.3.21)

[Egge98a] where the Doppler separation of beams is

$$\Delta f_d = \frac{f_s}{NB} = \frac{1}{d \cdot NB}$$
(3.3.22)

and normalisation with respect to λ is assumed.

Note that the TSUNAMI II case of K_a equal to 8 and NB equal to 21 has a DFT complexity of 168, Table 3.3.5 [Egge98a]. Using a FFT, one can either lower the complexity by 35 % using a 24 beam FFT, or increase the number of effective beams N_{eff} from 21 to 27 with the same complexity [Egge98a].

Note also that the FFT complexity remains fixed for array sizes up to K_a equal to NB, where the DFT complexity increase linearly with increasing array size. For typical array sizes K_a and number of beams NB the saving in complexity over the DFT approach is in order around a factor of two [Egge98a].

Table 3.3.5: Effective number of Doppler beams N_{eff} and FFT/DFT (i.e. $(NB \cdot \log(NB))/(K_a \cdot N_{eff})$) numerical complexity for $K_a = 8$, 12 and 16 ULA with different element spacing and number of FFT output beams.

K_a	$NB = 8$	$NB = 12$	$NB = 16$	$NB = 24$	$NB = 32$	$NB = 48$
8	7	11	13	21	27	41
	24/56	43/88	64/104	110/168	160/216	268/328
12	–	11	13	21	27	41
		43/144	64/192	110/288	160/384	268/576
16	–	–	13	21	27	41
			64/256	110/384	160/512	268/768

3.3.4.3.4. Fixed-beam antennas

In most algorithms and system concepts with adaptive antennas at the BS, the adaptivity of the signal processing can be considered as a dynamical optimisation of the antenna pattern. Systems with high adaptivity improve the performance considerably, while the BS cost and complexity increase. In [DFFR99] a method to determine the 'optimum' fixed radiation patterns for the BS antennas in urban micro-cellular environment is proposed. The basic idea is to tailor the antenna pattern to the particular BS site, i.e. to the surrounding urban topology. The final aim is to determine the optimum antenna pattern for each micro-cellular BS site class, such as rectilinear street-canyon, T-crossing, X-crossing etc. The multipath channel is sounded with a RT tool and then the BS antenna pattern is optimised in order to minimise the time DS and maximise the coverage in the selected service area. For instance, for a given test environment the DS exceeded with a probability of 10 % is reduced from 33 ns with an omnidirectional antenna pattern to 15 ns with the site-specific antenna pattern.

3.3.4.3.5. Real-time implementation of adaptive antenna arrays

In [KTBT99], the implementation of an 8-element real-time adaptive antenna array in a GSM 1800 BS, by using the run-time optimised adaptive antenna array processor **A³P,** is presented, Figure 3.3.6 [KTBT99].

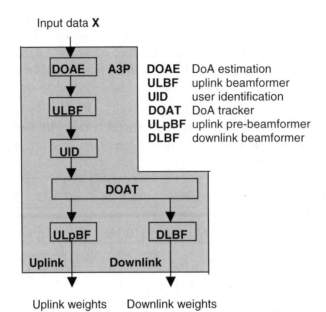

Figure 3.3.6: Adaptive antenna array processor A³P (© 1999 IEEE).

With a run-time of about 1 ms the processor allows an adaptation of the beam pattern in every GSM frame. **A³P** works within the GSM standard and is compatible with frequency hopping. All paths relevant for the intended user are resolved, and separate tracking for uplink and downlink is included. The smart antenna is used to suppress co-channel interference, i.e. Spatial Filtering for Interference Reduction (SFIR) is applied. The array processing is based on DoA estimation in the uplink applying, e.g. Unitary ESPRIT [HaNo95], see Section 3.3.3.2. Since in a FDD system the Rayleigh fading is uncorrelated in the up- and downlinks, using uplink weights for downlink beamforming can lead to erroneous results. However, as it was shown in [Czyl00], channel parameters like the DoAs and the average path loss are the same in uplink and downlink. While the uplink tracker is optimised for the instantaneous uplink channel situation, the downlink tracker is based on averaged uplink channel parameters. The **A³P**, Figure 3.3.6, uses the calibrated digital baseband signal **X** in the uplink to calculate the antenna weights for uplink and downlink beamforming. The processing is based on DoA estimation. A spatial pre-filter, the Uplink Beamformer (ULBF), and a User Identification (UID) decide, whether a DoA belongs to a user or to an interferer. The tracker averages the user DoAs. A Chebychev-based beamformer and an iterative beam cancellation scheme for placing broad nulls into the directions of the interferers are applied. The mean SINR improvement, which is evaluated by means of link level simulations with a synthetic channel model [FuMB98] for a macro-cell environment, is about 17 dB.

The design and implementation of an electronically steerable phased array antenna, which performs beamforming at RF level at 1.5 GHz in the uplink, is presented in [FPJF99]. An eight element linear array was developed using coupled microstrip patches, 0.4λ uniformly spaced. Each element array is also composed by a sub-array of two microstrip patches fed in phase in order to get some directivity in the E-plane. The amplitudes and phases of the received signals at the different antenna elements are controlled by DC signals coming from a control board connected to a digital processor. An eight-way power combiner/divider combines the RF signals of each array element. Outdoor measurements of the received power at the BS were performed while a mobile unit transmitting the desired signal was driving along a street in a suburban area and a second unit transmitting an interfering signal was kept fixed. During the measurements the antenna system first performs a fast search aiming to determine the DoA of the desired signal and the interference. Then, the main beam is pointed towards the mobile and enters in tracking mode. According to the received power level by the antenna system, a level detector generates a DC voltage, which is then acquired to update the weighting vector. The measurement results show the interference suppression capability and a high reception sensitivity of the implemented antenna system.

3.3.4.4. Space-time signal processing

3.3.4.4.1. Consideration of signal correlation properties

Joint signal processing in the time and space domain is called space-time signal processing, Figure 3.3.1. This type of processing allows us to optimise the RX in time and space signal processing. Multiuser detection schemes easily lend themselves to space-time signal processing, because the structure of the algorithms and the mathematical representation remain the same for multi-antenna RXs as for single antenna RXs and spatial and temporal correlation properties of the interference can be considered [WePH99], [WePS99], [WePa99a], [WePa99b], [AnMF99]. Due to the fact that JD is the preferred detection scheme in the UTRA TDD mode TD-CDMA, it is a promising candidate for space-time signal processing [WePH99], [AnMF99]. Using the discrete time lowpass model of a TD-CDMA RX [WePa99a] with an antenna array, the received signal can be described by a vector **e**, which is a concatenation of all received signals at all K_a antenna elements. The system matrix **A** includes all ChIRs of the K users to the K_a antenna elements convolved with the CDMA-codes. Setting out from $\mathbf{A}$, from the total data vector $\mathbf{d}$, which includes the data sequences of the K users, and from the total intercell MAI vector $\mathbf{n}$ one can represent the received signal $\mathbf{e}$ by

$$\mathbf{e} = \mathbf{A}\mathbf{d} + \mathbf{n} .$$ (3.3.23)

For the covariance matrix of the interference we obtain

$$\mathbf{R}_n = E\{\mathbf{n}\mathbf{n}^{*T}\},$$ (3.3.24)

and for the covariance matrix of the data we obtain

$$\mathbf{R}_d = E\{\mathbf{d}\mathbf{d}^{*T}\}.$$ (3.3.25)

To perform JD, a linear estimate

$$\hat{\mathbf{d}} = \mathbf{M}\mathbf{e}$$ (3.3.26)

of $\mathbf{d}$ can be obtained from (3.3.26). The choice of the matrix $\mathbf{M}$ determines the equaliser type:

- Zero-Forcing Block Linear Equaliser (ZF-BLE) [Whal71], [WePa99a]:

$$\mathbf{M} = \left(\mathbf{A}^{*T}\mathbf{R}_n^{-1}\mathbf{A}\right)^{-1}\mathbf{A}^{*T}\mathbf{R}_n^{-1}$$ (3.3.27)

- Minimum Mean Square Error Block Linear Equaliser (MMSE-BLE) [Whal71], [AnMF99]:

$$\mathbf{M} = \left(\mathbf{A}^{*T}\mathbf{R}_n^{-1}\mathbf{A} + \mathbf{R}_d^{-1}\right)^{-1}\mathbf{A}^{*T}\mathbf{R}_n^{-1}.$$ (3.3.28)

Whereas the matrix $\mathbf{A}$ is known at the RX with an accuracy of the channel estimates, $\mathbf{R}_n$ can be estimated from the chip-level defined training sequences in the burst midamble along with the received snapshots [AnMF99]. A second possibility to estimate $\mathbf{R}_n$ is based on an estimation of the interference vector [WePa99a], [WePa99b]

$$\hat{\mathbf{n}} = \mathbf{e} - \mathbf{A}\hat{\mathbf{d}}$$ (3.3.29)

To reduce the computational complexity only the spatial covariance matrices $\mathbf{R}_I$ are estimated [AnMF99], [WePa99a], [WePa99b], see also Section 3.3.4.3.1. Note that $\mathbf{R}_I$ contains only the spatial correlation properties of the intercell interference, due to the fact that intracell interference is eliminated by multiuser detection. Since it is assumed that interfering signals with different DoAs are uncorrelated and that they are essentially white in the temporal domain [AnMF99] or have the spectral form of the desired signals

[WePa99a], [WePa99b], then, together with the matrix $\mathbf{R}_l$ and the temporal covariance matrix $\mathbf{R}_t$, an expression for the matrix $\mathbf{R}_n$ is given by

$$\mathbf{R}_n = \mathbf{R}_l \otimes \mathbf{R}_t . \tag{3.3.30}$$

In [AnMF99] $\mathbf{R}_t$ equals the identity matrix $\mathbf{I}$ and in [WePa99a], [WePa99b] $\mathbf{R}_t$ is a priori known at the RX, since the spectral form of the interference is known. Table 3.3.6 summarises the obtained gains of E_b/N_0 and C/I at a BER of 10^{-3} under consideration of $\mathbf{R}_l$ instead of an identity matrix $\mathbf{I}$ in (3.3.27), (3.3.28) and (3.3.30), depending on the scenario and number K_a of antenna elements. To compare the performance of space-time signal processing with a beamforming scheme such as that presented in Section 3.3.4.3, the corresponding gains by utilising a MMSE detection scheme with beamforming (BF-MMSE) are also given in Table 3.3.6.

Table 3.3.6:   Gains of E_b/N_0 and C/I at a BER of 10^{-3} dependent on the scenario, K_a and under consideration of $\mathbf{R}_l$ compared to a beamforming approach.

Scenario	Indoor			Pedestrian	Vehicular
K_a	2	4	8	8	8
MMSE-BLE [AnMF99]	1 dB	4 dB	16 dB	6 dB	4 dB
BF-MMSE [AnMF99]	*	*	5 dB	0 dB	-2 dB
ZF-BLE [WePa99a]	*	4 dB	*	*	*

The improvements of the spectrum efficiency η, see (3.3.19), and of the spectrum capacity κ [PaHW99] in the uplink of a TD-CDMA system by performing space-time signal processing in the data detection under consideration of adaptive antenna arrays and the spatial covariance matrix $\mathbf{R}_s$ in (3.3.30) compared to single antenna RX are presented in [PaHW99]. Whereas the spectrum efficiency η is already defined in (3.3.19), the spectrum capacity κ of a TD-CDMA system can be determined by taking into account the offered traffic per cell A, the re-use factor r, the user bandwidth B and the area of a single cell A_c in km^2 [PaHW99]:

$$\kappa = \frac{A}{r \cdot B \cdot A_c} . \tag{3.3.31}$$

The spectrum capacity κ is measured in Erl / MHz km^2. The directional channel models used in the simulations are the UKL 2 (University of Kaiserslautern) directional channel models [ScPa98b]. The maximum number K of users is equal to 8, which constitutes a restriction. All other system parameters can be found in Table 3.3.7 and in [Blan98]. For determining the spectrum efficiency η the QoS criterion is specified by choosing the maximum coded BER P_b^M and the maximum outage probability P_o^M to be equal to 10^{-3} and $5 \cdot 10^{-2}$, respectively. Table 3.3.7 shows the improvement factors of the spectrum efficiency η and the spectrum capacity κ for the case of a cross array, with K_a equal to 8 antenna elements. The investigated system operation situations are rural, urban and dense urban, and for the blocking probability P_B is chosen equal to 10^{-2}.

Table 3.3.7:   Improvement factors of the spectrum efficiency η and the spectrum capacity κ between single and adaptive antenna operation, relative to single antenna operation for the case of a cross for three investigated system operation situations; cross array; K_a= 8 antenna elements; blocking probability $P_B = 10^{-2}$.

Channel model	User velocity	Improvement factor for η	Improvement factor for κ
UKL 2 rural	$v = 90$ km/h	10.7	12.6
UKL 2 urban	$v = 30$ km/h	6.4	6.9
UKL 2 dense urban	$v = 30$ km/h	6.0	6.7

3.3.4.4.2. Receive diversity for uplink and downlink

For WCDMA, which is the FDD mode of UTRA, several smart antenna concepts are presented in [JaDC99]. Receive diversity approaches for the uplink and downlink without DoA estimation are considered. To characterise the performance improvements of the different algorithms in terms of the uncoded BER, the conventional RX (CR) for the demodulation of uncoded signals is used as a reference system. The approaches comprise the schemes Linear Minimum Mean-Square Error (LMMSE), Principal Component (PC) MMSE, Maximum Likelihood (ML) and SAGE, see Section 3.3.4.2 [DJFH97], for channel parameter estimation and symbol detection. Except for CR, all other approaches can be seen as space-time signal processing, because spatial and temporal information of the signals is simultaneously exploited. In the uplink the PC-MMSE equaliser and the SAGE algorithm is employed to estimate the values of a parametric model of the received signal that leads to an iterative interference cancellation scheme for channel parameter estimation and symbol detection [JaDC99], [DJFH97]. Whereas both schemes CR and LMMSE employ single user demodulation, the SAGE algorithm is a scheme for joint channel parameter estimation and joint data detection. Figure 3.3.7 [JaDC99] shows the uncoded BER performance of the uplink for the CR, the PC-MMSE approach and the SAGE algorithm, where K equal to eight users are active and K_a equal to three antenna elements are considered at the RX. In the SAGE scheme, two iteration cycles are carried out for the channel estimation using the reference symbols in the DPCCH, and one cycle using the total slot signal, while four iteration cycles are carried out in a multistage detector [DJFH97]. The average SNR $\overline{\gamma}_k$, $k = 1, ..., K$, is the same for each user, the channelisation code length $n_{c,k}$, $\overline{\gamma}_k$, $k = 1, ..., K$, of each user k is 16, and $n_{ref,k}$ equal to 32 symbols is used for channel estimation.

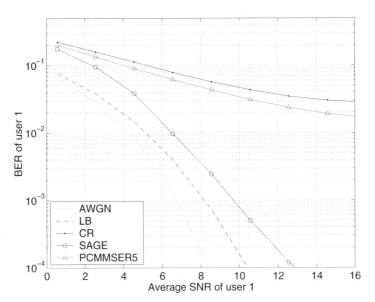

Figure 3.3.7: Uncoded bit error rates for uplink transmission with $K = 8$, $n_{c,k} = 16$, $K_a = 3$, $n_{ref,k} = 32$, $\overline{\gamma}_1 = ... = \overline{\gamma}_8$ and estimated channel parameters.

The other investigated approaches show a performance inferior to that of the SAGE algorithm in the uplink. In the downlink, the ML approach, under consideration of coloured noise, shows the best performance [JaDC99]. Further results show that the BER depends critically on the quality of the channel parameter estimation. Therefore, accurate channel parameter estimation seems to be the most challenging task in the design of advanced RXs with smart antennas.

3.3.4.4.3. Space-time codes

The concept of space-time code was introduced by Tarokh *et al.* [TaSC98] as a generalisation of delay diversity. For space-time coding, at the very least multi-port TX antennas are required. Very often multiple RX antennas are also considered to have a MIMO radio channel. The encoder provides multiple parallel outputs, which are fed to different antennas. In the convolutional case this could be a shift register with multiple combinations of modulo-2 adders. This approach compares with delay diversity, in which the signal taps of the shift register are fed to the antennas [Burr00b]. With such simple space-time coding approaches a diversity gain is achievable, but no coding gain. Nevertheless, the diversity gain can be obtained with only one RX antenna, and does not require knowledge of the channel at the TX, and thus allows diversity to be implemented, where multiple RX antennas are not feasible [Burr00b]. A system architecture called Bell Layered Space-Time, BLAST, is described in [Fosc96], [Burr00b]. The architecture is illustrated in Figure 3.3.8 [Burr00b]. The data is de-multiplexed into n_T parallel streams, which are referred to as 'layers'. These streams are separately encoded using a set of conventional scalar encoders. The encoded streams are then fed to the antennas. However, the connections between streams and antenna are cyclically permuted every few code symbols. At the RX the signals are decoded using techniques taken from multi-user systems like JD as recommended in UMTS. The BLAST approach gives both diversity and coding gain, as well as capacity increase.

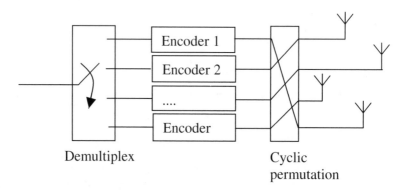

Figure 3.3.8: Architecture of BLAST system.

3.4. Millimetre-wave Propagation

José Fernandes

3.4.1. Path loss modelling

José Fernandes

The distance dependent path loss, at 27–29 GHz, is obtained by analysis of the spatial local mean of each measurement location [KaEm97]. The local means of the isotropic received power relative to the 1 m reference for LoS, OLoS and NLoS measurements in Malmö and Lund are plotted in Figure 3.4.1. The solid line in the figure shows the distance dependent Free Space Path Loss (FSPL) and it is obvious that the local mean of the 18 measurement locations in LoS are very close to the theoretical values. The path loss in OLoS and NLoS is much higher as a consequence of additional attenuation, diffraction and reflection losses at this high carrier frequency.

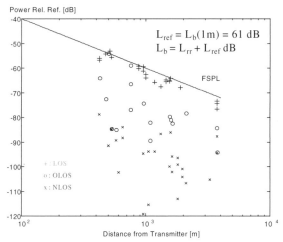

Figure 3.4.1: Received power relative to the 1 meter reference for LoS (+), OLoS (o) and NLoS (x) vs. TX – RX separation.

We have tried to model the additional path loss in OLoS and NLoS as external losses due to the specific objects that disturb the direct propagation path. In OLoS, measurement locations behind trees and in NLoS, locations behind houses, and where a reflected path is dominating, have been used to estimate the losses. The proposed path loss model is then given by:

$$L = L_0 + 20\log(d) + L_{add} \ [\text{dB}]$$

$$(3.4.1)$$

where L_0 is the free space loss at 1 m, d is the separation between TX and RX. The term L_{add} is obtained by statistical regression of the difference between measured local means and the free space loss:

$$L_{add} = L_{meas} - [L_0 + 20\log(d)] \text{ [dB]}$$ (3.4.2)

The values of L_{add} are given in Table 3.4.1, according to the specific measurement location. The locations are divided into a few categories, where the amount of trees in typical OLoS environments and the blocking structure in NLoS is considered.

Table 3.4.1: Path loss parameter values.

Blockage category	Number measur.	Additional path loss [dB]	Error [dB]
LoS	26	-	2.64
2-3 trees	12	17.8	8.21
>3 trees	10	29.2	8.79
House + tree	5	31.9	2.14
Reflections	15	32.4	8.65

Analysis of XPD is made by comparison of spatial local means received by the two antennas with vertical (co-) and horizontal (cross-) polarisation. In LoS, the antenna performance limits the XPD, since measurement results are similar to the antenna specifications, i.e. XPD = 25 dB. In OLoS and NLoS reflected and diffracted components give rise to depolarisation. Accordingly, the smallest XPD is found at locations where the additional path loss is high. The XPD can be as low as 4–8 dB [KaLJ98].

The co-polarised RX antenna was rotated 120° in azimuth for analysis of Angle of Arrival (AoA) and angular spread. The received power vs. angle for a typical LoS, OLoS and NLoS locations are shown in Figure3.4.2. There is no significant power outside the main beam in the case of LoS, and the response is close to the antenna calibration diagram made by the manufacturer Flann Microwave. It is obvious that the direct path dominates in LoS, which is in agreement with the path loss results.

To quantify the power for different AoA shown in Figure3.4.2, the angular spread, S, defined below, is used:

$$S = \sqrt{\int_{-\pi}^{\pi}(\xi - D)^2 P(\xi)d\xi \Big/ \int_{-\pi}^{\pi}P(\xi)d\xi},$$ (3.4.3)

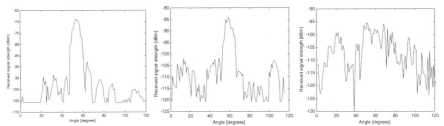

Figure 3.4.2: Typical received angular power in LoS (left) OLoS (middle) and NLoS (right) situations.

where D is the mean angle, defined as

$$D = \sqrt{\int_{-\pi}^{\pi} \xi P(\xi) d\xi \left/ \int_{-\pi}^{\pi} P(\xi) d\xi \right.}} \ . \tag{3.4.4}$$

The integration range should be 2π, for a complete analysis of the AS. But as described above, the maximum range of our measurement system is 120°. It has been assumed that the power from the angular range outside this span can be neglected for a comparison of the AS in LoS, OLoS and NLoS environments. The AS values have been calculated, according to the definitions for some of the measurements, in all three environments. In Figure 3.4.3 the CDFs of the AS for these different environments are displayed. The mean level of S increases from 15° in LoS to 22° and 29° in OLoS and NLoS, respectively.

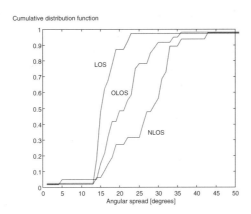

Figure 3.4.3: CDF of Angular spread in LoS, OLoS and NLoS environments.

Results from measured complex IR also show MPCs with excess delay between 20 and 40 ns. The time discrete power values show that such delayed components in LoS are typically 15–20 dB below the total received power, while the relative power of these components is higher in OLoS.

As the 60GHz band has been provisionally allocated to Mobile Broadband Systems (MBS) by ERC recommendation T/R 2203, there is a need for a path loss model that takes into account the specific propagation characteristics of this particular band, mainly the oxygen absorption and rain attenuation.

A model that estimates the average power dependency with distance and frequency at the 60GHz has already been established [CoFr94],

$$P_{r\,[\text{dBm}]} = -32.4 - 30\alpha + P_{t\,[\text{dBm}]} + G_{t\,[\text{dBi}]} + G_{r\,[\text{dBi}]} - 10\alpha\log(d_{\,[\text{km}]})$$
$$- 20\log(f_{\,[\text{GHz}]}) - \gamma_{r\,[\text{dB/km}]} d_{\,[\text{km}]} - \gamma_{o\,[\text{dB/km}]} d_{\,[\text{km}]} \qquad (3.4.5)$$

where P_r, P_t are the received and transmitted powers, G_t, G_r are the RX and TX antenna gains, d is the distance between RX and TX, γ_o, γ_r are the oxygen and rain attenuation coefficients. The crucial parameter of this model is α, since all the others are well known. Values for α have been presented in the literature, ranging from 1.4 to 2.5, see Table 1 in [CoRF97]; however, it is not clear how this parameter depends on the scenario, or on other factors.

Several measurement campaigns were performed in RACE-MBS in various street scenarios, using different TX and RX antennas. Space diversity was used, meaning that two different channels were registered in terms of distance, $P_1(d)$ and $P_2(d)$. The estimation of α was done by calculating the least-square linear fitting of the measured data (with the distances in a logarithmic scale). However, the result depends on the initial value that is considered for the distance, d_{min}, when the fitting is calculated between d_{min} and 130 m. Considering the raw data from the measurements, Figure 3.4.4 shows how α varies with this minimum distance for the two channels; it was observed that the increase of α with d_{min} only occurred until about 35 m.

Both channels show the same behaviour, with an almost constant difference between them (this may be due to problems of unbalanced gain behaviour between the amplifiers in the two channels). The range of variation is the same as the one found in the literature and it should not be expected to have the fourth-power law at this frequency band, since the break-point distance is far beyond the measurement distance [CoRF97]. The value of α depends very much on the measurement conditions, and the extrapolation of this value for a general purpose scenario must be done carefully.

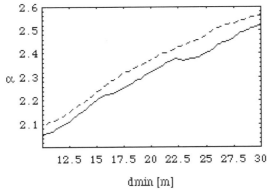

Figure 3.4.4: Variation of α with the minimum distance taken for the linear fitting (___ channel 1, _ _ _ channel 2) (© 1997 IEEE).

In many calculations of system design and performance evaluation (3.4.5) is not straightforward, since it does not present a unique term for the dependence on the distance, thus making it more difficult for application in certain models. A simpler model like those used in UHF, in which there is a global dependence of power with distance by intervals, would be preferable, for each interval $d \in [d_1, d_2]$

$$P_{r\,[dBm]} = C + P_{t\,[dBm]} + G_{t\,[dBi]} + G_{r\,[dBi]} - 10\zeta\log(d_{\,[km]})$$
$$- 20\log(f_{\,[GHz]}) \tag{3.4.6}$$

where C is a constant dependent on the interval, and ζ is the equivalent α parameter accounting for all attenuation with distance.

A similar approach can be considered at the 60 GHz band, by taking linear fittings (at the logarithmic scale) of (3.4.5) by distance intervals, i.e. ζ accounts also for the extra attenuation coming from oxygen and rain. Considering that rain is absent, and that $\alpha \in [2.0, 2.5]$, for a linear fitting of (3.4.5) (with a maximum absolute error of 1.5 dB) we obtain

$$\zeta \in \begin{cases} [2.30, 2.80] \,, & d \in [0, 250]\,\mathrm{m} \\ [3.93, 4.43] \,, & d \in [250, 1000]\,\mathrm{m} \\ [7.02, 7.52] \,, & d \in [1, 2]\,\mathrm{km} \end{cases}, \tag{3.4.7}$$

These values suggest that $\zeta \approx 2.5$ should be used for coverage calculations and $\zeta \approx 4.2$ or $\zeta \approx 7.3$ for interference calculations, due to the different distances involved. If rain is to be considered as well, these values will increase according to the rainfall rate and the respective attenuation. Of course, this approach can be extended to larger distances, if necessary.

The 40 GHz band is another band that has been considered for the MBS deployment, especially in the early phase. The specific frequencies are: [39.5, 43.5] GHz and [62, 66] GHz, with an interval of 2 GHz in between 1 GHz bands. The 40.5–42.5 GHz is also of interest, because it was allocated for Microwave Video Distribution Systems. A detailed study on this matter is presented in [HaJa99]. The propagation characteristics are not equal in these bands, with oxygen and rain presenting different attenuation coefficients values; moreover, these coefficients are not uniform within each of the bands. Since a larger attenuation leads to the possibility of reusing frequencies at a closer distance for approximately the same coverage, the usage of one or the other frequency band has significant consequences on system capacity, and on its cost/revenue performance. Thus, it is important to establish the correspondence between the maximum coverage and re-use distances, R and D, and the Carrier-to-Interference Ratio (CIR), for both bands, and to analyse the resulting consequences, in order to decide under which conditions it is preferable to use one band or another.

There is only a small difference for the free space path loss between both bands, 20 log(60/40) = 3.52 dB. Therefore, it is obvious that the difference between the two bands is not imposed by this parameter. For the oxygen absorption the difference is relevant.

Using the equations of ITU-R [ITUR94] for $f < 57$ GHz and the equations presented in [CoFr94] for $60 \leq f \leq 66$ GHz, one obtains the curves presented in Figure 3.4.5, where the frequency scale is normalised in order to superimpose the 40 and 60 GHz bands in the same graph (–2 GHz corresponds then to the lower limit of each band, 39.5 or 62 GHz respectively, and 2 GHz to the upper limit). In the 40 GHz band, γ_o is almost constant and negligible, less than 0.07 dB/km. However, in the 60 GHz band, the oxygen absorption has to be considered, decreasing from 14 dB/km (at 62 GHz) down to approximately 1 dB/km (at 66 GHz). In the case of the higher frequency band, the additional path loss caused by the oxygen absorption is negligible for short coverage distances, but it can present high values, larger than 10 dB, for typical re-use distances, which results in smaller values for D/R and a larger system capacity. This is, of course, the added value of using the 60 GHz band instead of the 40 GHz.

The rain attenuation has also to be considered. For a rain intensity of $I_r = 30$ mm/h, which occurs in Europe with a probability less than 0.03 %, the rain attenuation is approximately 8 dB/km at 40 GHz, and it slightly increases through the band. At the 60 GHz band, the behaviour is similar, with a value of the order of 12 dB/km. One may conclude that the presence of rain causes worst-case situations for the received power, which should be considered in the cellular design. Nevertheless, the difference between the two bands is not as significant as the one concerning oxygen [VeCo97]. In this reference, an analysis of the CIR can be found.

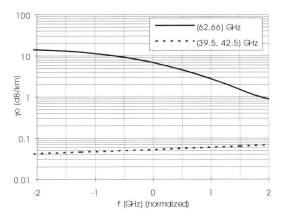

Figure 3.4.5: Oxygen attenuation coefficient as a function of normalised frequency for the 40 and 60 GHz bands (0 corresponds to the centre frequency of each band).

Path loss within buildings has also received attention. Attenuation distance dependence at 2, 5, 17 and 60GHz bands has been evaluated by measuring the received power within a room and across floors with different wall types. LoS and NLoS locations were considered for two antenna heights. Within a room the LoS path loss is typically less than that predicted for free space loss, with a greater scatter of results at 60 GHz [Nobl99].

Between rooms, the path loss increases with penetration through walls. At 60 GHz the signal barely passes through a plasterboard wall and was observed to spread out from a doorway. For an accurate channel model, the prediction of large-scale path loss combined with the small-scale effects of deep fades across both frequency and space will be required [Nobl99].

3.4.2. Wideband channel modelling

José Fernandes

Since the widespread use of high-speed data communications, requiring robust and reliable transmission, is envisioned in present and future millimetre-wave (mmw) broadband applications, an accurate analysis of the radio propagation and the development of adequate propagation models are essential.

3.4.2.1. Deterministic modelling

Dirk Didascalou and José Fernandes

Future Intelligent Transportation Systems (ITS) will deeply rely on mmw for communication and sensing purposes. The propagation mechanisms, reflection and diffraction of mmw at realistic road surfaces and geometries are investigated theoretically and validated by measurements at 77 GHz. The effect of guided waves underneath a vehicle's underbody is observed, which can be exploited in a favourable way for range extension in OLoS situations. It is also shown that the local curvature of road surfaces leads to focusing and defocusing effects of the transmitted energy, resulting in an increased signal fluctuation. Furthermore, the importance of diffraction phenomena at road crests is stressed, which may play an important role in security sensitive applications like collision avoidance systems, etc.

3.4.2.1.1. Specular reflections in mobile-to-mobile scenarios

Most of the available models for the inter-vehicle channel consider flat road geometry and approximate the reflectivity of road surfaces at grazing incidence equal to –1 [Schä93]. For mmw, asphalt layers act as slightly rough dielectric surfaces. Their polarisation dependent reflection coefficient (including slight surface roughness) can be computed using the so-called modified Fresnel coefficients as functions of the grazing angle with respect to tangential incidence [BeSp63]. They expand the well known Fresnel formulas, which deliver the reflection coefficients of an ideally smooth dielectric surface at horizontal (h) and vertical (v) polarisation. The value of the permittivity ε_r has been measured for a variety of dry asphalt surfaces at the frequency $f = 77$ GHz to a mean of $\varepsilon_r = 4.4 - j0.3$ [SWWN97], [Schn98].

Generally the assumption is made that an obstruction of the LoS path between two communicating vehicles by an intermediate third vehicle results in an interruption of the radio link. This is true if the obstacle (here: a vehicle) is non-permeable for the electromagnetic wave. In reality, however, the road surface and a car's underbody act as a waveguide: the electromagnetic wave can 'tunnel' between them via a reflection (road $\rightarrow$ underbody of the car $\rightarrow$ road, etc.). This effect has been experimentally verified by measurements with a high-resolution instrumentation radar at 77 GHz [Schn98], [ScDW00].

Considering the direct ray and one reflection on a flat road surface, it results in the well known two-path model. Applying GO rigorously, a more accurate modelling can be achieved. Consider the mobile-to-mobile scenario of Figure 3.4.6 a concave road course, built up by two flat parts and

an intermediate arc of the inner side of a cylinder. The receiving vehicle is
moved from the reference point at $d = 0$ m to the positive d-direction. Four
different types of rays can actually reach the RX: the direct path; a reflected
ray on the lower flat part (ray A); a reflected ray on the inclined flat part (ray
B); a reflected ray on the curved part (ray C).

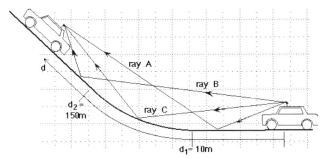

Figure 3.4.6: Mobile-to-mobile scenario with concave street course.

Not all of the four ray types are always present, but up to four rays can reach
the RX at a time. The complex superposition of all arriving rays at the RX
represents a simple propagation model for this mobile scenario. A
simulation with a cylinder radius of $R_0 = 791$ m and the transmitting and
receiving antennas located at 80 cm heights was performed at 60 GHz.
Figure 3.4.7 depicts the resulting overall attenuation at the receiving vehicle
as a function of distance d together with the free space attenuation.
Additionally, the number of rays for each reflection ray-type that occur at
the same distance, are also given.

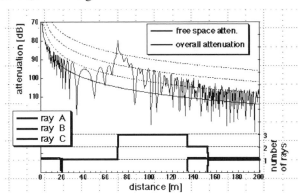

Figure 3.4.7: Simulation results for the mobile-to-mobile scenario of Figure
3.4.6 with parameters $h_{Tx} = h_{Rx} = 0.8$ m, $R_1 = 791$ m and
$f = 60$ GHz.

The results can be summarised as follows: if non-flat road geometry is considered, a simple two-path model is clearly insufficient and a more precise RT based on GO should be performed. Compared to a two-path model, an increased mean reception power results for a concave road course, the variation of the signal is higher for non-flat road geometries, which results in a greater standard deviation, the interference pattern changes more rapidly due to an increased number of received rays, resulting in shorter periods of high signal reception.

3.4.2.1.2. *Millimetre-wave scattering and attenuation in bounded vegetation*

Although in most cases the scenario geometry dominates the propagation mechanisms, vegetation, especially the trees, might influence the field strength in certain locations. A numerical algorithm based on the radiative transfer equation is presented, which allows us to calculate the attenuation of mmw propagating in bounded vegetation. A qualitative justification of the approach is given as well as a comparison with scattering measurements of ficus trees at 75 GHz.

A model based on the theory of radiative energy transfer, which is deduced from the power conservation law, that is formally expressed by the Radiative Transfer Equation (RTE), is proposed. In its differential form the RTE indicates the change of intensity I (power per unit area) along a differential path ds in the medium. The general vector form of RTE is given in [Ishi78].

If the vegetation layer is assumed to be statistical homogeneous, a method which is suitable for calculating the attenuation in isolated vegetation structures can be obtained if the RTE is discretised [DiYW99]. This will enable the RTE to be applied on an appropriate volume, and to be solved numerically for the unknown scattered intensity.

The vegetation structure is divided into several non-overlapping cubes of identical dimensions Δs. Each cube is an elementary volume localised by its discrete position (k,l,m) in space. The sum of all elementary volumes constitutes the vegetation structure and the RTE is discretised to apply for each of these elementary volumes [DiYW99].

The solution for a single elementary volume can be used to determine the mutual interaction between different elementary volumes and to calculate the attenuation in the vegetation. The vegetation structure is split into multiple parallel layers; each of the layers is normal to the direction of incidence of I_0, and consists of elementary volumes. The intensity I can be calculated successively for all layers 1 to K_{dim} according to the algorithm given in [DiYW99], [DiYW00].

It should be noted that the discrete iterative RTE algorithm is only used to determine the attenuation of a vegetation volume in the forward direction. Back or bistatic scattering cannot be treated by the algorithm because of the iterative structure of the approach, which is trimmed to the determination of the forward scattering. The simulations have been compared with measurements and the simulation results lie in between the limits of the standard deviation of the measurements, [DiYW99].

3.4.2.1.3. Sensitivity of ray tracing field prediction

Among the various models, 3-D Ray Tracing (RT) models, which operate on a site specific, detailed description of the environment, are now acknowledged as the most accurate and powerful [CDFR96]. In fact, the completely man-made indoor environment can be usually represented by a set of plane walls, and is thus very suitable for RT field prediction. However, the geometrical and electromagnetic characteristics of each environment element, as well as the characteristics of furniture, are often roughly known or even *a priori* unknown.

In the following, the effect of furniture is inspected by performing RT simulations with and without introducing as input a description of furniture into the simulator. The environment is a typical office building, see [DCFR97] for details. A comparison between field prediction performed with and without furniture is shown in Figure 3.4.8. It is evident that the introduction of furniture helps in decreasing the mean error from 5.2 to 2.2 dB. Also the standard deviation of the error decreases from 4.5 to 3.7 dB. Nevertheless, even prediction without furniture shows an acceptable accuracy if we consider that the adopted environment is a very challenging NLoS topology. Then, the sensitivity of RT prediction to inaccuracies in the electromagnetic characteristics of construction materials is analysed.

The effects of variations of ε_r and σ are shown in Figure 3.4.9, respectively. For simplicity, furniture is not considered here. Reference values are ε_r (relative permittivity) = 2 for internal walls, ε_r = 3 for floor, ceiling, external walls and glass windows; σ (conductivity) = 0.01 S/m for all walls. Note that the average error always increases with ε_r and decreases with σ. As expected the reference values, which are taken from the literature, are not exactly the best choice. However, even this 'blind choice' for ε_r and σ yields acceptable results. Since two or three walls are interposed between TX and RX the received power decreases with σ as transmission power loss increases. Probably, a different behaviour would show up in a LoS situation.

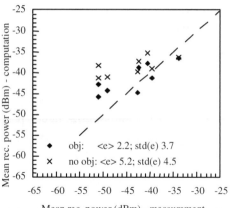

Figure 3.4.8: Mean received power: RT prediction vs. measurement. Dark dots: prediction with furniture (objects); light crosses: prediction without furniture.

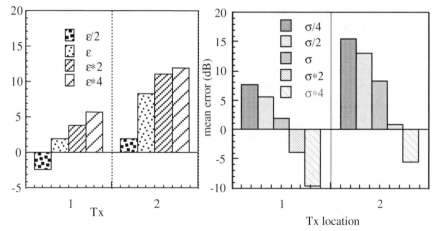

Figure 3.4.9: Mean received power error (prediction-measurement) for different ε_r and $\sigma = 0.01$ S/m (left), and for different σ and ε_r fixed to the reference value (right).

3.4.2.2. Combination of statistical and deterministic approaches

José Fernandes

A realistic ChIR model should be able to take into account the propagation environment and the antenna configuration. This part should be handled deterministically once it is not possible to establish a channel profile for all kinds of scenarios and for all kinds of antenna configurations, while the fast variations should be made statistically. This approach has the advantage of keeping the deterministic part simpler (only a simple description of the scenario is necessary) and versatile to use in system simulations.

A wideband site-specific radio propagation channel model, consisting of a combination of a deterministic and statistic parts is presented. The deterministic part generates the channel profile, taking into account the influence of the propagation environment and the antenna configuration once they have a strong impact on the channel IR, while the statistical part is used to include the small-scale variations of the amplitude and phase. The model is tapped-delay-line based, and it is shown that the Rice distribution is adequate to model the fast variations of the amplitude of each tap, the respective parameters being extracted from the deterministic modelling part. Moreover, a statistical phase distribution about the phase of the dominant MPC in each time bin proves to be more appropriate than the uniform one. Due to the lack of experimental measurements, the results of the model are assessed via a RT tool [FeNS95], previously validated with measurements, and have shown the validity of the model.

3.4.2.2.1. Deterministic modelling

[FeNe97] presents an analytical propagation channel model for the mmw band, based on high frequency ray theory approximation, where the propagation environment is described by its dimensions and reflection properties. This model consists of a set of equations derived from the propagation theory and is able to take into account the propagation environment characteristics, as well as the location of the transceiver antennas and their radiation patterns. It can be regarded as an extension of the Friis formula for multipath propagation. To include the radiation patterns, the model estimates the departure and the arriving angles to the transceiver antennas in order to evaluate the amplitude of the radiated electromagnetic field in each direction and the effective electric length of the receiving antenna. The amplitude of each MPC is evaluated through its travel distance between the transmitting and receiving antennas, which depend on their location, the physical dimensions and the electromagnetic properties of each surface. The excess time delay of each MPC is calculated from its travel distance and then the site-specific channel profile with an infinite time resolution is constructed. The validation of the model presented in [FeNe97] demonstrates that it can predict, with a good accuracy, the average signal level and the PDP in any position of the MS and with different antenna types.

Figure 3.4.10 depicts the PDP calculated by the analytical model in an empty room ($13 \times 9 \times 4$ m^3), made of low reflective materials. The BS and the MS use a biconical horn antenna with 10° beam width in the elevation plane and vertically polarised. The BS is centrally placed in the room at 2.6 m height and for this specific PDP the MS is 4.3 m away from the BS.

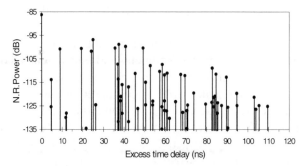

Figure 3.4.10: Typical PDP with the MS at 4.3 m from the BS (© 1998 IEEE).

3.4.2.2.2. Statistical modelling of envelope distribution

In a tapped-delay-line model [Pars92], each tap coefficient results from the vectorial sum of the MPCs arriving at the respective time bin. As a result of high phase variations of such components, due to small displacements of the MS, the amplitude and phase of each tap varies rapidly.

Rice distribution has been selected as being most appropriate to model the fast variations of the signal amplitude of each tap [MaFN98]. The parameters of the statistical distributions for all taps are derived from the output of the analytical model referred to above. The methodology consists in selecting the strongest path in each time bin and calculate the Rice factor based on its amplitude and on the mean power of the other components. This is a key issue of this approach because the statistical distribution of each tap is established according to the propagation environment and the antenna configuration expressed by the deterministic part. The derived parameters are valid in a small area where the mean power can be considered constant. For larger displacements of the MS, the procedure is to calculate a new channel profile by using the analytical model, and again calculating the parameters of the statistical distributions for all time bins. This brings the possibility to have a channel model for any scenario with any antenna configuration and in any location of the MS.

3.4.2.2.3. Phase modelling

It has been a common practice to consider the phase uniformly distributed in [0, 2π] [Hash93]. This is only true when several MPCs are present and none of them is dominant. On the other hand, the phase does not change abruptly as the MS moves, and it tends to follow the phase of the dominant MPC in case it exists. It is not important to predict the absolute phase value of a MPC at a fixed point in space; the modelling effort should rather be placed on the phase evolution as the mobile moves. A model for the phase evolution of each tap during a continuous movement of the MS is presented hereafter.

The phase of the dominant ray-path arriving in a specific time bin can be easily modelled through a deterministic increment between adjacent positions. Using the arriving angles (θ and ϕ) calculated by analytical model [FeNe97] the phase evolution can be written as

$$\varphi(j) = \varphi(j-1) + \Delta\varphi, \quad j = 1,2,3,\dots \; ,$$ (3.4.8)

with $\varphi(0) = 2\pi L/\lambda$ and

$$\Delta\varphi = \frac{2\pi}{\lambda} \Delta s \cos(\gamma - \phi) \cos(\theta) \; ,$$ (3.4.9)

where L is the ray-path length, Δs is a very small spatial displacement between the $(j-1)$th and jth positions, λ is the wavelength, θ and ϕ are, respectively, the elevation and the azimuth angles of the dominant ray-path and γ is the angle of the MS movement direction [Pars92].

Beyond the dominant MPC, there are usually other components that introduce random phase variations, which should be added on. Similarly to the amplitude, this part is modelled statistically, being the PDF of the phase in a Ricean fading channel given by [BiDI91]

$$pdf(\xi) = \frac{e^{-A^2/2\sigma^2}}{2\pi} \left[1 + \sqrt{\pi}\psi e^{\psi^2} \left[1 + erf(\psi) \right] \right]$$ (3.4.10)

with

$$\psi = A\cos(\xi)/\sigma\sqrt{2}, \quad |\xi| \leq \pi \; .$$ (3.4.11)

Note that when A (deterministic amplitude of the stronger path) tends to zero (Rice tends to Rayleigh distribution), $pdf(\xi)$ approaches $1/2\pi$, i.e. the phase distribution of that specific tap is uniform over the interval $[0, 2\pi]$.

3.4.2.2.4. Comparison with simulations

Due to the lack of experimental measurements, a RT tool [FeNS95] was used to validate the model. The accuracy of the model is very dependent on the accuracy of the deterministic part. Fortunately, this model has been widely validated with measurements being some of those results shown in [FeNe97]. The results of the model were extensively compared with simulations obtained with a RT tool, and a sample of those results is shown in Figure 3.4.11, see [MaFN98] for more details. Figure 3.4.11(left) depicts the CDF of the amplitude for the first, fourth and fifth taps. As it can be seen, the fading depth is quite different between them and is directly related with the Rice factor shown in Table 1 in [MaFN98]. There is a good agreement between the results obtained with the RT tool (Simulation in the figure's legend) and with the model.

These two examples show that the Rice factor can vary significantly from tap to tap; the first tap is not always the strongest one, and this depends on the position of the MS in the cell, which is related to the coverage provided by the antenna configuration. Other configurations have shown that a quite different channel profile and the corresponding parameters can be obtained for the same positions of the MS. This illustrates how important it is to use site-specific models for the design and system performance evaluation.

The phase of the taps with a high Rice factor have a low variation about the phase of the dominant component following its deterministic evolution, while for very low Rice factors, the distribution broadens having a higher variation range, tending to uniform distribution. This is illustrated in Figure 3.4.11(right) that depicts the phase evolution obtained with the RT tool, with (3.4.8) for the dominant ray and the result of the proposed model calculated by combining (3.4.8) and (3.4.10). Although we cannot rely on the absolute phase values obtained through RT in a fixed point in space, it provides an estimation of the phase evolution and it appears that the proposed model performs quite well. Moreover, it tends to be more realistic than the ones previously proposed once it uses information from propagation environment.

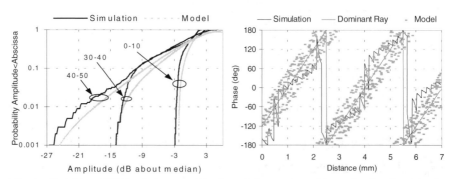

Figure 3.4.11: CDF of the amplitude of three time bins obtained with the model and with RT (left); phase evolution of the time bin 30–40 ns obtained with RT, with (3.4.8) for the dominant ray and with the proposed model (right) (© 1998 IEEE).

3.4.3. Impact of shaped lens antennas on the ChIR and cell coverage

José Fernandes

3.4.3.1. Lens antennas for mmw applications

Carlos Fernandes

Particularly at millimetre waves, antennas decisively influence the system performance. The power that is available from solid-state devices is quite limited and its distribution over the cell must be carefully considered when designing the antenna. On the other hand, the antenna should discriminate the MPCs to a certain extent, thus contributing to reduce the channel time dispersion. Simultaneously these antenna requirements should not entail any restriction of terminal mobility. Adaptive multibeam antennas are the current approach used to cope with this type of requirement at lower frequencies. However, especially for mmw systems, an efficient and inexpensive implementation of such an antenna is still not available.

An affordable, yet quite effective, alternative solution is provided by dielectric lens antennas, which can easily produce different types of highly shaped beam radiation patterns to significantly enhance the system performance. These antennas have been extensively studied and tested experimentally in the context of wireless broadband communications, with very good results [FeFB95], [FeFe99], [Fern99]. The dielectric lens antenna is rugged and simple to fabricate, since tolerances are acceptable, and moulding can be used [FBZF98].

The idea is to design the BS antenna to produce a $\sec^2 \theta$ pattern within the cell limits, while the mobile or portable terminal antenna is designed to produce a hemispherical pattern. This combination ideally compensates the free space attenuation at each direction θ, so that a uniform received power distribution is obtained over the cell. A remarkable characteristic is that the cell dimensions are scaled to the antenna height, which allows antennas with a single design to be used for different cell sizes according to the installation height. This characteristic further enables the control of the wall illumination at the cell edge to maintain a compromise between multipath effects and the need for alternative paths in the case of LoS blockage.

Figure 3.4.12 shows the basic antenna configuration: a homogeneous dielectric body with an embedded feed. Possible feeds are the aperture of a rectangular or circular waveguide, a coaxial feed, or a printed element. The polarisation can be linear or circular.

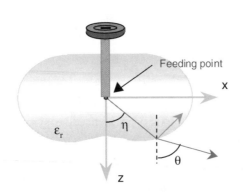

Figure 3.4.12: Dielectric lens antenna fed by a waveguide. Lens configuration (left); Lens prototype for $\sec^2\theta$ pattern (right) (© 1999 IEEE).

The lens surface is designed to transform the feed radiation pattern into the desired output pattern. Lens dimensions are large in terms of λ so that GO can be used for the design. This approach yields the best design in terms of operation bandwidth. Only circular symmetric lenses are treated here. The details of the formulation are addressed in [SaFY98]. Usually, a slight correction of the GO profile is required to take into account diffraction effects, and the fact that $\lambda \neq 0$ [LFSB96]. Calculation of the lens radiation pattern using the PO formulation allows the assessment of the lens correction prior to fabrication.

Circular polarisation is particularly useful when coverage is also required in the region directly below the BS [FBZF98]. For circular polarisation, lenses may be excited by the TE_{11} mode of the circular waveguide with circular polarisation. The resulting measured elevation pattern for a BS antenna is presented in Figure 3.4.13 (left), superimposed on the ideal $\sec^2\theta$ pattern. The horizon corresponds to $\theta = 90°$, and the nadir to $\theta = 0°$. The lens is designed for $\theta_M \approx 75°$ cut-off angle. The coincidence between the measured pattern and the desired $\sec^2\theta$ characteristic is quite acceptable, showing a sharp fall of the radiation for $\theta \approx \theta_M$. The pattern is circular symmetric, and the directivity calculated from the measured patterns is $D = 9.4$ dBi.

The same TE_{11} mode excitation with circular polarisation may be used for a mating MS antenna. The desired elevation characteristic in this case is flat top up to $\theta_M \approx 75°$ and null radiation outside this interval. This flat top characteristic and the pattern circular symmetry favour free movement of the mobile within the cell limits, and even some tilting of the MS. Figure 3.4.13 (right) shows the measured pattern of a 66 mm lens designed according to the above specification. For the MS antenna, $\theta = 0°$ corresponds to the zenith, and $\theta = 90°$ to the horizon.

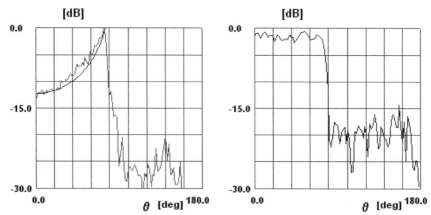

Figure 3.4.13: Elevation power patterns of BS and MS lenses measured at $f = 62.5$ GHz. $\mathrm{Sec}^2\theta$ pattern of BS lens (left); flat top pattern of MS lens (right).

Note that the product of the BS and MS patterns is at least 30 dB below maximum for any angle $\theta > \theta_M$, meaning that all the reflected paths falling in this interval are strongly attenuated, which minimises multipath fading. Note that these regions correspond to ground, ceiling, and part of the wall area at the edges of the cell.

The TM_{01} mode of the circular waveguide can be used to excite the BS lens to obtain a linearly polarised coverage. The resulting radiation pattern is similar to the one presented in Figure 3.4.13 (left) except that, intrinsic to the linear polarisation, a null appears for $\theta \approx 0°$ (direction of BS nadir).

For the MS, the hemispherical characteristic can be intentionally modified in this case to eliminate radiation towards the region bellow the BS antenna. This slightly improves gain, and rejects some multipath pick-up. Figure 3.4.14 (left) shows a MS antenna prototype designed to radiate a linearly polarised flat top characteristic only for $60° < \theta < 95°$ [FeFA97]. The lens is fed inside its body by the aperture of a coaxial line with extended central conductor. This particular pattern allows for a double lens coaxial mounting for diversity at the receive branch, without the risk of mutual blocking.

The top lens is shared by the *TX* and the RX, and the lower is used by the second diversity receiving channel. The radiation pattern is circular symmetric, and E_θ is the dominant component of the radiated field.

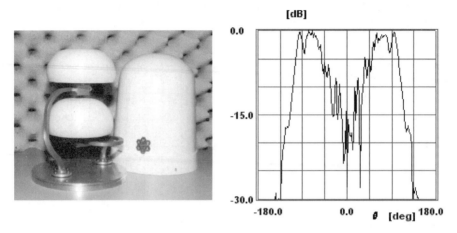

Figure 3.4.14: MS antenna assembly, and radiation pattern measured at 62.5 GHz. Antenna prototype (left) (© 1999 IEEE); elevation pattern of top lens (E_θ) (right).

It is stressed that the shaped beam concept presented here is not restricted to fixed beams. It can be conciliated with fast beam switching capability (5 ns) in a single dielectric lens for improved performance [Fern99]. Moreover, the same lens antenna configuration is compatible with electronic azimuth scanning of a beam shaped in the elevation plane.

3.4.3.2. Cell coverage

José Fernandes

The measured antennas' radiation patterns were included in a radio propagation simulation tool [FeNS95] in order to evaluate its performance in typical environments. Two main parameters are used to evaluate cell coverage:

- The Normalised Received Power (NRP = P_r / P_t) defined at the antenna input/output terminals. This parameter includes the gain of transmitting and receiving antennas and is of major importance for the budget link calculation and to define the cell boundaries;
- The Sliding Delay Window (SDW) containing 90 % of the energy of the ChIR as a measure of the channel time dispersion.

CW measurements were performed in a small room to evaluate the antenna configuration presented in Figure 3.4.13. The ceiling and sidewalls material is concrete, and the floor is covered with 1' thick ceramic tiles. An anechoic chamber with laminated wood external walls, tables, shelves and equipment racks are the main objects in this room.

Figure 3.4.15 shows the BS antenna location and the measured received power distribution along linear paths taken in the uncluttered part of the room, for $\Delta h = 1$ m. The circular symmetric nature of the cell is quite apparent. The received power level is reasonably constant within about a 3.5 m radius, and falls off rapidly outside this region. A quantitative measure of this behaviour is given in Figure 3.4.15 (right) that corresponds to the closest path to BS taken along the room length with simulation results superimposed, showing a good agreement. Despite the omnidirectional characteristic of BS and MS antennas, fading depth within the cell is negligible due to weak illumination of the walls.

Further simulations were performed in an empty but otherwise similar room for different BS antenna heights (h_b) and MS antenna tilting angles (γ_m), to demonstrate the performance of this antenna configuration as well as the allowed user movement freedom. The simulation uses measured antenna radiation pattern data. The MS antenna height (h_m) is always kept at 1.5 m. Figure 3.4.16 shows the distribution of the normalised received power, NRP, calculated for $h_b = 3$ m. As expected, the cell is larger than in Figure 3.4.15 because in this case Δh is increased by a factor of 1.5, but previous conclusions about the uniformity of power distribution and sharp cell boundary apply again.

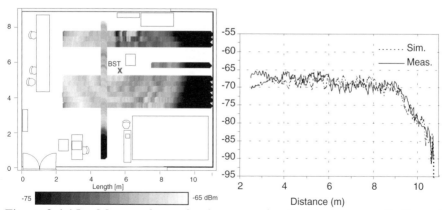

Figure 3.4.15: Measured received power in the test room (left) and comparison with simulations along a longitudinal path (right), $\Delta h = 1$ m, $h_b = 2.5$ m (© 1999 IEEE).

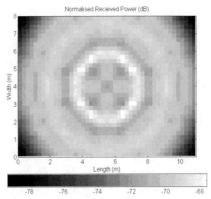

Figure 3.4.16:   Simulated NRP distribution along the room for, $h_b = 3$ m (© 1998 IEEE).

The corresponding CDF in Figure 3.4.17 reveals that in 80 % of the room NRP > –69 dB and only in 10 % is below –73 dB. For $h_b > 3.5$ m, the NRP distribution tends to be uniform all over the room since the circle defined by the BS antenna radiation pattern covers the whole area.   However, as h_b increases, the channel time dispersion (here represented by the sliding delay window, SDW 90%) also increases as the MS picks up more and more significant MPCs mainly reflected from the side walls. This is an important characteristic of the proposed antenna configuration, since alternative paths can be provided in a controlled way through h_b to cope with possible LoS blockage. The effect of LoS blockage, movement freedom of the MS and other characteristics were studied, and details can be found in [FeFe99].

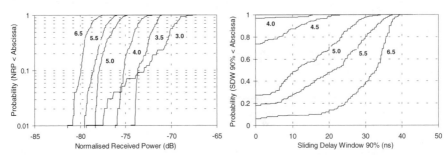

Figure 3.4.17:   CDF of the NRP and SDW for several h_b values, in meters (© 1998 IEEE).

Another antenna configuration, specially dedicated for outdoor and large indoor environments, is sketched in Figure 3.4.18 and is called a wide cell. The BS antenna alone provides the secant-squared characteristic, so the matching MS radiation pattern corresponds to a hemispherical pattern, like the one represented in Figure 3.4.14.

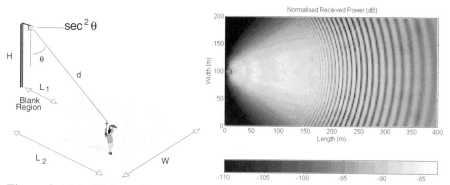

Figure 3.4.18: Wide cell antenna configuration (left) (© 1999 IEEE); NRP distribution in open space with the BS located at (0, 100, 12) m (L, W, H) and using MRC space diversity technique at the MS (right) (© 1998 IEEE).

According to typical values achieved with current technology for MBS operating at mmw frequencies, the NRP can be as low as –95 dB [FeFe98]. Figure 3.4.18 (right) shows the NRP distribution in an open area and it can be seen from that some fading starts to appear at 150 m away from the BS due to the interference between direct and ground reflected rays, increasing towards the end of the cell. For the above-mentioned threshold of –95 dB, the cell dimensions can be as high as 200 m by 50 m. Besides the good power distribution, this antenna configuration allows complete movement freedom of the MS in the horizontal plane, and also some tilting in the vertical plane. It can be used to cover city squares, pavilions, wide streets, etc. In order to test the tilting freedom of the MS in the elevation plane and the pointing sensitivity of the BS antenna in this configuration, several simulations were performed for different misalignment situations, [FeFe98].

'Lourenço Peixinho' is an avenue located in the central part of Aveiro, and depending on the time of the year, the trees' dimensions and their foliage density are susceptible to causing some propagation problems, due to signal attenuation, which causes some shadow regions. Hence, cellular planning should be performed in such a way that those problems can be minimised.

In [FeFe98] it was seen that with a wide cell antenna configuration in a 40 m wide city street, a cell length of 300 m by the total street width can be achieved. The larger cell length compared to that obtained in an open area, is due to the MPCs that originated on the lateral walls. As long as there are no objects within the street, there is no problem concerning either shadow fading or time dispersion. In order to evaluate the influence of trees on the signal propagation in this scenario, two different solutions were studied, considering the trees modelled as two cylinders [FeGa99], [FeMG99]. One solution uses one BS located in the central region of the avenue, as can be seen in Figure 3.4.19 (a). The shadow regions imposed by trees are very significant, mainly in the regions further away from the BS, leading to an inefficient coverage of the avenue from the point of view of the NRP distribution. The larger shadow regions located further away from the BS are mainly due to the attenuation caused by the trees. The other solution uses two BSs located near the side buildings, and the MS will select one of them according to the received power level, with the results being presented in Figure 3.4.19 (b). This brings us to the conclusion that the use of two BSs almost permits us to overcome the shadow regions imposed by the trees. From the results, it was observed that the SDW values are very low, making it possible to achieve very high bit rates with a low equalisation effort.

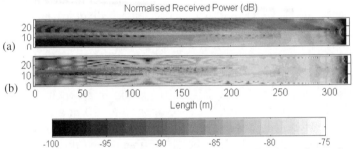

Figure 3.4.19: (a) BS located at (320, 14, 12); (b) Combination of two BSs, one located at (320, 25.7, 10) and the other at (320, 2.3, 10).

The previous results show that multiple-coverage is an efficient technique to cope with signal attenuation and shadow fading, but the inherent need of different frequencies at the BSs is an overhead, which can be critical in system deployment. The use of equal frequencies at the BSs is advantageous, especially in terms of system capacity and the overhead caused by the need of inter-frequency HO, and was studied in indoor and outdoor scenarios by extensive simulation; see [FeGa99] for details.

The interference of the additional MPCs from the second BS causes fading and an increase of time dispersion. The increase in time dispersion is not critical, because the buildings confine the MPCs and consequently the time dispersion can be related to the street width, and hence is considerably lower than the system's limits for typical street widths. Therefore, the main problem is to reduce the fading depth in the power distribution. Figure 3.4.20 (left) depicts the power distributions along the central path of the avenue, obtained using different frequencies at the BSs. The spatial diversity reduces the fading depth in about 10 %. Using equal frequencies at both BSs, the fading depth can increase up to 26 % when compared with the case where two different frequencies are used, Figure 3.4.20 (right). In this case the space diversity reduces the fading depth in the order of 15 %, which makes the penalty in the power distribution caused by the use of equal frequencies at the BSs not so severe. The net result is that we have up to 10 % degradation when compared to the use of different frequencies [FeGa99].

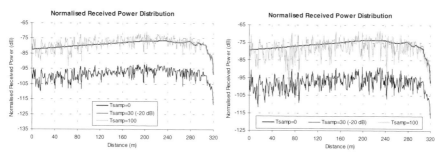

Figure 3.4.20: NRP distribution along the central path for different Tsamp with diversity, using different frequencies at the BSTs (left) and equal frequencies (right).

It is clear that along the avenue's axis, the distances between both BSs and the MS are similar and hence the amplitudes of the direct rays are identical when there are no obstacles in the scenario, resulting in severe destructive interference. The decorrelation of those components can be achieved by shifting the position of one BS, and therefore the two direct rays suffer different path losses and their amplitudes are no longer identical, resulting in lower fading depth. In this new configuration, one BS was shifted horizontally along the avenue by 45 m. The shifting distance cannot be too high to keep the time dispersion to a controlled level. Results obtained by simulation are shown in Figure 3.4.21 (left), and comparing them with the previous ones it can be concluded that by shifting one BS an additional reduction of the fading depth up to 13 % can be achieved. Resorting to space diversity, the fading depth can be further reduced up to 8 %, hence through the combination of BS shifting and space diversity a reduction of the fading depth in the order of 16–21 % can be achieved. As expected, shifting one BS and using spatial diversity at the MS leads to a more uniform power distribution, similar to that obtained when different frequencies are used at the BSs. The better uniformity in the power distribution reflects the lower fading deepness obtained, 3–4% smaller than that obtained with different frequencies at the BST.

Figure 3.4.21 (right) summarises the results and main conclusions drawn from the study of multiple-coverage techniques using equal frequencies at the BST applied to 'Lourenço Peixinho' avenue. Identical improvement results were obtained in a large sports pavilion [FeGa99].

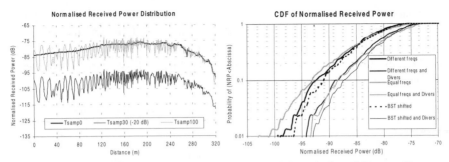

Figure 3.4.21: NRP distribution along the central path for different Tsamp, using equal frequencies at the BSs and asymmetric positioning of the BS (left); CDF of the NRP considering Tsamp = 30 ns.

3.4.3.3. Impact on transmission performance

José Fernandes

With the antenna configuration presented Figure 3.4.13 and using a moderate equalisation effort in the RX, e.g. a DFE with seven taps, it is possible to achieve a Carrier Bit Rate (CBR) of 170 Mbit/s (SDW < 40 ns) or even higher, as explained in [FeFe99]. On the other hand, a significant high bit rate can be achieved without equalisation: using the well known rule of thumb that the RMS DS cannot exceed 10 % of symbol time Ts, a symbol rate of 10 Msym/s can be achieved in a LoS condition.

Some results obtained from the field trials performed in the framework of the SAMBA project are presented to highlight the impact of the wide cell antenna configuration on system performance. A simplified trial platform composed of a BS and MS and a Control & Monitoring system able to collect the measurement data of various parameters, namely the RSSI, BER and real-time position indication of the MS, was used. The trial platform was installed in different locations in Aveiro City, Portugal, and measured data was acquired for processing [DFPH99]. The MS was mounted on a van for outdoor operation at speeds of up to 50 km/h. The infrastructure comprising cables, a BS tower, the electrical power supply, communication equipment and a van are shown in Figure 3.4.22.

The wide cell antenna configuration was used in 'Nova' Street with the BS located at 11.3 m in height. Figure 3.4.23 depicts the received power in both channels (Ch.1 and Ch.2), the combination of the signals (diversity reception), using MRC, and the BER as a function of the distance between the BS and MS measured along a horizontal path. The received power level was sampled once per frame [DLPZ97], [FDPH98], and the BER was calculated as the ratio of the number of bit errors detected in one frame to the number of transmitted bits. The MRC and Ch.2 curves are shifted by +10 and –20 dB, respectively, for better legibility. The results shown correspond to a central path along the lane, where the van is shown in Figure 3.4.22, which is a central path along the cell.

Figure 3.4.22:   Van and BS tower in 'Nova' street.

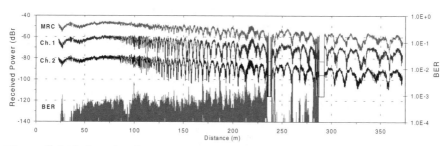

Figure 3.4.23: Received power and BER per frame vs. distance measured along a path. Ch.2 is shifted –20 dB and the MRC +10 dB.

The average received power level dynamic range is rather low due to the shaped radiation pattern of the BS antenna. Looking to the curves, we can conclude that the cell length is larger than 200 m, which complies quite well with the predictions in [FeFe98]. The fading depth increases significantly for distances above 100 m as the MS moves away from the BS. When the mobile is at 235 m and 285 m, the radio link breaks down due to a deep fade, and it will not recover after the second break. This is the reason why there are no errors for distances above 285 m.

The CDF of both channels and the respective MRC, calculated for two short path intervals, are shown in Figure 3.4.24. The theoretical Rice distribution calculated for Ch.1 and the MRC is superimposed. For Ch.1 the Rice factor is 19.7 and 3.7 dB for the 45–55 m and 100–110 m intervals while the Rice factor corresponding to the MRC is 21.7 and 12.9 dB, which complies with the fading behaviour of the curves in Figure 3.4.23. The Rice factor increases more for the MRC curve in the interval 100–110 m, which is evident in Figure 3.4.23 and Figure 3.4.24. A very good match was obtained between the theoretical Rice distribution and the corresponding measured data, leading again to an almost complete overlap of the curves.

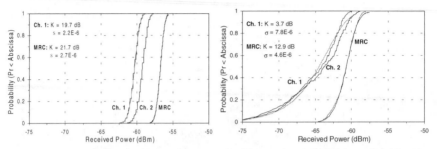

Figure 3.4.24: CDF of received power in intervals along the path: 45–55 m (left) and 100–110 m (right).

Figure 3.4.25 depicts the BER as a function of the received power after MRC. The BER does not decrease with the increase of the receiving power because the power level is higher for distances below 150 m where channel time dispersion tends to be higher. As a general remark, we can say that the performance of the trial platform is quite good and the measured results closely follow the ones obtained via simulation [DFPH99].

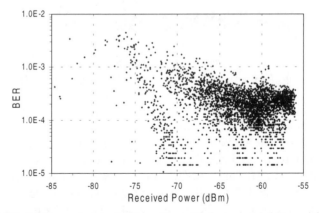

Figure 3.4.25: BER per frame (left) and per 10 frames (right) as a function of MRC received power.

3.5. Antennas for Mobile Phones

Gert Frølund Pedersen

A major proportion of all telephones in the future will be portable, according to market predictions. With existing portable phones, antennas have not been a major issue, often more of an add-on component, which was necessary, but not the focus of major R&D. This is likely to change, since investigations presented under COST259 have shown a large difference in the performance of antennas indicating a high return from better antennas, which makes the antenna an important component. Integration of the antenna into the handset is an interesting area of research as well, promising not only convenience from a user point of view, but also reducing the absorption in the user, which is an unfortunate use of valuable electromagnetic energy. The biological issues related to this absorption are to some extent still unresolved, although recommendations exist. In this section, the emphasis is on the telecommunication performance including investigations of absorption.

In the first subsection the single antenna performance is described, together with relevant parameters, which are necessary for a description of the antenna performance in a fading environment, including the effect on the user. In Section 3.5.2, test set-ups for measuring handset communication performance investigated under COST259 are described.

3.5.1. Communication performance

Gert Frølund Pedersen

3.5.1.1. Received signal strength by phone

A study on the communication performance of typical antennas on handheld phones, including the mobile environment and a large number of 'test users' using the same phone, was reported in [PNOK98]. The measurements were carried out at 1 880 MHz having 200 test people using a handset in a normal speaking position. The mock-up handheld consists of a commercially available GSM handheld equipped with a retractable whip/helical antenna that was modified to also include a back-mounted patch antenna. Two cables were used to connect the antennas to the receiving equipment.

The measurements were carried out by asking each test person to hold the handheld in what he or she felt was a natural speaking position. Then the person was asked to follow a path marked on the floor. The path was a square of some 2×4 m^2, and each recording of data lasts one minute corresponding to 3–4 rounds, Figure 3.5.1.

Figure 3.5.1: One of the test persons holding the handheld in what he feels as natural speaking position during measurement.

To record all three antennas, each person had to follow the path for one minute, change the whip antenna to the helix and walk the path once again. Hence, first the whip antenna and the patch were recorded, and next the helix antenna was recorded together with the patch (the patch is recorded twice).

Altogether, four locations were measured, one path on each floor and 50 test persons were used on each floor. The windows on level 3, level 2 and the ground level were facing towards the TX, but there was no LoS due to higher buildings in-between. On the first floor the windows were facing opposite the TX. The handset was connected to a correlation sounder capable of measure two antennas at a time. A bandwidth of 20 MHz was used to suppress the fast fading.

To match a typical urban GSM1800 small cell the TX antenna was located approximately 700 m away on the sixteenth level of a high building in an urban environment. The transmit antenna was a high gain sector antenna. Figure 3.5.2 shows the transmit antenna, together with the view of the environment. Other buildings on the picture hide the building in which the measurements were performed, and therefore no LoS exists between the TX and the handheld RX.

Figure 3.5.2: Picture of the BS antenna and view of the urban area.

As an example, the measured received power for 50 people on one floor is shown in Figure 3.5.3. The average results for the Mean Effective Gain (MEG), in all four levels are summarised in Table 3.5.1. From the results it is clear that the helix performs worse than the two other antennas concerning both the average and the spread values. In Figure 3.5.3 a peak difference between the highest received power and the lowest received power of 10 dB for the helix antenna can be seen. For the whip and patch antenna, peak difference in the order of 5–7 dB can be found. These values are surprisingly high.

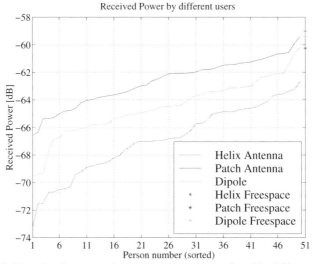

Figure 3.5.3: Received power by a mobile phone for 50 different people in an indoor environment. The BS is located outdoor some 700 m from the mobile in an urban environment.

Table 3.5.1: Measured average MEG with a tilted whip on a handheld as reference, in dB. On each level of the building the MEG is measured for 50 users to obtain the average values. The average body-effect is found for each antenna as being the difference with and without the user present. The inclination angle from vertical of the handheld is 60° without the user present, whereas the user holds the handheld in whatever position he or she feels as natural.

Antenna	Level 1	Level 2	Level 3	Level 4	Average	Body-effect
Whip / no user	0	0	0	0	0	
Helical / no user	0.70	0.97	0.60	0.41	0.67	
Patch / no user	-3.12	-3.46	-2.69	-0.9	-2.42	
Whip / with user	-6.61	-7.09	-7.04	-5.08	-6.37	-6.37
Helical / with user	-9.61	-10.29	-9.54	-7.48	-9.09	-9.76
Patch / with user	-6.71	-8.34	-6.24	-3.43	-5.80	-3.38

All the findings in the investigation can be summarised as:

- The variations in MEG from one person to another can vary up to 10 dB.
- The body loss defined as the difference between 'no person present' and a person present is on average 10 dB for a helical antenna, 6 dB for a whip antenna and 3 dB for a directive patch antenna.
- The average MEG for all test persons with respect to the dipole antenna is +0.5 dB for the patch and –3 dB for the helical antenna.
- The influence on the MEG from difference in peoples' heights and from those who wear glasses is small (less than one dB).
- The influence on the MEG from people who use the left or the right hand can be large depending on the antenna, 5 dB for the patch antenna, 3 dB for the helix antenna and 1 dB for the whip antenna.
- It is possible to reproduce the MEG results within ±0.5 dB.

3.5.1.2. Whole body absorption

To find why the loss in the communication link varies significantly from one user to another and to find where the power disappears, an investigation in an anechoic room with full control of all factors comprising the body loss was conducted [PeOL99].

An interesting question is what causes the apparent loss in the link budget. The absorption by the human body is one contribution to the body loss, but altogether four factors comprise the body loss: absorption; antenna mis-tuning; change of radiation pattern; and change of the polarisation state.

To investigate the contribution from each factor comprising the body loss for real live people a set-up was built in an anechoic room. By measuring the field strength on a sphere in the far field of the transmitting phone held by a test person, the absorption can be found by simply integrating the radiated power and comparing it to the power delivered to the antenna. By using a directional coupler before the antenna on the phone, both the forward and reflected power can be measured, and the losses due to antenna mis-tuning can be found. The 'loss' due to change in the direction of the radiation and polarisation of the fields radiated by the antenna and test person can be found from measurements of the radiation patterns for both polarisations of the phone held by a test person and information of the incoming multipath fields in an mobile environments [Taga90], [PeAn99], [SuVa99].

The anechoic room is a rectangular room with dimensions $7 \times 7 \times 10$ m^3 with absorbers on all sides, Figure 3.5.4. The suppression of reflections is better than 40 dB at 2 GHz. To overcome the problem of moving the person in two dimensions, the probe antennas are used to scan the sphere around the test person in elevation whereby the test person and phone only needs to rotate in azimuth. A construction for moving the two probe antennas (one for each polarisation) was made.

Figure 3.5.4: The set-up in the anechoic room for measurements of three-dimensional radiation patterns of a person using a phone.

The distance from the antenna on the phone to the probe is 2.1 m, the former being placed in the centre of rotation by moving the person while seated, but before the measurements started, in (X, Y, Z) directions. The measurements were conducted at 1.89 GHz and measuring one person took 22 minutes.

The reflected power was used for three purposes: to estimate the loss due to mis-tuning, in the calculation of the absorption, and to monitor if the test person was changing the way he or she holds the phone against the head. Just moving a finger on the back of the phone clearly shows in the reflected power. The electric field, $E_\theta(\phi,\theta)$ and $E_\phi(\phi,\theta)$, for each person, and both antennas on the phone, was measured and the transmitted power calculated by a spherical integration.

From a test with repeated measurements, it was found that the overall uncertainty was less than 1 dB, and as the transmitted power by one person compared to another can change by 10 dB the set-up was concluded to be sufficiently accurate. The loss in the power delivered to the antenna due to mis-tuning of the antenna was less than 2.0 dB for both antennas for all persons. In Figure 3.5.5 the transmitted power for the test people are shown. The test people are sorted according to the transmitted power, except for the free space measurement, which is located as the last. The average overall loss for the helix and patch is 9.7 and 3.4 dB, respectively. For the free space measurements, the antenna with the lowest loss is the patch whereas the helix has 1 dB higher loss in free space.

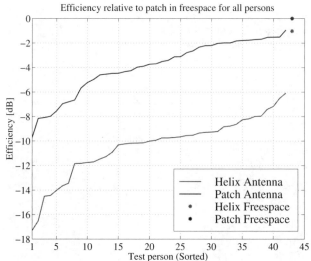

Figure 3.5.5: Transmitted power by the mobile phone held by 43 different people measured in an anechoic room.

3.5.1.3. Absorption in phantom hand and head

From the investigation in the anechoic room, it was clear that the primary parameter causing the high body loss is the absorption. To find out why the absorption is strongly user-dependent, an investigation in the anechoic room using a phantom instead of real people was performed [PeTK00].

The interesting question is what causes the large variation in the absorption and thereby the body loss? The following parameters are identified:

- Position of hand on the phone.
- Distance between head and phone.
- Tilt angle of the phone.
- The shape of the head and hand.
- The size of the person.
- If the skin of the person is dry or not.
- And many possibly minor parameters as age, sex, amalgam in the teeth, if the person wears glasses, amount of hair, etc.

Work investigating the first three items was reported in [PeTK00]. The investigation was conducted by measurements using a simple head plus hand phantom and the test phone used in the previously described measurements.

The set-up in the anechoic room is the same as described earlier, but using a phantom instead of a live person to hold the phone. The investigation is conducted for two antennas on the phone (a normal mode helix and an integrated patch). The frequency used is 1.89 GHz. The phantom consists of a plastic head filled with brain/muscle simulating liquid [Tart99] and a thin rubber glove filled with the same liquid to simulate the hand holding the phone. To retain the shape of the hand some thin transparent tape was wrapped around each finger and around the bulk of the hand, Figure 3.5.6.

To measure if the tilt angle of the phone changes the loss, nine measurements were made where the tilt angle was changed in steps of 10°. The reported variation in the total loss for both antennas due to tilt angle is small, less than ±0.6 dB for each antenna. A figure as small as ±0.6 dB may be close to the measuring accuracy.

Some people may hold the phone a small distance from the head, whereas others may press the phone firmly towards the head. If the phone is pressed firmly to the head, the tissue will reshape and the loss may increase. Using a hard plastic head phantom, it is not possible to press it towards the head in a way that the phantom head changes its shape. The loss as a function of the distance to between phone and phantom is shown in Figure 3.5.7.

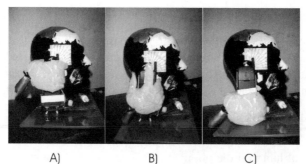

A)   B)   C)

Figure 3.5.6: The simple head plus hand phantom used in the study of change in the loss as a function of the tilt angle. Note the scale of tilt angles attached to the head. A) shows the hand in 'top' position, B) the hand in 'near top', C) the hand in bottom.

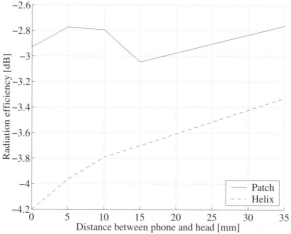

Figure 3.5.7: Plot of the loss as a function of the distance between phone and head phantom.

Although the distance is very important for the local Specific Absorption Rate (SAR) value, it is not an important issue for the total absorption. In [ToHA93] simulations of absorption vs. distance for a monopole antenna on a conducting box close to a head plus hand phantom of brain tissue are performed. The dimensions of the box used in the simulations are similar to the size of the phone used in these measurements. The result also shows a small drop in the loss of 0.9 dB (for a 35 mm distance), at the same frequency of 1.89 GHz. This is in good agreement with the results for the helix antenna (0.87 dB drop at 35 mm distance) and the curves are very similar, although the total absorption is a few percent higher for the shaped phantom and helix antenna.

In [ToHA93] simulations of the loss in the simple head plus hand phantom indicates that the loss due to the hand is very small compared to the loss in the head. Also measurements reported in [OlLa98] using an artificial hand holding the phone at the bottom of the phone showed a small excessive loss due to the hand of less than 2 dB. In the simulations the hand covers all of the handset except for 50 mm at the top. In order to see how losses in the hand depend on the position of the hand, three positions were measured, Figure 3.5.6.

The results are shown in Figure 3.5.8. As can be seen, the absorption is practically the same for the hand holding the phone at the bottom and for the case of no hand. This is in good agreements with the findings in [ToHA93], [OlLa98]. However, if the hand is close to the top of the phone, and thereby the antenna, this changes completely. If the hand covers most of the top, the absorption increases to 8 dB for both antennas. This shows that the hand is the single most important part when considering the loss for this simple head plus hand phantom.

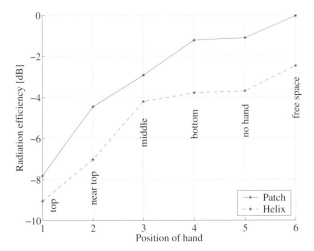

Figure 3.5.8: Loss due to the simple head plus hand phantom as a function of the position of the hand.

Summarising the results for the measurements of the phone together with the simple head plus hand phantom, only the position of the hand can explain why the absorption from one person to another can vary by more than 10 dB. If the position of the hand is dominating for the absorption, it should be possible to link the absorption with the position of the hand for the live people measured in the anechoic room.

3.5.1.4. Absorption in a live person's hand

During the measurements with live people described earlier [PeOL99], each person was recorded on video, and at the back of the phone a thin tape with black and white squares of 10×10 mm^2 was attached. Due to the black and white squares, it is possible to locate the position of the hand very precisely just from the video [PeTK00].

Each person is, of course, holding the phone in a unique way, and in order to ease the investigation, positions are quantified by the number of centimetres that are not covered by at least one finger, in the height of the phone.

The results are shown in Figure 3.5.9. As can be seen, losses do correlate strongly with the position of the hand, the correlation coefficient between the loss and the position of the hand being 0.7 for the patch and 0.67 for the helix. It should be noted that only one person was holding the phone in positions 3, 4, 8 and 12 cm; all other positions were used by at least three people. The loss seems constant when the hand covers up to 10 cm, the critical part is when the finger enters the top 3 cm. Of the people tested, 40 % entered this region and more than 2/3 entered the top 4 cm.

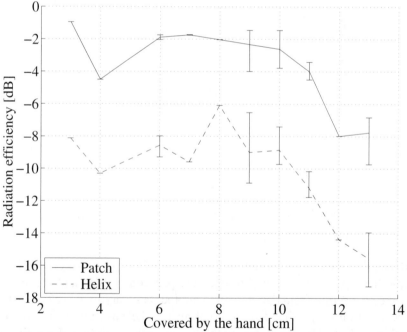

Figure 3.5.9: Losses due to live people using the phone sorted according to the way they hold the phone. Upper curve is Patch, lower curve is Helix antenna.

The loss due to the position of the hand on the phone is found to be the primary parameter, whereas the tilt angle and distance relative to the head are minor parameters. For this rather large phone (height of 13 cm), it causes high losses for 1/6 of the people tested, but for the future (much smaller) phones, losses can be expected to increase significantly for the majority of users, if nothing is done to prevent this. More than 2/3 of the people tested entered the top 4 cm on the phone, whose length is 13 cm. It may be that this is the percentage that will enter the top 2 cm for a 10 cm long phone. This is the length of most new phones on the market today.

3.5.1.5. Local absorption

As described, using a phone close to the human body results in significant absorption. The concern is not only about the waste of energy, but also on the public concern about the possible health hazards connected with mobile phones. There are safety limits on the amount of energy that is allowed to be dissipated into the head of the user. These safety limits are based on local heating, and the limits range from 1.6 to 2.0 mW/g averaged over 1 or 10 g of tissue. Work investigating the local heating and SAR values (which define the local heating) from different phones using a new heterogeneous head phantom as opposed to the usual homogenous one has been reported [HaGa99]. Measurements using the new head model have shown that the SAR was more than four times below the strictest limit [HaGa99]. These values are very different from values found from the standard SAR test with a homogenous head model. The investigation was conducted at 900 MHz, with 0.25 W of output power corresponding to the most common handheld phones in GSM900. From the comparison, it is clear that SAR depends strongly on the antenna type, and on the location of the antenna. Measured values ranging from 0.048 to 0.366 mW/g in the brain were reported. Further, it was suggested that for a fair comparison, the antenna efficiency should be included in the SAR measure. In this way, the power transmitted and not the power delivered to the antenna is the important figure.

3.5.2. Standardised phone measurements

Gert F. Pedersen

3.5.2.1. Initial considerations

Gert F. Pedersen

New types of antennas, especially those focusing on integration of the antenna into the handset, is an interesting area of research as well, promising not only convenience from a user's point of view, but also reductions in the absorption in the user, which is an unfortunate use of valuable electromagnetic energy. The basic requirements for an ideal antenna from a user's and manufacturer's point of view are the following: invisible, small, sufficient bandwidth, lossless, and well matched. A number of constraints of a theoretical and practical nature make this an impossible goal, but the art of engineering is of course to approach goals as near as possible. Small size antennas are in high demand, but it is difficult to design antennas that are small and have sufficient bandwidth without a strong coupling to the phone body. An investigation of a printed Bowtie antenna for GSM1800 reported in [Vazq00] shows a relative bandwidth of only 0.48 %, which is a factor of 20 from that required, and illustrates the difficulties in designing small integrated antennas fulfilling the basic requirements.

As described above, it is a characteristic of antennas in mobile systems that their performance depends on the environment and not just on the basic parameters. The environment for the portables consists of the user, which in some respects can be considered as part of the antenna, and the external environment. The multiplicity of waves usually incident on the user creates a Rayleigh fading channel, which distribution cannot be changed by a single, one-port antenna. Instead, it is necessary to look at averaged environments, including the polarisation properties.

An expression for the received power at a matched antenna port was first given in [Jake74]

$$P_{rec} = \oint \{P_1 P_\theta(\Omega)G_\theta(\Omega) + P_2 P_\varphi(\Omega)G_\varphi(\Omega)\}d\Omega \ , \qquad (3.5.1)$$

where P_1 is the power in the θ-polarisation, $\oint P_\theta(\Omega)d\Omega = 1$, and similarly for P_2 and the φ-polarisation.

By normalizing the powers, [Taga90] arrived at the very useful definition of the Mean Effective Gain (MEG) as

$$MEG = \oint \left\{ \frac{XPD}{1+XPD} P_\theta(\Omega)G_\theta(\Omega) + \frac{1}{1+XPD} P_\varphi(\Omega)G_\varphi(\Omega) \right\}d\Omega \quad (3.5.2)$$

using the ratio between the two polarisations, and defining XPD as

$$XPD = \frac{<P_\theta>}{<P_\varphi>} \ . \qquad (3.5.3)$$

An optimum handset antenna will be one that maximises the MEG for the relevant environmental scenarios in addition to the basic requirements listed.

All that is needed to design a good antenna for a handset is the radiation pattern and distribution of the incoming power. The relevant radiation pattern is the one measured in the typical user positions [BrEd99], which may be standardised to a few typical positions. In [BrEd99] the relevant positions are defined as:

- Telephone in talk position.
- In waist position.
- In handheld position.
- On table.
- Free space.

A simpler measure than the MEG is the Pattern Averaging Gain (PAG), where only the antenna gain in the azimuth plane is considered, and the incoming power is assumed to be uniform distributed over azimuth. Moreover, an investigation of the MEG vs. the PAG concludes that the MEG is a better measure [SuVa99].

In the first subsection below, test methods including the environment, and thereby giving the MEG directly, are described. Next, measurements of radiation patterns for mobile phones are discussed, and the last subsection describes measurements of the incoming multipath.

3.5.2.2. Set-up including environment

Bo Olsson

The increasing interest in handset antenna performance has led to a variety of methods to measure such performance, depending on the specific need. Network operators need to test commercially available handsets' performance in real mobile networks or in simulated environments that are really close to the mobile environment. Test results can then be used to improve network planning, and possibly to avoid those handsets with the worst performance. Handset manufacturers are interested in methods to evaluate their own and their competitor's products. Antenna designers are also interested in early development, i.e. to test different properties in the early design phase.

A summary of provisional test condition requirements, based on discussions within the Ectel/MoU-TWG group, are given in [Edva98]:

1. Test conditions should include the possibility to include a phantom, if possible with a hand (and possibility to use real subjects).
2. Measurements should be done in scattered field conditions, similar to real conditions.

3. Both uplink and downlink should be measured, typically with a GSM tester.
4. A quick establishment of a method that can be implemented with quite simple equipment is preferred.

The scattered field condition can be either a drive around test or a walk around test in real environments with a fixed BS, and comparing the average antenna performance for different antenna configurations with some reference antenna on the same routes. Another method is to establish some indoor or 'in box' scattering environments, that have spatial properties simulating real mobile environments, and to perform the movement in a reproducible way. See [Olss99a] for a summary of proposed requirements.

Drive around tests

This measurement can be performed in both downlinks and uplinks, measuring the signal strength in the MS or BS. This method is the simplest of all methods and will, properly performed, give a measure of the real efficiency in a real cellular network. The drawback is of course that the results are limited to the environments in which the tests are performed.

These tests have mainly been drive around tests with mobile handsets, in live GSM networks, tested by the front passenger in 'talk position' at the ear closest to the centre of the car. This method has been proposed by the MoU TWG Antenna Sub Group. The signal strength has been measured using the standard GSM Rx-Lev (full) measurements in MS and BS, reported to the GSM system. MS Power Control and DTX has been disabled, and the MSs reported Rx-Lev values have been equalised using calibrated cable sensitivity measurements at –60, –80 and –100 dBm signal levels.

A mobile phone with calibrated output power, RX sensitivity and cable has been used with a roof top magnetic mounted quarter wave monopole antenna as a performance reference. The handset antenna performance is then given as the mean signal difference over the whole test route. Even though there are no official reported results available from tests, the method has been used by Telia Mobile, Telstra, Telecom Italia and Vodafone.

Scattered field measurements

Telia has developed a method called 'Antenna Test Method Performed in Scattered Field Environment' [OlLa98]. 72 GSM900 and 47 GSM1800 mobile handsets (including different versions and prototypes of the same handset) have been tested by Telia Mobile using this method. The method measures the efficiency of the handset antenna in transmit mode at 'talk position', including the influence of a simulated human, relative to a reference antenna. The indoor propagation environment with multipath fading propagation is intended to simulate the spatial properties of an indoor MS environment served by an outdoor BS.

The method uses an ordinary large office room with a metal net to obstruct LoS between transmit and receive antennas to create a multipath, and thus fading, Figures 3.5.10 and 3.5.11. The handset tested is fixed to a phantom in 'talk position', which simulates the influence of the human as shown in Figure 3.5.12. A torso phantom V2.2 from Schmid & Partner Engineering AG filled with 22.2 l of water with a salt concentration of 1.49 g/l has been used. Tests are normally performed both with and without a simulated hand. The simulated hand is a rubber glove loosely filled with the same liquid as the phantom. The glove is thus very flexible, and is fixed at the lower part of the handset, with no fingers between the handset and the phantom. The position of the glove is regarded as the 'good position' of a hand, with low additional loss. The additional measured median loss is 2.6 dB for 72 GSM900 handsets and 2.4 dB for 47 GSM1800 handsets.

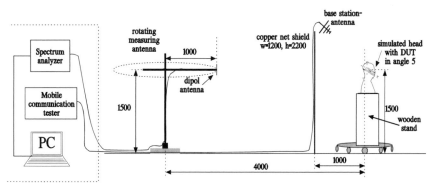

Figure 3.5.10: Arrangement of test equipment. Handset speaker centred at ear and above centre of turntable (© 1998 Telia).

When the handset is transmitting at nominal maximum power, the received power is sampled with a rotating RX antenna (vertical half wave dipole antenna). One hundred samples of the received power are collected during one revolution and the median power is calculated. This measurement is repeated for eight different phantom azimuthal angles, 45° separated as shown in Figure 3.5.11, and the average of the eight measured median received powers is then compared with the received reference power. The result is expressed as a (body) loss in dB relative to the received reference power.

Received reference power is obtained with a half wave dipole transmit antenna in free space (reference antenna; no phantom) fed with maximum nominal MS power (i.e. 33 dBm for GSM900 and 30 dBm for GSM1800). The reference antenna is tilted 60° from vertical and with its centre at the ear point (Figure 3.5.12), but without the phantom. The loss also includes handset output power deviation from nominal power.

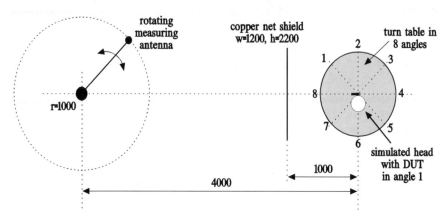

Figure 3.5.11: Arrangement of test equipment, top view (© 1998 Telia).

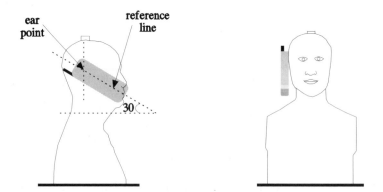

Figure 3.5.12: Simulated head. Note that the handset in practice is fixed tightly to the cheek of the phantom (not vertical as shown in the right hand part of the figure) (© 1998 Telia).

A HP 8594E spectrum analyser with special GSM measurement software has been used to measure the received power and a Racal 6103 GSM MS tester to control the mobile telephone.

Measurement results

The trend during the last years towards decreased size handsets with built in antennas has also decreased the antenna performance. This is shown in Figure 3.5.13, where the average antenna loss has increased over two years.

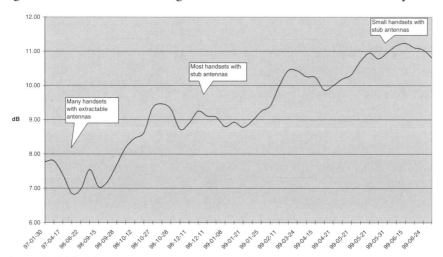

Figure 3.5.13: The average antenna loss for the 10 latest tested handsets relative to 60° tilted dipole from 97-01-30 until 99-06-24. Measured at 900 MHz with phantom and simulated hand.

Figure 3.5.14 shows the measured handset antenna loss relative to a 60° tilted dipole with nominal power for 72 different GSM900 and 47 different GSM1800 handsets with simulated head and hand. Note that the results include multiple versions and prototypes for some handset models. It is interesting to note the large variations 3 to 16 dB for GSM900 and −1 to 9 dB for GSM1800 handsets. In the case where the handsets have extractable antennas, they are extracted in these measurements. The median loss is 3.15 dB higher for GSM900 than for GSM1800 handsets.

Measurements of MEG for body carried phones have also been reported [Olss00]. In this extraction from these measurements, 15 GSM900 handsets were tested using the described method, but with the following exceptions:

1. Simulating head that is a 200 mm tube filled with saline water. Height of the tube is 300 mm. 7.3 kg water is used.
2. Simulated body that is a 300 mm pipe (height 1.7 m) filled with saline water to simulate paging position, vertical, at the waist. A 200 mm internal pipe is used to reduce the weight (59 kg water).
3. Free space: this is the same position as simulated head, but without the saline water inside the thin plastic cylinder.

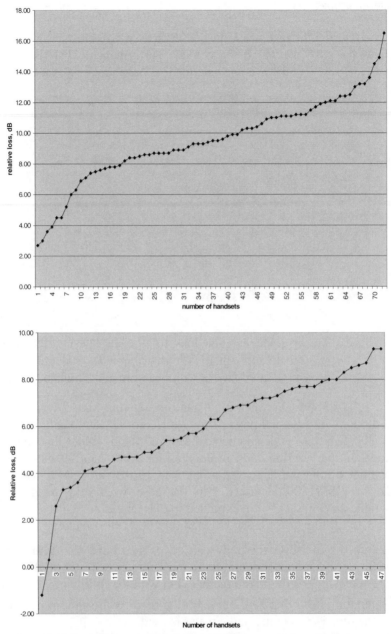

Figure 3.5.14: GSM900 (top) and GSM 1800 (bottom) handset antenna loss, including phantom and simulated hand, relative to 60° tilted dipole. Extractable antennas are extracted. Median values are 9.45 and 6.30 dB (© 2000 Telia).

4. All phones with retractable antenna (11 out of 15) are shown with retracted antenna, since this is the most probable waist position.

The results are illustrated by Figure 3.5.15 showing the attenuation below the received reference power for the 15 phones tested. By comparing with free space readings, partly different curves are obtained in Figure 3.5.16. In the most typical position, 45° tilted phone at the simulating head plus hand, 50 % of the phones have an attenuation within the interval 9–11 dB as compared to the REF DUT. In free space, 50 % of the phones have 3.5 ± 1.5 dB loss. This loss is typical efficiency compared with the reference $\lambda/2$-antenna, that has 33 dBm input. Paging position (at the waist of a simulated body) gives 3 dB more attenuation than the most typical position.

Scattered field properties

Measurements have been performed in an attempt to characterise the propagation properties of the scattered field measurement environment [Olss99c]. The investigations of the properties have been performed using reference half wave dipoles at 900 MHz. The analysis is based on the assumption that the electromagnetic (EM) field distributions at the antennas could be modelled as uniform in azimuth with vertical Gaussian distributions with zero mean.

The results indicate that the scattered field measurement environment, including the method to average over both transmit and receive positions/angles, will give the same measured results as an environment with uniform field distributions in azimuth and Gaussian elevation distribution at both transmit and receive antennas and for both polarisations.

It must be noted that this does not imply that the test set-up have the assumed vertical EM field distributions, only that the measured coupling gains would be the same using half wave dipoles.

With the above assumptions, it is possible to calculate the properties of the set-up as a function of the vertical field distribution standard deviation of the environment. In Figure 3.5.17, XPD (P_θ/P_ϕ) at the RX antenna is shown for different transmit antennas.

The estimated standard deviation of the vertical field distribution (both VP and HP) is less than 25°. This is due to the fact that XPD for the vertical TX antenna will pass infinity and turn negative, which is of course not possible. Also, the cross-polarisation power ratios at the RX antenna for vertical, horizontal or 60° tilted dipole transmit antennas are estimated to > 8, < –8.5 and –2 to –3 dB, respectively. Further investigations, including 3D measurements are necessary to validate these results.

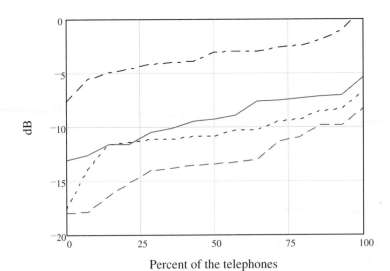

Percent of the telephones

Figure 3.5.15: Received power relative to reference (REF-DUT). From above to below: free space (da-dot), simulated head (solid), simulated head plus hand (dotted) and simulated body (dashed) (© 2000 Telia).

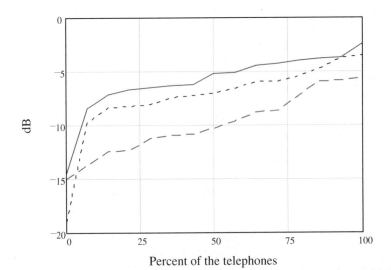

Percent of the telephones

Figure 3.5.16: Loss below free space reading for the same telephone. From above to below: simulated head (solid), simulated head plus hand (dotted) and simulated body (dashed) (© 2000 Telia).

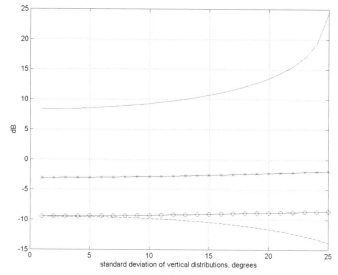

Figure 3.5.17: Calculated XPD at the RX for vertical(-)/horizontal(--) polarised and horizontal (-o)/60° tilted(-x) dipole transmitted signals vs. standard deviation of the vertical field distribution.

Compact Shield Box Measurements

A compact shield box, $80 \times 80 \times 120$ cm^3, to be used for handset antenna tests has been described by [Eric98]. The idea is to use a reverberating chamber, i.e. a metallic box with no absorbents. The box is intended for use in 890–960 MHz and 1710–1990 MHz frequency bands. The box has a 55 cm diameter turntable in one end, where the handset is placed under test and a 45° tilted antenna to the MS test set in the other end. The handset is first placed on a Kuster head/torso with a 45° vertical inclination, and then without the phantom 45° tilted on the turntable for a reference measurement. A metal board, obstructing the direct path between handset and MS test set antenna, is suggested. A call will thus be established between the MS test set and the handset, and reported received signal levels will be logged during the rotation of the turntable. Test results have not been reported.

3.5.2.3. Radiation measurements

Cristophe Grangeat

Measuring the radiation patterns from a mobile phone seams, at first glance, trivial, but measuring on a phone is different from measuring on 'normal' antennas due to:

1. The presence of the human body.
2. The phone is small, causing problems with cables.

Several set-ups for measuring the radiation in two or three dimensions have been reported [vaSc99], [VVMO99], [Baro99]. To find set-ups with good accuracy and low measuring time, it is important to compare all candidates. This was one of the important tasks of SWG 2.2.

One test bench developed for the measurement of the transmitted power and reception sensitivity of mobile phones in anechoic chambers is reported in [Gran99], Figure 3.5.18. As radiation from the mobile phone under test is not directional, measurements must be taken in three dimensions throughout the sphere surrounding the terminal. Measurements are performed for the terminal alone, or for the terminal associated with the phantom, to determine performance in normal use.

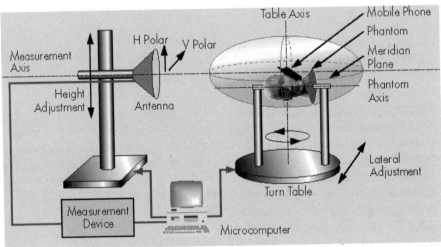

Figure 3.5.18: Measurement set-up for measuring 3-dimensional radiation pattern including a phantom.

The interaction between the mobile phone and the human body is not limited to the vicinity of the head, and it is found that the phantom must describe the head and the upper part of the chest. The influence of the hand must also be considered. Measurements are currently carried out using the generic phantom or a left hand. The terminal is placed in the normal position of use on the left or right side of the phantom. The EM parameters of the phantom are those used for SAR measurements.

The measuring rig comprises two positioning systems. The probe antenna, installed on a horizontal axis, measures both vertical and horizontal polarisation. The mobile phone and the phantom (or the terminal alone) are mounted on a turntable. The axis of the phantom is parallel to the axis of the probe antenna. One complete revolution of the turntable enables the flow of radiated power or the RX sensitivity to be measured in a meridian plane. All that then remains is to pivot the phantom on the horizontal axis to obtain the measurements for the other meridians. A set of 6 or 12 meridians is sufficient to determine the transmitted power or RX sensitivity averaged over the whole sphere surrounding the phantom. The RX sensitivity is measured with a radio communication tester. The test bench is driven entirely by computer. The total capture sequence takes from 5 to 30 minutes, depending on the nature of the measurements. Post-processing and visualisation of the directivity diagrams, radiation patterns and averaged parameters of the phone under test are performed with the computer.

3.5.2.4. Mobile environment

Mikael B. Knudsen

To calculate the MEG, the antenna radiation pattern of a phone and the distribution of the incoming power at the phone are required. For standardisation of phone performance including the antenna a model of the incoming multipath power is needed. A model of the environment resulting in an accurate average MEG is desired. To find one or even a number of models describing the environment as, for example delay profiles in GSM, information of the incoming fields is needed in both azimuth and elevation.

In [Taga90] a 3D outdoor propagation model was proposed, which was validated by measurements in a land mobile environment. The proposed model assumes Gaussian distributed power in vertical and uniform in azimuth. Measurements of 3D incoming fields for indoors with the TX outdoor showing a more complex distribution [PeAn99], but it is concluded that even though the distributions look different, the resultant MEG for outdoor to indoor remains similar for the locations measured. To test the method of measuring the radiation and environment separately, in practise, an investigation in a more controlled environment was used.

In [KaLa99], measured statistical distributions of incident waves to a mobile in an urban micro-cellular environment and for indoor environments at 2.15 GHz were reported. The power spectra of azimuth angle, elevation angle and Doppler frequency were included. The measurement was performed using the spherical 32-element dual-polarised antenna array shown in Figure 3.5.19 [KaLa99] and one channel.

The azimuth and elevation angles of both the vertically polarised and horizontally polarised components of the incident waves were measured.

Both an indoor and outdoor mobile route were measured with a single BS site. The collected data corresponds to a restricted environment, and therefore no profound conclusion was made. However, the assumption of Gaussian distribution in the elevation seems to be validated by the results, Figure 3.5.20 [Taga90], but not the uniform distribution in azimuth.

Figure 3.5.19: Spherical array of 32 dual-polarised microstrip patch elements (© 2000 IEEE).

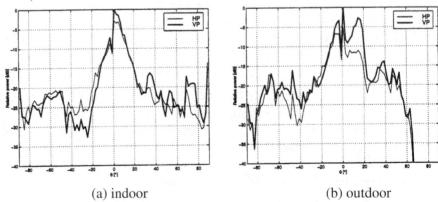

(a) indoor   (b) outdoor

Figure 3.5.20:   Measured average power distribution in elevation. (© 2000 ESA).

Characterisation of scattered field test room

In [KnPO00] investigations of the test method for validating mobile handset proposed in [OlLa98], [Olss99c] have been reported. As described in Section 3.5.2.2, the proposed measuring set-up fits in a room with metallic reflectors to mimic the fading and distribution of incoming fields in a real radio environment. Test results of 10 commercially GSM handsets measured in the set-up have been compared with measurements of power, based on 3D radiation pattern measurements performed in an anechoic room, including a phantom head [Tart99]. Comparing the efficiency with results including the emulated radio environment shows a rather large difference indicating the importance of including the environment [KnPO00]. The emulated radio environment has been investigated by performing 3D radio environment measurements of the incoming power in the set-up [KnPO00]. The set-up is shown in Figure 3.5.10. In the set-up, the transmitted signal of the phone is received by a rotating measuring antenna (only the uplink is measured). The received signal power from the rotating measuring antenna is sampled 100 times for eight different angles of the phone.

To measure the incoming multipath fields from each direction at the phone, a dual polarised horn antenna was placed at the phone location. The horn is mounted on a pedestal, which can rotate the horn in azimuth and in elevation from –45° to 65°. Therefore, measurements of received angular distributions of multipath fields at the phone are obtained. The horn antenna was connected to a correlation sounder capable of measuring both polarisations at the same time. The carrier frequency was 1 890 MHz and a bandwidth of 20 MHz was used. With the pedestal replacing the phone, channel measurements were performed for 8 equally spaced positions of the rotating measuring antenna. For each measurement the TX antenna was fixed.

A drawing of the scattered field room is shown in Figure 3.5.21 (a). The room contains a lot of corrugated sheets of metal, copper net shields and metal shelves, which together with concrete walls and window openings, may create a multipath environment. A net shield blocks the direct path between the phone and the rotating antenna. In Figure 3.5.21 (b) the measured vertical polarised power is averaged for all eight measured positions. The plot consists of a number of rectangles; the rectangles close to the centre of the plot represent the measurements recorded at the highest elevation angle. When the horn antenna moves in azimuth it corresponds to moving on a constant radius in the plot. A change in elevation corresponds to moving closer to the edge of the plot. The power values are in dB.

From the vertical polarised mean power distribution, it is concluded that the shield in front of the phone blocks the direct path, and that the major part of the energy comes from two strong reflections at the walls. The horizontal polarised mean power distribution is more diffuse, since the transmitting antenna is vertically polarised, and the energy in the horizontal polarisation is cross-coupled by means of reflections, diffraction and scattering.

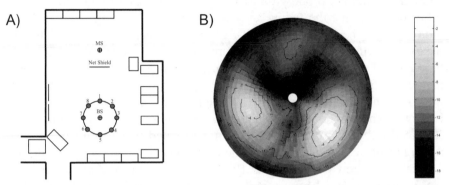

Figure 3.5.21: Measurements of the mean power distribution for the vertical polarisation. A) Drawing of the room. Dots indicate positions of the TX antenna. B) Plot of the mean power distribution.

The power transmitted by 10 GSM phones has been measured at 900 MHz using the same phones and the same phantom head as tested in the set-up [Tart99]. Figure 3.5.22 shows the transmitted power relative to the best phone and the results obtained in the set-up.

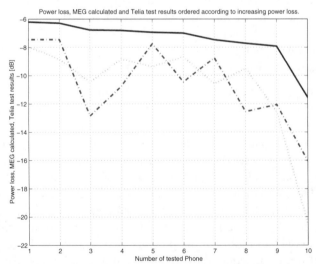

Figure 3.5.22: Test of 10 GSM handsets. The solid line shows the power loss. The dot-dash line is the MEG measured in the set-up. The dotted line is the calculated MEG.

In Figure 3.5.22 the handsets are sorted relative to the phone with the highest total transmitted power; the dotted line is the calculated MEG from the measured 3D power distribution of the room and the measured 3D radiation patterns of the handset antennas. A difference of up to 6.1 dB with a standard deviation of 1.87 dB between the results obtained in the set-up and the radiated power is found. By including the measured radio environment to the measured 3D radiation the difference in MEG is maximum 4.5 dB with a standard deviation of 1.18 dB.

It is concluded that the remaining difference could be due to different measuring frequencies for the handset and radio environment, or too simple model of the incoming power due to too few measurements.

3.6. References

[AaAP98]   Aanvik,F., Antonsen,E. and Petersen,M., "An Experimental Performance Comparison Between Antenna Diversity Schemes at 1800MHz", *COST 259*, TD(98)016, Bern, Switzerland, Feb. 1998.

[AaAP98]   Aanvik,F., Antonsen,E. and Pettersen,M., "An experimental performance comparison between antenna diversity schemes at 1800 MHz", in *Proc. of PIMRC'98 – 9th IEEE International Symposium on Personal, Indoor and Mobile Radio Communications*, Boston, Mass, USA, Sep. 1998 [also available as TD (98)016].

[AlBe99a]   Alayon-Glazunov,A. and Berg,J.-E., "Statistical analysis of measured Short-Term Impulse Response Functions of 1.88 GHz radio channels in Stockholm with corresponding channel model", in *Proc. of VTC'99 Fall – 50th IEEE Vehicular Technology Conference*, Amsterdam, The Netherlands, Sep. 1999 [also available as TD(99)009].

[AlBe99b]   Alayon-Glazunov,A. and Berg,J.-E., "CDMA Rake Receivers Performance Evaluation using a Wideband Channel Model", *COST 259*, TD(99)010, Thessaloniki, Greece, Jan. 1999.

[AlRC98]   Alves,B., Ribeiro,C. and Correia,L.M., "A Comparison of Propagation Models for Urban Micro-Cells in GSM", *COST 259*, TD(98)069, Bradford, UK, Apr. 1998.

[Ande94]   Andersen,J.B., "Transition zone diffraction by multiple edges", *IEE Proceedings Microwaves, Antennas and Propagation*, Vol. 141, No. 5, Oct. 1994, pp. 382–384.

[Ande98a]   Andersen,J.B., "High Gain Antennas in a Random Environment", in *Proc. of PIMRC'98 – 9th IEEE International Conference on Personal, Indoor and Mobile Radio Communications*, Boston, Mass., USA, Sep. 1998.

[Ande98b]   Andersen,J.B., "Intelligent Antennas in a Scattering Environment—An Overview", in *Proc. of GLOBECOM'98*, Sydney, Australia, Nov. 1998.

[Ande98c]   Andersen,J.B., "Intelligent Antennas in a Scattering Environment", in *Proc. of COST 252/259 Joint Workshop*, Bradford, UK, Apr. 1998.

[Ande98d]   Andersen,J.B., "Transmit-Receive Diversity in a Scattering Environment", *COST 259*, TD(98)011, Bern, Switzerland, Feb. 1998.

[Ande98e]   Andersen,J.B., *Angular and Temporal Spread versus Distance for Different Outdoor Environments*, Center for PersonKommunikation, Aalborg University, Internal Report, Denmark, 1998.

[Ande99a]   Andersen,J.B., "Intelligent Antennas in Personal Communications", in *Proc. IEEE International Conference on Personal Wireless Communications*, Jaipur, India, Feb. 1999.

[Ande99b]   Andersen,J.B., "Antenna Arrays in Mobile Communications—Gain, Diversity and Channel Capacity", *Radio Science Bulletin*, Vol. 290, 1999, pp. 4–7.

[AnMF99]   Anton-Haro,C., Mestre,X. and Fonollosa,J.R., "Adaptive Antennas for the TDD mode of UTRA", *COST 259*, TD(99)040, Vienna, Austria, Apr. 1999.

[ÂnNC98]   Ângelo,G.C., Neto.,I. and Correia,L.M., "Health and Penetration Issues in Buildings with GSM Base Station Antennas on Top", in *Proc. of VTC'98 – 48th IEEE Vehicular Technology Conference*, Ottawa, Canada, May 1998 [also available as TD(97)072].

[ArTP94]   Arawojolu,A.A., Turkmani,A.M.D. and Parsons,J.D., "Time dispersion measurements in urban microcellular environments", in *Proc. of VTC'94 – 44th IEEE Vehicular Technology Conference*, Stockholm, Sweden, May 1994.

[AsBe97]   Asplund,H. and Berg,J-E., "An investigation of measured wideband channels at 1880 MHz with applications to 1.25 MHz and 5 MHz CDMA systems", *COST 259*, TD(97)026, Turin, Italy, May 1997.

[AsBe98]   Asplund,H. and Berg, J.-E., "Estimation of Scatterer Locations from Wideband Array Channel Measurements at 1800 MHz in an Urban Environment", *COST 259*, TD(98)087, Duisburg, Germany, Sep. 1998.

[AsBe99a] Asplund,H. and Berg,J.-E., "An empirical model for the probability of line of sight in an urban macrocell", *COST 259*, TD(99)107, Leidschendam, The Netherlands, Sep. 1999.

[AsBe99b] Asplund,H. and Berg,J.-E., "Parameter distributions for the COST 259 directional channel model", *COST 259*, TD(99)108, Leidschendam, The Netherlands, Sep. 1999.

[AsLB98] Asplund,H., Lundqvist,P. and Berg,J.-E., "A Channel Model for Positioning", *COST 259*, TD(98)020, Bern, Switzerland, Feb. 1998.

[Aspl00a] Asplund,H., "Bandwidth-independent fading characterization", *COST 259*, TD(00)013, Valencia, Spain, Jan. 2000.

[Aspl00b] Asplund,H., Private Communication, Ericsson Radio, Stockholm, Sweden / TU Vienna, Vienna, Apr. 2000.

[BaGi98] Barbiroli,M. and Giannetti,C., "A Methodology for Comparison of Propagation Models", *COST 259*, TD(98)019, Bern, Switzerland, Feb. 1998.

[BaPa82] Bajwa,A.S. and Parsons,J.D., "Small-area characterisation of UHF urban and suburban mobile radio propagation", *IEE Proceedings*, Vol. 129, Pt. F, No. 2, Apr. 1982, pp. 102–109.

[Baro99] Baro,J., "Assessment of Handheld Mobile Stations by Spherical Weighting Based on Elevation Measurements", *COST 259*, TD(99)126, Stockholm, Sweden, June 1999.

[BaSa97] Bachelier,T. and Sante,J.F., "Influence of Mobility on Capacity of DCS Network using Switched-Beam Antenna", in *Proc. of IEE Symposium on Intelligent Antennas*, Guildford, UK, Aug. 1997 [also available as TD(97)048].

[Basi99] Basile,M., "Obtaining accurate prediction with an untuned semi-deterministic model used with an advanced statistical database", *COST 259*, TD(99)031, Vienna, Austria, Apr. 1999.

[BeKa98] Bergljung,C. and Karlsson,P., "Propagation Characteristics for Indoor Broadband Radio Access Networks in the 5 GHz Band", in *Proc. of PIMRC'98 – 9th IEEE International Conference on Personal, Indoor and Mobile Radio Communications*, Boston, Mass., USA, Sep. 1998.

[BeKB99] Bergljung,C., Karlsson,P. and Börjesson,H., "Penetration Loss and Spatial Propagation Characteristics in the 5 GHz Band", *COST 259*, TD(99)097, Leidschendam, The Netherlands, Sep. 1999.

[Bell63]   Bello,P.A., "Characterization of Randomly Time-Variant Linear Channels", *IEEE Trans. Commun. Systems*, Vol. CS-11, No. 12, Dec. 1963, pp. 360–393.

[Bere94]   Berenger,J.P., "A perfectly matched layer for the absorption of electromagnetic waves", *J. Comp. Phys.*, Vol. 114, No. 2, Oct. 1994, pp. 185-200.

[Berg94]   Berg,J.-E., "A macrocell model based on the parabolic differential equation", in *Proc. of Virginia Tech's 4th Symposium on Wireless Personal Communications*, Blacksburg, Virgin., USA, June 1994.

[Berg99]   Bergljung,C., "On the diffraction by two staggered half planes", *COST 259*, TD(99)066, Vienna, Austria, Apr. 1999.

[BeSp63]   Beckmann,P. and Spizzichino,A., *The Scattering of Electromagnetic Waves from Rough Surfaces*, Pergamon Press, London, UK, 1963.

[BiDI91]   Bic,J.C., Duponteil,D. and Imbeaux,J.C., *Elements of Digital Communication*, John Wiley, New York, New York, USA, 1991.

[Blan98]   Blanz,J.J., *Receiver Antenna Diversity in Mobile Radio systems Using Joint Detection of User Signals* (in German), VDI Fortschrittberichte, Vol. 10, No. 535, VDI-Verlag, Düsseldorf, Germany, 1998.

[BlJu98]   Blanz,J.J. and Jung,P., "A flexibly configurable spatial model for mobile radio channels", *IEEE Trans. Comm.*, Vol. COM-46, No. 3, Mar. 1998, pp. 367–371.

[BrCu97]   Brennan,C. and Cullen,P., "Efficient techniques for the computation of UHF grazing incidence terrain scattering", *COST 259*, TD(97)091, Lisbon, Portugal, Sep. 1997.

[BrCu98a]   Brennan,C. and Cullen,P., "Application of the Fast Far-Field Approximation to the computation of UHF path loss over irregular terrain", *IEEE Trans. Antennas Propagat.*, Vol. 46, No. 6, June 1998, pp. 881–890.

[BrCu98b]   Brennan,C. and Cullen,P., "Tabulated Interaction Method for UHF terrain propagation problems", *IEEE Trans. Antennas Propagat.*, Vol. 46, No. 5, May 1998, pp. 738–739 [also available as TD(97)005].

[BrCu98c]   Brennan,C. and Cullen,P., "An integral equation approach to long distance UHF terrain propagation", *COST 259*, TD(98)064, Bradford, UK, Apr. 1998.

[BrDe91]   Braun,W.R. and Dersch,U., "A Physical Mobile Radio Channel Model", *IEEE Trans. Veh. Technol.*, Vol. 40, No. 2, May 1991, pp. 472–482.

[BrEd99]   Braun,Ch. and Edvardsson,O., "On intended use positions for wireless terminals", *COST 259*, TD(99)098, Leidschendam, The Netherlands, Sep. 1999.

[BrNM99a] Braun,C., Nilsson,M. and Murch,R.D., "Measurement of the Interference Rejection Capability of Smart Antennas on Mobile Telephones", in *Proc. of VTC'99 Spring – 49th IEEE Vehicular Technology Conference*, Houston, Texas, USA, May 1999.

[BrNM99b] Braun,C., Nilsson,M. and Murch,R.D., "Simulation and Measurement of the Interference Rejection Capability of Handset Antenna Diversity", *COST 259*, TD(99)033, Vienna, Austria, Apr. 1999.

[Burr00a]   Burr,A., "Channel Capacity Evaluation of Multi-Element Antenna Systems using a Spatial Channel Model", *COST 259*, TD(00)006, Valencia, Spain, Jan. 2000.

[Burr00b]   Burr,A., "Evaluation of Space-Time Codes", *COST 259*, TD(00)007, Valencia, Spain, Jan. 2000.

[Burr97]   Burr,A.G., "Power-delay profile of spatial channel model", *COST 259*, TD(97)066, Lisbon, Portugal, Sep. 1997.

[Burr98]   Burr,A.G., "Wide-band channel modelling using a spatial model", in *Proc. of ISSSTA'98 – IEEE International Symposium on Spread Spectrum Techniques and Applications*, Sun City, South Africa, Sep. 1998 [also available as TD(98)013].

[CaFr98]   Cardona,N. and Fraile,R., "Propagation Path Loss Prediction in GSM Macro- and Microcells Using Neural Networks", *COST 259*, TD(98)105, Duisburg, Germany, Sep. 1998.

[CaMA95]   Cardona,N., Möller,P. and Alonso,F., "Application of Walfish-type urban propagation models", *IEE Electronics Letters*, Vol. 31, No. 23, Nov. 1995, pp. 1971–1972.

[Cann90a]   Canning,F.X., "The Impedance Matrix Localisation (IML) Method for Moment Method Calculations", *IEEE Antennas Propagat. Mag.*, Vol. 32, No. 5, Oct. 1990, pp. 18–30.

[Cann90b]   Canning,F.X., "On the application of some radiation boundary conditions", *IEEE Trans. Antennas Propagat.*, Vol. 38, No. 5, May 1990, pp. 740–745.

[Capo69]   Capon,J., "High-Resolution Frequency-Wavenumber Spectrum Analysis", *Proc. IEEE*, Vol. 57, No. 8, Aug. 1969, pp. 1408–1418.

[CDFR96]   Corazza,G.E., Degli-Esposti,V., Frullone,M. and Riva,G., "A Characterization of Indoor Space and Frequency Diversity by Ray-Tracing Modelling", *IEEE J. Select. Areas Commun.*, Vol. 14, No. 3, Apr. 1996, pp. 411–419.

[ChOC00]   Chaves,J., Oliveira,P. and Correia,L.M., "Comparison of Propagation Models for Small Urban Cells in GSM 900 and 1800", *COST 259*, TD(00)008, Valencia, Spain, Jan. 2000.

[Chua87]   Chuang,J.C.-I., "The Effects of Time Delay Spread on Portable Radio Communications Channels with Digital Modulation", *IEEE J. Select. Areas Commun.*, Vol. SAC-5, No. 5, June 1987, pp. 879–889.

[ClFC99]   Claro,A.R., Ferreira,J.M. and Correia,L.M., "Assessment of a propagation model for crossroads in micro-cells", in *Proc. of VTC'99 Fall – 50th IEEE Vehicular Technology Conference*, Amsterdam, The Netherlands, Sep. 1999 [also available as TD(99)035].

[CoCe97]   Correia,A.M.C. and Cercas,F.B., "On Channel Coding for CDMA over Multipath Rayleigh Fading Channels", *COST 259*, TD(97)057, Lisbon, Portugal, Sep. 1997.

[CoFr94]   Correia,L.M. and Francês,P.O., "A propagation model for the estimation of the average received power in an outdoor environment at the millimetre waveband", in *Proc. of VTC'94 – 44th IEEE Vehicular Technology Conference*, Stockholm, Sweden, July 1994.

[CoLe75]   Cox,D.C. and Leck,R.P., "Correlation Bandwidth and Delay Spread Multipath Propagation Statistics for 910-MHz Urban Mobile Radio Channels", *IEEE Trans. Commun.*, Vol. COM-23, No. 11, Nov. 1975, pp. 1271–1280.

[CoRF97]   Correia,L.M., Reis,J.J. and Francês,P.O., "Analysis of the Average Power to Distance Decay Rate at the 60GHz", in *Proc. of VTC'97 – 47th IEEE Vehicular Technology Conference*, Phoenix, Ariz., USA May 1997 [also available as TD(97)010].

[CoRW93]   Coifman,R., Rokhlin,V. and Wandzura,S., "The Fast Multipole Method for the Wave Equation: A Pedestrian Prescription", *IEEE Antennas Propagat. Mag.*, Vol. 35, No. 3, June 1993, pp. 7–12.

[COST89]   COST 207, *Digital Land Mobile Radio Communications*, Final Report, COST Telecom Secretariat, European Commission, Brussels, Belgium, 1989.

[COST99]   COST 231, *Digital mobile radio towards future generation systems,* Final report, COST Telecom Secretariat, European Commission, Brussels, Belgium, 1999.

[Cox72]   Cox,D.C., "Time- and frequency-domain characterizations of multipath propagation at 910 MHz in a suburban mobile-radio environment", *Radio Science*, Vol. 7, No. 12, Dec. 1972, pp. 1069–1077.

[Cox73]   Cox,D.C., "910 MHz Urban Mobile Radio Propagation: Multipath Characteristics in New York City", *IEEE Trans. Commun.*, Vol. COM-21, No. 11, Nov. 1973, pp. 1188–1194.

[Czyl00]   Czylwik,A., "Downlink Beamforming for Systems with Frequency Division Duplex (FDD)", *COST 259*, TD(00)034, Valencia, Spain, Jan. 2000.

[DCFR97]   Degli-Esposti,V., Carciofi,C., Frullone,M., and Riva,G., "Sensitivity of Ray Tracing Indoor Field Prediction to Environment Modelling", in *Proc. of ICEAA'97 – International Conference on Electromagnetics in Advanced Applications*, Turin, Italy, Sep. 1997 [also available as TD(97)049].

[DDZW99]   Didascalou,D., Döttling,M., Zwick,T. and Wiesbeck,W., "A Novel Ray-Optical Approach to Model Wave Propagation in Curved Tunnels", in *Proc. of VTC'99 Fall – 50th IEEE Vehicular Technology Conference*, Amsterdam, The Netherlands, Sep. 1999 [also available as TD(99)088].

[deBa97]   de Backer,B., "An integral equation approach to the prediction of indoor wave propagation", *COST 259*, TD(97)089, Lisbon, Portugal, Sep. 1997.

[DeVV98]   De Coster,I., Van Lil,E. and Van de Capelle,A., "A PO approach for a more accurate computation of the reflected field in indoor propagation", in *Proc. of VTC'98 – 48th IEEE Vehicular Technology Conference*, Ottawa, Canada, May 1998.

[DFFR99]   Degli-Esposti,V., Falciasecca,G., Frullone,M. and Riva,G., "Antenna Pattern Optimization for Base Stations in Urban Microcellular Systems", *COST 259*, TD(99)060, Vienna, Austria, Apr. 1999.

[DFPH99]   Dinis,M., Fernandes,J., Prögler,M., Herzig,W. and Zubrzyck,J., "The SAMBA Trial Platform in the Field", in *Proc. of ACTS Mobile Communication Summit*, Sorrento, Italy, June 1999.

[Dida00]   Didascalou,D., *Ray-Optical Wave Propagation Modelling in Arbitrarily Shaped Tunnels,* Ph.D. Thesis, Universität Karlsruhe (TH), Karlsruhe, Germany, Feb. 2000.

[Diet00]   Dietert,J.E., Private Communication, TU Aachen, Aachen, Germany / TU Vienna, Vienna, Austria, Jan. 2000.

[DiKR00]   Dietert,J.E., Karger,S. and Rembold,B., "Statistical Channel Modelling based on Raytrace Simulations", *COST 259*, TD(00)004, Valencia, Spain, Jan. 2000.

[DiMW00]   Didascalou,D., Maurer,J. and Wiesbeck,W., "Natural wave propagation in subway tunnels at mobile communications frequencies", in *Proc. of VTC'2000 Spring – 51st IEEE*

Vehicular Technology Conferencee, Tokyo, Japan, May 2000 [also available as TD(00)011].

[DiRe00] Dietert,J.E. and Rembold,B., "Stochastic Channel Model for Outdoor Applications based on Raytrace Simulations", *COST 259*, TD(00)005, Valencia, Spain, Jan. 2000.

[DiYW00] Didascalou,D., Younis,M. and Wiesbeck,W., "Millimeter-Wave Scattering and Penetration in Isolated Vegetation Structures", submitted to *IEEE Trans. Geoscience Remote Sensing* (Special Issue), 2000.

[DiYW99] Didascalou,D., Younis,M. and Wiesbeck,W., "Millimeter wave scattering and attenuation in limited vegetation structures", in *Proc. of IGARSS'99*, Hamburg, Germany, June 1999.

[DJFH97] Dahlhaus,D., Jarosch,A., Fleury,B. and Heddergott,R., "Joint Demodulation in DS/CDMA Systems Exploiting the Space and Time Diversity of the Mobile Radio Channel", in *Proc. of PIMRC'97 – 8th IEEE International Symposium on Personal, Indoor and Mobile Radio Communications*, Helsinki, Finland, Sep. 1997 [also available as TD(97)007].

[DLPZ97] Dinis,M., Lagarto,V., Prögler,M. and Zubrzycki,J., "SAMBA: a Step to Bring MBS to the People", in *Proc. of ACTS Mobile Communications Summit*, Aalborg, Denmark, Oct. 1997.

[DMSV97a] De Coster,I., Madariaga,G.A., Souto,B.P., Van Lil,E. and Pérez-Fontan,F., "An Extended Propagation Software Package For Indoor Communication Systems", in *Proc.of ICAP'97 – International Conference on Antennas and Propagation*, Edinburgh, UK, Apr. 1997.

[DMSV97b] De Coster,I., Madariaga,G.A., Souto,B.P., Van Lil,E. and Pérez-Fontan,F., "Development of an indoor communication system software package (EPICS)", *COST 259*, TD(97)027, Turin, Italy, May 1997.

[DRFF97] Degli-Esposti,V., Riva,G., Falciasecca,G., Frullone,M. and Corazza,G.E., "Performance evaluation of space and frequency diversity for 60 GHz wireless LAN using a ray model", *COST 259*, TD(97)021, Turin, Italy, May 1997.

[DVLD98] De Coster,I., Van Lil,E., Loureiro,B.G., Delmotte,P. and Van de Capelle,A., "Validation measurements for the Fresnel approximation of the PO model", in *Proc. of EuMC'98 – 28th European Microwave Conference*, Amsterdam, The Netherlands, Oct. 1998.

[DVLP97] De Coster,I., Van Lil,E., Loureiro,B.G. and Pérez-Fontan,F., "EPICS: Illustration of Penetration and Comparison Between

GO and PO Results", *COST 259*, TD(97)059, Lisbon, Portugal, Sep. 1997.

[DVLV98]   De Coster,I., Van Lil,E., Loureiro,B.G. and Van de Capelle,A., "Validation Measurements for a PO Reflection Model", *COST 259*, TD(98)003, Bern, Switzerland, Feb. 1998.

[Edva98]   Edvardsson,O., "In Search for a Standardized Test Method for Mobile Phone Antennas", *COST 259*, TD(98)109, Duisburg, Germany, Sep. 1998.

[Egge98]   Eggers,P.C.F., "Generation of base station DOA distributions by Jacobi transformation of scattering areas", *Electronic Letters*, Vol. 34, No. 1, Jan. 1998, pp. 24–26.

[Egge98a]   Eggers,P.C.F., "Base Station Virtual Doppler Beams used in Beam Orientated SDMA", *COST 259*, TD(98)024, Bern, Switzerland, Feb. 1998.

[EgKO98]   Eggers,P.C.F., Kóvacs,I.Z. and Olesen,K., "Penetration effects on XPD with GSM1800 handset antennas, relevant for BS polarisation diversity for indoor coverage", in *Proc. of VTC'98 – 48th IEEE Vehicular Technology Conference*, Ottawa, Canada, May 1998 [also available as TD (98)022].

[EGTL92]   Erceg,V., Ghassemzadeh,S., Taylor,M., Li,D. and Schilling,D., "Urban/suburban out of sight propagation modelling", *IEEE Commun. Mag.*, Vol. 30, No. 6, June 1992, pp. 56–61.

[EnKF99]   Englert,T., Kattenbach,R. and Früchting,H., "Investigation of the Characteristics of Measured Delay Power Density Spectra for the 5.2 GHz Indoor Radio Channel", in *Proc. of European Wireless'99*, Munich, Germany, Oct. 1999 [also available as TD(00)009].

[EnMa77]   Engquist,B. and Majda,A., "Absorbing Boundary Conditions for the numerical simulation of waves", *Math. Comp.*, Vol. 31, No. 139, July 1977, pp. 629–651.

[Eric98]   ERICSSON, Internal Report, Tdoc SMG2 WPB 81/98, 1998.

[FBRS94]   Feuerstein,M.J., Blackard,K.L., Rappaport,T.S., Seidel,S.Y. and Xia,H.H., "Path loss, delay spread, and outage models as functions of antenna height for microcellular system design", *IEEE Trans. Veh. Technol.*, Vol. VT-43, No. 3, Aug. 1994, pp. 487–498.

[FBZF98]   Fernandes,C., Brankovic,V., Zimmermann,S., Filipe,M. and Anunciada,L., "Dielectric lens antennas for wireless broadband communications", *Wireless Personal Communications*, Vol. 10, No. 1, June 1998, pp. 19–32.

[FCGB98]   Frullone,M., Carciofi,C., Grazioso,P., Barbiroli,M. and De Bernardi,R., "Identification of the Validity Domains of

Different Prediction Models in Urban Environment", *COST 259*, TD(98)021, Bern, Switzerland, Feb. 1998.

[FDPH98] Fernandes,J., Dinis,M., Prögler,M., Herzig,W. and Zubrzyck,J., "The SAMBA Trial Platform: Initial Results", in *Proc. of ACTS Mobile Communication Summit*, Rhodes, Greece, June 1998.

[FDSS94] Foster,H.,   Dehghan,S.,   Steele,R.,   Stefanov,J.   and Strelouhov,H., "Microcellular measurements and their prediction", in *Proc. of IEE Coloquium on Role of Site Shielding in Prediction Models for Urban Radiowave Propagation*, London, UK, 1994.

[FeFA97] Fernandes,C., Filipe,M. and Anunciada,L., "Lens Antennas For The SAMBA Mobile Terminal", in *Proc. of ACTS Mobile Telecommunications Summit,* Aalborg, Denmark, Oct. 1997.

[FeFB95] Fernandes,C., Francês,P. and Barbosa,A., "Shaped coverage of elongated cells at millimetrewaves using a dielectric lens antenna", in *Proc. of EuMC'95 – 25th European Microwave Conference*, Bologna, Italy, Oct. 1995.

[FeFe98] Fernandes,J. and Fernandes,C., "Impact of Shaped Lens Antennas on MBS Systems", in *Proc. of PIMRC'98 – 9th IEEE International Conference on Personal, Indoor and Mobile Radio Communications*, Boston, Mass., USA, Sep. 1998.

[FeFe99] Fernandes,C. and Fernandes,J., "Performance of Lens Antennas in Wireless Indoor Millimeter-Wave Applications", *IEEE Trans. Microwaves Theory Tech.*, Vol. 47, No. 6, Part 1, June 1999, pp. 732-737.

[FeGa99] Fernandes,J. and Garcia,J., "Cell Coverage for MBS Environments", *COST 259*, TD(99)099, Leidschendam, The Netherlands, Sep. 1999.

[FeHe94] Fessler,J.A. and Hero,A.O., "Space-alternating generalized expectation-maximization algorithm", *IEEE Trans. Signal Process.*, Vol. 42, No. 10, Oct. 1994, pp. 2664–2677.

[FeMG99] Fernandes,J., Marques,A. and Garcia,J., "Cellular Coverage for MBS Using the Millimetre-Wave Band", in *Proc. of ACTS Mobile Communications Summit*, Sorrento, Italy, June 1999.

[FeNe97] Fernandes,J. and Neves,J., "New Wide-Band Propagation Channel Model for the mm-Wave Band", in *Proc. of 2nd International Symposium on Mathematical Modelling*, Vienna, Austria, Feb. 1997.

[FeNS95] Fernandes,J., Neves,J. and Smulders,P., "Mm-Wave Indoor Radio Channel Modelling vs. Measurements", *Wireless Personal Communications*, Vol. 1, No. 3, 1995, pp.211–219.

[Fern99]   Fernandes,C., "Shaped Dielectric Lenses for Wireless Millimeter-Wave Communications", *IEEE Antennas Propagat. Mag.*, Vol. 41, No. 5, Oct. 1999, pp. 141–150.

[FlBH96]   Fleury,B.H., Bernhard,U.P. and Heddergott,R., "Advanced radio channel model for Magic WAND", in *Proc. of the ACTS Mobile Communication Summit*, Granada, Spain, Nov. 1996.

[Fosc96]   Foschini,G.J., "Layered space–time architecture wireless communication in a fading environment when using multi-element antennas", *Bell Labs Tech. Journal*, Vol. 1, No. 2, Autumn 1996, pp.41–59.

[FPJF99]   Fernandes,J., Pereira,J.R., Jesus,P. and Ferreira,J., "Adaptive Array Antenna with Electronically Steerable Beam", in *Proc. of COST259/260 Joint Workshop on Spatial Channel Models and Adaptive Antennas*, Vienna, Austria, Apr. 1999.

[FrCa97]   Fraile,R. and Cardona,N., "Fast Neural Network Method for Propagation Loss Prediction in Urban Environments", *IEE Electronics Letters,* Vol. 33, No. 24., Nov. 1997, pp. 2056–2058 [also available as TD(98)006].

[FrCa98a]   Fraile,R. and Cardona,N., "Coverage predictin in microcells using neural networks" (in Spanish), in *Proc. of Spanish National Symposium of URSI*, Pamplona, Spain, Sep. 1998.

[FrCa98b]   Fraile,R. and Cardona,N., "Macrocellular Coverage Prediction for all Ranges of Antenna Height using Neural Networks", in *Proc. of ICUPC'98 – IEEE International Conference on Universal Personal Communications*, Florence, Italy, Oct. 1998.

[FuMB98]   Fuhl,J., Molisch,A.F. and Bonek,E., "Unified channel model for mobile radio systems with smart antennas", *IEE Proceedings – Radar Sonar Navigation*, Vol. 145, No. 1, Feb. 1998, pp. 32–41 [also available as TD(97)014].

[GaCo98]   Gaggioli,F. and Correia,L.M., "Analysis of Space Diversity in a Canyon-Like Street at 60 GHz", in *Proc. of VTC'98 – 48th IEEE Vehicular Technology Conference*, Ottawa, Canada, May 1998 [also available as TD(97)079].

[GDVP98]   Gavilanes-Loureiro,B., De Coster,I., Van Lil,E. and Perez-Fontan,F., "Comparison of Antenna Diversity Schemes", *COST 259*, TD(98)059, Bradford, UK, Apr. 1998.

[GeWi98]   Geng,N. and Wiesbeck,N., *Radio Propagation Modeling for Wireless Communications System Design* (in German), Springer, Berlin, Germany, 1998.

[GEYC97]   Greenstein,L.J., Erceg,V., Yeh,Y.S. and Clark,M.V., "A new path-gain/delay-spread propagation model for digital cellular channels", *IEEE Trans. Veh. Technol.*, Vol. 46, No. 2, May 1997, pp. 477–485.

[GoCo98] Goncalves,N.C. and Correia,L.M., "Propagation Model for Urban Microcellular Systems at the UHF Band", in *Proc. of PIMRC'98 – IEEE 9ᵗʰ International Symposium on Personal, Indoor and Mobile Radio Communications*, Boston, Mass., USA, Sep. 1998 [also available as TD(98)079].

[Goll94] Gollreiter,R. (ed.), *Channel Models Issue 2*, RACE ATDMA Deliverable R2084/ESG/CC3/DS/P/029/b1, Brussels, Belgium, May 1994.

[Gran99] Grangeat,C., "Test methods for handset antennas", *COST 259*, TD(99)124, Vienna, Austria, Apr. 1999.

[HaGa99] Haider,H. and Garn,H., *Basic Investigation of Mobile Phone Antennas with regard to Power Efficiency and Radiation Safety.* Report No. P10412-Tec, Technical University of Vienna, Vienna, Austria, 1999.

[HaJa99] Hayn,A. and Jakoby,R., "Propagation and Standardisation Issues for 42 GHz Digital Microwave Video Distribution System (MVDS)", *COST 259*, TD(99)021, Thessaloniki, Greece, Jan. 1999.

[HaNo95] Haardt,M. and Nossek,J.A., "Unitary ESPRIT: How to Obtain Increased Estimation Accuracy with Reduced Computational Burden", *IEEE Trans. Signal Process.*, Vol. 43, No. 1, May 1995, pp. 95–107.

[Hash93] Hashemi,H., "The Indoor Radio Propagation Channel", *Proc. IEEE*, Vol. 81, No. 7, July 1993, pp. 943–968.

[HeBF97] Hedergott,R., Bernhard,U.P. and Fleury,B.H., "Stochastic Radio Channel Model for Advanced Indoor Mobile Communication Systems", in *Proc. of PIMRC'97 – IEEE 8ᵗʰ International Symposium on Personal, Indoor and Mobile Radio Communications*, Helsinki, Finland, Sep. 1997 [also available as TD(98)057].

[HeKa97] Heiska,K. and Kalliola,K., "Wideband Propagation Modelling in Urban Microcells by Using Ray-Tracing", *COST 259*, TD(97)076, Lisbon, Portugal, Sep. 1997.

[HeTr00] Heddergott,R. and Truffer,P., "Statistical characteristics of indoor radio propagation in NLoS scenarios", *COST 259*, TD(00)024, Valencia, Spain, Jan. 2000.

[HeTr99] Heddergott,R. and Truffer,P., "Results of Indoor Wideband Delay-Azimuth-Elevation Measurements for Stochastic Radio Channel Modeling", *COST 259*, TD(99)083, Leidschendam, The Netherlands, Sep. 1999.

[HNBK00]   Hansen,J., Nold,M., Benedickter,H. and Kossel,M., "A Model for Quick Calculation of Power Delay Profiles in Geometrically Simple Environments", *COST 259*, TD(00)015, Valencia, Spain, Jan. 2000.

[Holm96]   Holm,P.D., "UTD-Diffraction Coefficients for Higher Order Wedge Diffracted Fields", *IEEE Trans. Antennas Propagat.*, Vol. 44, No. 6, June 1996, pp. 879–888.

[HoWL99]   Hoppe,R., Wölfle,G. and Landstorfer,F.M., "Fast 3-D Ray Tracing for the Planning of Microcells by Intelligent Preprocessing of the Data Base", in *Proc. of EPMCC'99 – 3rd European Personal and Mobile Communications Conference*, Paris, France, Mar. 1999 [also available as TD(99)006].

[HRSS99]   Hampicke,D., Richter,A., Scheider,A., Sommerkorn,G. and Thomä,R., "Measurement-based characterization of time-variant directional radio channels", *COST 259*, TD(99)039, Vienna, Austria, Apr. 1999.

[HTTN99]   Heddergott,R., Truffer,P., Tschudin,M. and Nold,M., "Comparison of High Resolution Channel Parameter Measurements with Ray Tracing Simulations in a Multipath Environment", in *Proc. of EPMCC'99 – 3rd European Personal and Mobile Communications Conference*, Paris, France, Mar. 1999 [also available as TD(98)098].

[Hugl99]   Hugl,K., "Frequency transformation based downlink beamforming", in *Proc. of COST259/260 Joint Workshop on Spatial Channel Models and Adaptive Antennas*, Vienna, Austria, Apr. 1999.

[HuLB99]   Hugl,K., Laurila,J., and Bonek,E., "Downlink beamforming for frequency division duplex systems", in *Proc. of GLOBECOM'99*, Rio de Janeiro, Brazil, Dec. 1999.

[IHAP99]   Isasi,F., Hernando,J.M., Aguado,F. and Pagel,S., "Validation and Adaptation of Models of Urban Propagation to Land and Irregular Construction", *COST 259*, TD(99)041, Vienna, Austria, Apr. 1999.

[Ishi78]   Ishimaru,A., *Wave Propagation and Scattering in Random Media*, Vol. 1, Academic Press, New York, New York, USA, 1978.

[ITUR94]   ITU-R, *Attenuation by Atmospheric Gases*, Recommendations and Reports of the ITU-R, Report 719-2, Vol. V (Propagation in Non-ionized Media), International Telecommunication Union, Geneva, Switzerland, 1994.

[IwKa93]   Iwai,H. and Karasawa,Y., "Wideband propagation model for the analysis of the effect of the multipath fading on the near-far problem in CDMA mobile radio systems", *IEICE Trans. Communications*, Vol. 76-B, No. 2, Feb. 1993, pp. 342–354.

[IYTU84]   Ikegami,F., Yoshida,S., Takeuchi,T. and Umehira,M., "Propagation factors controlling mean signal strength on urban streets", *IEEE Trans. Antennas Propagat.*, Vol. 32, No. 8, Aug. 1984, pp. 531–537.

[JaAn00]   Janaswamy,R. and Andersen,J.B., "Path Loss Predictions in Urban Areas with Irregular Terrain Topography", *Wireless Personal Communications*, Vol. 12, No. 3, Mar. 2000, pp. 255–268 [also available as TD(98)060].

[JaDC99]   Jarosch,A., Dahlhaus,D. and Cheng,Z., "Smart Antenna Concepts for the UMTS Terrestrial Radio Access", in *Proc. of Smart Antenna Workshop of the European Microwave Week*, Munich, Germany, Oct. 1999 [also available as TD(00)017].

[Jake74]   Jakes,W.C., *Microwave Mobile Communications*, IEEE Press, Piscataway, New Jersey, USA, 1974.

[Jana97]   Janaswamy,R., "Propagation Predictions Over Rural Terrain by the Split-Step Parabolic Equation Method", *COST 259*, TD(97)045, Lisbon, Portugal, Sep. 1997.

[Jana98]   Janaswamy,R., "Path loss predictions in urban areas with irregular terrain topography", *COST 259*, TD(98)060, Bradford, UK, Apr. 1998.

[KaEM92]   Kalivas,G.A., El-Tanany,M. and Mahmoud,S.A., "Millimeter-Wave Channel Measurements for Indoor Wireless Communications", in *Proc. of VTC'92 – 42nd IEEE Vehicular Technology Conference*, Denver, Colorado, USA, May 1992.

[KaEm97]   Karlsson,P. and Emanuelsson,P., "Propagation measurements and models in the 27-29 GHz band", *COST 259*, TD(97)009, Turin, Italy, 1997.

[KaEn97]   Kattenbach,R. and Englert,T., "Wideband Statistical Modeling of Indoor Radio Channels based on the Time-Variant Transfer Function", *COST 259*, TD(97)071, Lisbon, Portugal, Sep. 1997.

[KaEn98]   Kattenbach,R. and Englert,T., "Investigation of Short Term Statistical Distributions for Path Amplitudes and Phases in Indoor Environment", in *Proc. of VTC'98 – 48th IEEE Vehicular Technology Conference*, Ottawa, Canada, May 1998.

[KaLa99]   Kalliola,K. and Laitinen,H., "Statistical distribution of incident waves to mobile antenna in microcellular environment at 2.15 GHz", *COST 259*, TD(99)045, Vienna, Austria, Apr. 1999.

[KaLJ98]   Karlsson,P., Löwendahl,N. and Jordana,J., "Narrowband and Wideband Propagation Measurements and Models in the 27–29

GHz Band", *COST 259*, TD(98)017, Bern, Switzerland, Feb. 1998.

[Kall98]   Kalliola,K., "Examples of Dynamic Wideband Mobile Radio Channel Measurements with an Antenna Array", *COST 259*, TD(98)034, Bern, Switzerland, Feb. 1998.

[Kall99]   Kalliola,K., "Directional 3D Real-Time Dual-polarized Measurement of Wideband Mobile Radio Channel", *COST 259*, TD(99)013, Thessaloniki, Greece, Jan. 1999.

[KaPa99]   Karlsson,P. and Pamp,J., "Propagation Measurements and Spatial Channels for HIPERLAN 2", *COST 259*, TD(99)067, Vienna, Austria, Apr. 1999.

[Katt97a]   Kattenbach,R., *Characterization of Time-Variant Indoor Radio Channels by Means of their System and Correlation Functions* (in German), Ph.D. Thesis, University of Kassel, published by Shaker Verlag, Aachen, Germany, 1997.

[Katt97b]   Kattenbach,R., "Considerations about the Validity of WSSUS for Indoor Radio Channels", *COST 259*, TD(97)070, Lisbon, Portugal, Sep. 1997.

[Katt98]   Kattenbach,R., "Statistical Distribution of Path Interarrivaltimes in Indoor Environment", in *Proc. of VTC'98 – 48th IEEE Vehicular Technology Conference*, Ottawa, Canada, May 1998.

[Katt99]   Kattenbach,R., "Statistical Modeling of Short-Term Fading Effects for Directional Radio Channels", *COST 259*, TD(99)112, Leidschendam, The Netherlands, Sep. 1999.

[KaWe00]   Kattenbach,R. and Weitzel,D., "Wideband Channel Sounder for Time-Variant Indoor Radio Channels", *in Proc. of AP2000 – Millennium Conference on Antennas & Propagation*, Davos, Switzerland, Apr. 2000 [also available as TD(00)018].

[KBMH98]   Karlsson,P., Bergljung,C., Medbo,J. and Hallenberg,H., "Outdoor Spatio-Temporal Propagation Measurements for Evaluation of Smart Antennas in the 5 GHz Band", *COST 259*, TD(98)082, Duisburg, Germany, Sep. 1998.

[KBTB99]   Karlsson,P., Bergljung,C., Thomsen,E. and Börjesson,H., "Wideband Measurement and Analysis of Penetration Loss in the 5 GHz Band", in *Proc. of VTC'99 Fall – 50th IEEE Vehicular Technology Conference*, Amsterdam, The Netherlands, Sep. 1999.

[KeMo90]   Keenan,J.L. and Motley,A.J., "Radio coverage in buildings", *British Telecom Technology Journal*, Vol. 8, No. 1, Jan. 1990, pp. 19–24.

[KeOB99]   Kemp,A.H., Orriss,J. and Barton,S.K., "The Impact of Base Station Diversity on CDMA System Capacity", *COST 259*, TD(99)089, Leidschendam, The Netherlands, Sep. 1999.

[KiVa97]   Kivinen,J. and Vainikainen,P., "Wideband Indoor Radio Channel Measurements at 5.3 GHz", in *Proc. of EuMC'97 – 27th European Microwave Conference*, Jerusalem, Israel, Sep. 1997 [also available as TD(97)054].

[KiVa98a]   Kivinen,J. and Vainikainen,P., "Analysis of Wideband Indoor Propagation Measurements at 5.3 GHz", *COST 259*, TD(98)067, Bradford, UK, Apr. 1998.

[KiVa98b]   Kivinen,J. and Vainikainen,P., "Wideband propagation measurements in corridors at 5.3GHz", in *Proc. of ISSSTA'98 – 5th International Symposium on Spread Sepctrum Techniques and Applicatins*, Sun City, South Africa, Sep. 1998.

[KiVa99]   Kivinen,J. and Vainikainen,P., "Indoor Propagation Measurements at 5 GHz Band", *COST 259*, TD(99)064, Vienna, Austria, Apr. 1999.

[KiZV99]   Kivinen,J., Zhao,X. and Vainikainen,P., "Wideband Indoor Radio Channel Measurements with Direction of Arrival Estimations in the 5 GHz Band", in *Proc. of VTC'99 Fall – 50th IEEE Vehicular Technology Conference*, Amsterdam, The Netherlands, Sep.1999 [also available as TD(99)024].

[KlAn97]   Kloch,C. and Andersen,J.B., "Scattering by a small urban environment", *COST 259*, TD(97)056, Lisbon, Portugal, Sep. 1997.

[KLTH00]   Kalliola,K., Laurila,J., Toeltsch,M., Hugl,K., Vainikainen,P. and Bonek,E., "3-D Directional Wideband Dual-Polarized Measurement of Urban Mobile Radio Channel with Synthetic Aperture Technique", in *Proc. of AP2000 – Millennium Conference on Antennas & Propagation*, Davos, Switzerland, Apr. 2000.

[KnPO00]   Knudsen,M.B., Pedersen,G.F. and Olsson,B., "Characterisation of Telia Scattered Field Test Room by use of 3D Radio Environment Measurements", *COST 259*, TD(00)021, Valencia, Spain, Jan. 2000.

[KoEO98]   Kovács,I.Z., Eggers,P.C.F. and Olesen,K., "Comparison of mean effective gains of different DCS1800 handset antennas in urban and suburban environment", in *Proc. VTC'98 – 48th IEEE Vehicular Technology Conference*, Ottawa, Canada, May 1998.

[KTBT99]   Kuchar,A., Taferner,M., Bonek,E., Tangemann,M. and Hoek,C., "A Run-time Optimized Adaptive Antenna Array Processor for GSM", in *Proc. of EPMCC'99 – 3rd European*

Personal Mobile Communications Conference, Paris, France, Mar. 1999 [also available as TD(99)046].

[KüFa97]   Kürner,T. and Fauß,R., "Impact of Digital Terrain Databases on the Prediction Accuracy in Urban Areas – A Comparative Study", in *Proc. of EPMCC'97 – 2nd European Personal Mobile Communications Conference*, Bonn, Germany, Sep. 1997 [also available as TD(97)060].

[Kürn99]   Kürner,T., "A Run-Time Efficient 3D Propagation Model for Urban Areas Including Vegetation and Terrain Effects", in *Proc. of VTC'99 Spring – 49th IEEE Vehicular Technology Conference*, Houston, Texas, USA, May 1999 [also available as TD(99)047].

[LaFr99a]   Layer,F. and Früchting,H., "Impacts of movement on the directional properties of indoor radio channels based on a 3D ray-tracing model", in *Proc. of COST259/260 Joint Workshop on Spatial Channel Models and Adaptive Antennas*, Vienna, Austria, Apr. 1999.

[LaFr99b]   Layer,F. and Früchting,H., "A UTD-based Model of the Time-Variant Indoor Radio Channel in the Vicinity of a Human", *COST 259*, TD(99)081, Leidschendam, The Netherlands, Sep. 1999.

[Lars98]   Larsson,M., "Spatio-Temporal Channel Measurements at 1800 MHz for Adaptive Antennas", *COST 259*, TD(98)107, Duisburg, Germany, Sep. 1998.

[Lee82]   Lee,W.C.Y., *Mobile Communications Engineering*, McGraw-Hill, New York, New York, USA, 1982.

[Lehn98]   Lehne,P.H.(ed.), *Review of existing channel sounder measurement setups and applied calibration methods*, SMT-METAMORP Deliverable META/D-1/TR/D-1/1/b1, Brussels, Belgium, June 1998.

[LeMo98]   Leth-Espensen,P. and Mogensen,P., "Uplink Combining Performance in a GSM 1/3 Reuse Network using Smart Antennas", *COST 259*, TD(98)039, Bern, Switzerland, Feb. 1998.

[LFSB96]   Lemaire,D., Fernandes,C., Sobieski,P. and Barbosa,A., "A method to overcome the limitations of G.O. in the design of axis-symmetrical lenses", *J. Infrared and Millimetrewaves*, Vol. 17, No. 8, Ago. 1996, pp. 1377–1390.

[LHTB99]   Laurila,J., Hugl,K., Toeltsch,M., Bonek,E., Kalliola,K. and Vainikainen,P., "Directional Wideband 3-D Measurements of Mobile Radio Channel in Urban Environment", *COST 259*, TD(99)092, Leidschendam, The Netherlands, Sep. 1999.

[LiBD00]   Liénard,M., Betrencourt,S. and Degauque,P., "Propagation in Road Tunnels / Influence of the Traffic Conditions and Channel Characterisation for Adaptive Antennas", *COST 259*, TD(00)001, Valencia, Spain, Jan. 2000.

[LiDe99]   Lienard,M. and Degauque,P., "Channel Modelling for the Propagation in Tunnel", *COST 259*, TD(99)049, Vienna, Austria, Apr. 1999.

[LiKo99]   Liberti,J.C. and Koshy,B.J., "Spatial channel measurements and modeling for smart antenna systems", in *Proc. of COST 259/260 Joint Workshop on Spatial Channel Models and Adaptive Antennas*, Vienna, Austria, Apr. 1999.

[Lind75]   Lindman,E.L., "Free-space boundary conditions for the time dependent wave equation", *J. Comp. Phys.*, Vol. 18, No. 1, May 1975, pp. 67–78.

[LiWL97]   Li,J., Wagen,J.F. and Lachat,E., "Propagation over rooftop and in the horizontal plane for small and micro-cell coverage problems", in *Proc. of VTC'97 – 47th IEEE Vehicular Technology Conference*, Phoenix, Ariz., USA, May 1997 [also available as TD(97)020].

[LKTH00]   Laurila,J., Kalliola,K., Toeltsch,M., Hugl,K., Vainikainen,P. and Bonek,E., "Wideband 3-D Characterization of Mobile Radio Channels in Urban Environment", submitted to *IEEE Trans. Antennas Propagat.*, 2000.

[LlCa97]   Llacer,L.J. and Cardona,N., "UTD solution for the multiple building diffraction attenuation function for mobile radio propagation", *IEE Electronics Letters*, Vol. 33, No. 1, Jan. 1997, pp. 92–93 [also available as TD(97)041].

[LOAN99]   Larsson,M., Olsson,B., Arkner,T., Nordin,S. and Wallberg,J., "Time Dispersion Measurements at NMT 450 Frequency Band", *COST 259*, TD(99)080, Leidschendam, The Netherlands, Sep. 1999.

[Lohs98]   Lohse,N., "Mobile Channel Analysis with Smart Antennas", *COST 259*, TD(98)112, Duisburg, Germany, Sep. 1998.

[LSJB99]   Lohse,N., Stege,M., Jelitto,J., Bronzel,M. and Fettweis,G., "Parameter validation for a space-time channel model", *COST 259*, TD(99)109, Leidschendam, The Netherlands, Sep. 1999.

[LTDV99]   Loredo,S., Torres,R.P., Domingo,M. and Valle,L., "Measurements and Predictions of the Local Mean Power and Fast Fading Statistics in Indoor Wireless Environments", *Proc. of 7th International Symposium on Recent Advances in Microwave Technology*, Málaga, Spain, Dec. 1999 [also available as TD(00)030].

[LTVD00] Loredo,S., Torres,R.P., Valle,L. and Domingo,M., "Measurements and predictions of the local mean power and small scale fading statistics in indoor wireless environments", accepted for publication in *Micro. Opt. Tech. Lett.*, 2000.

[LuJi96] Lu,N. and Jin,J.M., "Application of Fast Multipole Method to finite-element boundary-integral solution of scattering problems", *IEEE Trans. Antennas Propagat.*, Vol. 44, No. 6, June 1996, pp. 781–786.

[MaBT81] Maestrello,L., Bayliss,A. and Turkel,E., "On the interaction of a sound pulse with the shear layer of an axisymmetrical jet", *Journal of Sound and Vibration.*, Vol. 74, No. 2, Jan. 1981, pp. 281–301.

[MaBX93] Maciel,L., Bertoni,H. and Xia,H., "Unified approach to prediction of propagation over buildings for all ranges of base station antenna height", *IEEE Trans. Veh. Technol.*, Vol. 42, No 1, Feb. 1993, pp. 41–45.

[MaFN98] Marques,P., Fernandes,J. and Neves,J., "Complex Impulse Response Modeling for Wideband Radio Channels", in *Proc. of VTC'98 – 48th IEEE Vehicular Technology Conference*, Ottawa, Ontario, Canada, May 1998.

[MAHS00] Molisch,A.F., Asplund,H., Heddergott,R., Steinbauer,M. and Zwick,T., "The COST 259 directional channel model I. Philosophy and general aspects". Asplund,H., Glazunov,A.A., Molisch,A.F., Pedersen,K.I. and Steinbauer,M., "II. Macrocells". Dietert,J.E., Heddergott,R., Kattenbach,R., Molisch,A.F., Steinbauer,M. and Zwick,T., "III Micro- and picocells". To be submitted.

[Mart96] Martin,U., "Statistical mobile radio channel simulator for multiple-antenna reception", in *Proc. of ISAP'96 – International Symposium on Antennas and Propagation*, Chiba, Japan, Sep. 1996.

[MASA98] Medbo,J., Andersson,H., Schramm,P., Asplund,H. and Berg,J.-E., "Channel models for HIPERLAN/2 in different indoor scenarios", *COST 259*, TD(98)070, Bradford, UK, Apr. 1998.

[MFDO96] Mogensen,P., Frederiksen,F., Dam,H., Olesen,K. and Larsen,S., "TSUNAMI II Stand-alone Testbed", in *Proc. of ACTS Mobile Communications Summit*, Granada, Spain, Nov. 1996.

[MKLH99] Molisch,A.F., Kuchar,A., Laurila,J., Hugl,K. and Bonek,E., "Efficient implementation of a geometry-based directional model for mobile radio channels", in *Proc. of VTC'99 Fall – 50th IEEE Vehicular Technology Conference*, Amsterdam, The Netherlands, Sep.1999.

[MoCu95]   Moroney,D. and Cullen,P., "A Fast Integral Equation Approach to UHF Coverage Estimation", in del Re,E.(ed.), *Mobile and Personal Communications*, Elsevier Press, Amsterdam, The Netherlands, 1995.

[MoFP96]   Molisch,A.F., Fuhl,J. and Proksch,P., "Error Floor of MSK Modulation in a Mobile-Radio Channel with Two Independently Fading Paths", *IEEE Trans. Veh. Technol.*, Vol. 45, No. 2, May 1996, pp. 303–309.

[MoLK98]   Molisch,A.F., Laurila,J. and Kuchar,A., "Geometry-base stochastic model for mobile radio channels with directional component", in *Proc. of 2nd Intelligent Antenna Symp.*, Univ. Surrey, Surrey, UK, July 1998 [also available as TD(98)076].

[MoMi80]   Monzingo,R.A. and Miller,W.T., *Introduction to Adaptive Antennas*, John Wiley, New York, New York, USA, 1980.

[Morg99]   Morgado,A., "Spatial Filtering in UTRA", *COST 259*, TD(99)101, Leidschendam, The Netherlands, Sep. 1999.

[MPFF98]   Mogensen,P.E., Pedersen,K.I., Fleury,B., Frederiksen,F., Leth-Espensen,P., Olesen,K. and Leth-Larsen,S., *TSUNAMI-II Final Report*, ACTS-TSUNAMI, Brussels, Belgium, Sep. 1998.

[MPFF99]   Mogensen,P., Pedersen,K.I., Frederiksen,F. and Fleury,B., "Measurements, channel statistics, and performance of adaptive base station antennas", in *Proc. of COST 259/260 Joint Workshop on Spatial Channel Models and Adaptive Antennas*, Vienna, Austria, Apr. 1999.

[MSAB00]   Molisch,A.F., Steinbauer,M., Asplund,H. and Bergner,A., "Compatibility of the COST259 directional channel model with previous channel models", to be submitted.

[Mur81]   Mur,G., "Absorbing Boundary Conditions for the finite difference approximation of the time domain electromagnetic field equations", *IEEE Trans. Electromagn. Compat.*, Vol. EMC-23, No. 4, Nov. 1981, pp. 377–382.

[Naka60]   Nakagami,M., "The m-Distribution – A General Formula of Intensity Distribution of Rapid Fading", in Hoffman,W.C.(ed.), *Statistical Methods in Radio Wave Propagation*, Pergamon Press, London, UK, 1960.

[NeEg99]   Neubauer,T. and Eggers,P.C.F., "Simultaneous Characterization of Polarisation matrix Components in Pico Cells", in *Proc. of VTC'99 Fall – 50th IEEE Vehicular Technology Conference*, Amsterdam, The Netherlands, Sep. 1999 [also available as TD (99)004].

[NeMe64]   Nelder,J.A. and Mead,R., "A simplex method for function minimization", *Computer Journal*, Vol. 7, 1964/65, pp. 308–313.

[NiWP99]   Nijs,J., Witrisal,K. and Prasad,R., "Characterization and Simulation of the 18 GHz Radio Channel", *COST 259*, TD(99)020, Thessaloniki, Greece, Jan. 1999.

[NLAL99]   Nilsson,M., Lindmark,B., Ahlberg,M., Larsson,M. and Beckman,C., "Measurements of the spatio-temporal polarization characteristics at the 1800 MHz band", in *Proc. of COST 259/260 Joint Workshop on Spatial Channel Models and Adaptive Antennas*, Vienna, Austria, Apr. 1999.

[Nobl99]   Nobles,P., "A comparison of indoor pathloss measurements at 2 GHz, 5 GHz, 17 GHz and 60 GHz", *COST 259*, TD(99)100, Leidschendam, The Netherlands, Sep. 1999.

[OBKC98]   O´Brian,W.M., Kenny,E. and Cullen,P.J., "A snapshot of an evolving three-dimensional micro-cell propagation tool for indoor and outdoor urban environment", *COST 259*, TD(98)103, Duisburg, Germany, Sep. 1998.

[OBKC99]   O'Brien,W.M., Kenny,E. and Cullen,P., "An efficient implementation of a 3D microcell propagation tool for indoor and outdoor urban environments", submitted to *IEEE Trans. Veh. Technol.*, 1999.

[OlLa98]   Olsson,B. and Larsson,S., "Description of Antenna Test Method Performed in Scattered Field for GSM MS", *COST 259*, TD(98)106, Duisburg, Germany, Sep. 1998.

[Olss00]   Olsson,B., "Telia Scattered Field Measurements of Mobile Phone Antennas", in *Proc. of COST 259 Workshop on the Mobile Terminal and Human Body Interaction*, Bergen, Norway, Apr. 2000.

[Olss99a]   Olsson,B., "On the Basics Requirements for Test Methods for Handset Antennas", *COST 259*, TD(99)026, Thessaloniki, Greece, Jan. 1999.

[Olss99b]   Olsson,B., "Analysis of Scattered Field Handset Antenna Measurements", *COST 259*, TD(99)052, Vienna, Austria, Apr. 1999.

[Olss99c]   Olsson,B., "Test of the Electromagnetic Distribution Properties for Scattered Field Handset Antenna Measurements", *COST 259*, TD(99)087, Leidschendam, The Netherlands, Sep. 1999.

[Ortg99]   Ortgies,G., "Applicability of semi-empirical propagation models for path loss prediction in urban microcells", *COST 259*, TD(99)007, Thessaloniki, Greece, Jan. 1999.

[PaBa82]   Parsons,J.D. and Bajwa,A.S., "Wideband characterisation of fading mobile radio channels", *IEE Proceedings*, Vol. 129, Part F, No. 2, Apr. 1982, pp. 95–101.

[PaFB97]   Papathanassiou,A., Furio,I. and Blanz,J.J., "Link Level Performance in the Uplink of a Joint Detection CDMA Mobile Radio System Employing Multi-Antenna Array Configurations", *COST 259*, TD(97)051, Lisbon, Portugal, Sep. 1997.

[PaHW99]   Papathanassiou,A., Hartmann,C. and Weber,T., "Uplink Spectrum Efficiency and Capacity of TD-CDMA with Adaptive Antennas", *COST 259*, TD(99)113, Leidschendam, The Netherlands, Sep. 1999.

[Paju98]   Pajusco,P., "Experimental Characterization of D.O.A. at the Base Station in Rural and Urban Area", in *Proc. of VTC'98 – 48th IEEE Vehicular Technology Conference*, Ottawa, Canada, May 1998 [also available as TD(98)002].

[Pars92]   Parsons,J.D., *The Mobile Radio Propagation Channel*, Pentech Press, London, UK, 1992.

[Pätz99]   Pätzold,M., *Mobile Radio Channels* (in German), Vieweg Verlag, Braunschweig/Wiesbaden, Germany, 1999.

[PeAn99]   Pedersen,G.F. and Andersen,J.B., "Handset Antennas for Mobile Communications – Integration, Diversity, and Performance", in *Review of Radio Science 1996–1999*, Oxford University Press, Oxford, UK, 1999.

[PéJi94]   Pérez,V. and Jiménez,J.(eds.), *Final Propagation Model*, RACE-CoDiT Deliverable R2020/TDE/PS/DS/P/040/a1 Brussels, Belgium, June 1994.

[PeMF00]   Pedersen,K.I., Mogensen,P.E. and Fleury,B.H., "A Stochastic Model of Temporal and Azimuthal Dispersion seen at the Base Station in Outdoor Propagation Environments", accepted for publication at *IEEE Trans. Veh. Technol.*, 2000.

[PeMF97]   Pedersen,K.I., Mogensen,P.E. and Fleury,B.H., "Power Azimuth Spectrum in Outdoor Environments", *IEE Electronics Letters*, Vol. 33, No. 18, Aug. 1997, pp. 1583–1584.

[PeMF98]   Pedersen,K.I, Mogensen,P.E. and Fleury,B.H., "Spatial Channel Characteristics in Outdoor Environments and their Impact on BS Antenna System Performance", in *Proc. of VTC'98 – 48th IEEE Vehicular Technology Conference*, Ottawa, Canada, May 1998 [also available as TD(98)040].

[PeMF99a]   Pedersen,K.I., Mogensen,P.E. and Frederiksen,F., "Joint Directional Properties of Uplink and Downlink Channel in

Mobile Communications", *IEE Electronics Letters*, Vol 35, No. 16, Aug. 1999, pp. 1311–1312.

[PeMF99b]   Pedersen,K.I., Mogensen,P.E. and Fleury,B.H., "Dual-Polarized Model of Outdoor Propagation Environments for Adaptive Antennas", in *Proc. of VTC'99 Spring – 49th IEEE Vehicular Technology Conference*, Houston, Texas, USA, May 1999 [also available as TD (98)059].

[PeNo97a]   Pensel,K. and Nossek,J.A., "Uplink and Downlink Calibration of an Antenna Array in a Mobile Communication System", *COST 259*, TD(97)055, Lisbon, Portugal, Sep. 1997.

[PeNo97b]   Pensel,K. and Nossek,J.A., "DoA Estimation for SDMA Systems", *COST 259*, TD(97)024, Turin, Italy, May 1997.

[PeOL99]   Pedersen,G.F., Olesen,K. and Larsen,S.L., "Bodyloss for Handheld Phones", in *Proc. of VTC'99 Spring – 49th IEEE Vehicular Technology Conference*, Houston, Texas, USA, May 1999 [also available as TD(99)069].

[PeTK00]   Pedersen,G.F., Tartiere,M. and Knudsen,M.B., "Radiation efficiency of Handheld Phones", *COST 259*, TD(00)033, Valencia, Spain, Jan. 2000.

[PFBB99]   Papathanassiou,A., Furio,I., Blanz,J.J. and Baier,P.W., "Smart Antennas with Two-Dimensional Array Configurations for Performance Enhancement of a Joint Detection CDMA Mobile Radio System", *Wireless Personal Communications*, Vol. 11, No. 1, Oct. 1999, pp. 89–108.

[PFBH97]   Papathanassiou,A., Furio,I., Blanz,J.J., Haardt,M. and Schmalenberger,R., "Suboptimum Combined Direction of Arrival and Channel Estimation for Time-Slotted CDMA with Joint Detection", *COST 259*, TD(97)025, Turin, Italy, May 1997.

[PLNR99]   Pettersen,M., Lehne,P.H., Noll,J., Rostbakken,O., Antonsen,E. and Eckhoff,R., "Directional wideband channel measurements in urban and suburban areas", *COST 259*, TD(99)053, Vienna, Austria, Apr. 1999.

[PMFF98]   Pedersen,K.I., Mogensen,P.E., Fleury,B.H., Frederiksen,F., Olesen,K. and Larsen,S.L., "Analysis of Time, Azimuth, and Doppler Dispersion in Outdoor Radio Channels", in *Proc. of ACTS Mobile Communication Summit*, Aalborg, Denmark, Oct. 1997 [also available as TD(98)041].

[PNOK98]   Pedersen,G.F., Nielsen,J.Ø., Olesen,K. and Kovacs,I.Z., "Measured Variation in performance of handheld antennas for a large number of test persons", in *Proc. of VTC'98 – 48th IEEE*

Vehicular Technology Conference, Ottawa, Canada, May 1998 [also available as TD(98)025].

[PSWB98] Papathanassiou,A., Schmalenberger,R., Weckerle,M. and Baier,P.W., "User Angular Separation Considerations on the Uplink Performance of a TD-CDMA Mobile Radio System", in *Proc. of ISSSTA'98 – 5th IEEE International Symposium on Spread Spectrum Techniques and Applications*, Sun City, South Africa, Sep. 1998 [also available as TD(98)055].

[Rapp89] Rappaport,T.S., "Characterization of UHF multipath radio channels in factory buildings", *IEEE Trans. Antennas Propagat.*, Vol. AP-37, No. 8, Aug 1989, pp. 1058–1069.

[Rapp96] Rappaport,T.S., *Wireless Communications – Principles and Practice*, Prentice Hall, New Jersey, USA, 1996.

[RoCB99] Rossi,L., Cullen,P. and Brennan,C., "An analytical approximation of the TIM look-up table", *COST 259*, TD(99)025, Thessaloniki, Greece, Jan. 1999.

[RoKa89] Roy,R. and Kailath,T., "ESPRIT—Estimation of Signal Parameters via Rotational Invariance Techniques", *IEEE Trans. Acoust., Speech, Signal Process.*, Vol. 37, No. 7, July 1989, pp. 984–995.

[Rokh90] Rokhlin,V., "Rapid Solution of integral equations of scattering theory in two dimensions", *J. Comp. Phys.*, Vol. 86, No. 2, Feb. 1990, pp. 414–439.

[RuCO98] Ruiz-Boque,S., Covarrubias,D. and Olmos,J., "Impact of real channel characterisation on wireless packet networks with S-ALOHA channel access", *COST 259*, TD(98)096, Duisburg, Germany, Sep. 1998.

[RuSA99] Ruiz-Boque,S., Samper,Y. and Agusti,R., "Radiowave Propagation and Coverage in a Construction Site", *COST 259*, TD(99)110, Leidschendam, The Netherlands, Sep. 1999.

[RVCG98] Rizk,K., Valenzuela,R., Chizhik,D. and Gardiol,F., "Application of the slope diffraction method for urban microcellular systems at the UHF band", *COST 259*, TD(98)071, Bradford, UK, Apr. 1998.

[RVFC98] Rizk,K., Valenzuela,R., Fortune,S., Chizhik,D. and Gardiol,F., "Lateral, Full-3D and Vertical Plane Propagation in Microcells and Small Cells", *COST 259*, TD(98)047, Bern, Switzerland, Feb. 1998.

[RWLG97] Rizk,K., Wagen,J.F., Li,J. and Gardiol,F., "Lamppost and panel scattering compared to building reflection and diffraction", annexed to Ph.D. Thesis, K. Rizk, *Propagation in microcellular and small cell urban environment*, EPFL, Lausanne, Switzerland, 1997 [also available as TD(97)033].

[SaBo94]   Saunders,S.R. and Bonar,F.R., "Prediction of mobile radio wave propagation over buildings of irregular heights and spacings", *IEEE Trans. Antennas Propagat.*, Vol. 42, No. 2, Feb. 1994, pp. 137–144.

[SACH99]   Steinbauer,M., Asplund,H., de Coster,I., Hampicke,D., Heddergott,R., Lohse,N. and Molisch,A., "SWG 2.1 Modelling Unification Workshop", *COST 259*, TD(99)061, Vienna, Austria, Apr. 1999.

[SaFY98]   Salema,C., Fernandes,C. and Yha,R., *Solid Dielectric Horn Antennas*, Artech House, New York, New York, USA, 1998.

[SaVa87]   Saleh,A. and Valenzuela,R.A., "A statistical model for indoor multipath propagation", *IEEE J. Select. Areas Commun.*, Vol. SAC-5, No. 2, Feb. 1987, pp. 128–137.

[ScDW00]   Schneider,R., Didascalou,D. and Wiesbeck,W., "Impact of road surfaces on millimeter wave propagation", accepted for publication at *IEEE Trans. Veh. Technol.*, 2000.

[Schä93]   Schäfer,W., "A new deterministic/stochastic approach to model the intervehicle channel at 60 GHz", in *Proc. of VTC'93 – 43rd IEEE Vehicular Technology Conference*, Secaucus, New Jersey, USA, May 1993.

[Schn98]   Schneider,R., "Wave propagation modelling for vehicular-based radar imaging" (in German), Ph.D. Thesis, Universität Karlsruhe (TH), Karlsruhe, Germany, 1998.

[ScPa98a]   Schmalenberger,R. and Papathanassiou,A., "Downlink Spectrum Efficiency of a JD-CDMA Mobile Radio System with Array Transmit Antennas", *COST 259*, TD(98)094, Duisburg, Germany, Sep. 1998.

[ScPa98b]   Schmalenberger,R. and Papathanassiou,A., "Two compatible channel models for system and link level simulations of mobile radio systems", *COST 259*, TD(98)055, Bradford, UK, Apr. 1998.

[SeLS99]   Segovia,D., Lorenzo,R. and Sierra,M., "Robust DOA Estimation using a Polarisation Sensitive Array for Mobile Communications", *COST 259*, TD(99)065, Vienna, Austria, Apr. 1999.

[SHSS00]   Steinbauer,M., Hampicke,D., Sommerkorn,G., Schneider,A., Molisch,A.F., Thomä,R. and Bonek,E., "Array-measurement of the double-directional mobile radio channel", in *Proc. of VTC'2000 Spring – 51st IEEE Vehicular Technology Conference*, Tokyo, Japan, May 2000.

[Sibi00]   Sibille,A., "Sector and polarisation dependent wideband indoor propagation channel measurements at 5.1 GHz", *COST 259*, TD(00)010, Valencia, Spain, Jan. 2000.

[SoCh95]   Song,J.M. and Chew,W.C., "Multilevel fast multipole algorithm for solving combined field integral equations of electromagnetic scattering", *Micro. Opt. Tech. Lett.*, Vol. 10, No. 1, Sep. 1995, pp. 14–19.

[Sore98]   Sorensen,T.B., "Correlation model for slow fading in a small urban macro cell", in *Proc. of PIMRC'98 – IEEE 9th International Symposium on Personal, Indoor and Mobile Radio Communications*, Boston, Mass., USA, Sep. 1998.

[Stee92]   Steele,R.(ed.), *Mobile Radio Communications,* Pentech Press, London, UK, 1992.

[Stei98]   Steinbauer,M., "A comprehensive transmission and channel model for directional radio channels", *COST 259*, TD(98)027, Bern, Switzerland, Feb. 1998.

[SuVa99]   Sulonen,K. and Vainikainen,P., "Evaluation of Handset Antenna Configurations", *COST 259*, TD(99)055, Vienna, Austria, Apr. 1999.

[SWWN97] Schneider,R., Wenger,J., Wanielik,G. and Neef,H., "Millimetre wave images of traffic scenes and their automatic interpretation", in *Proc. of Microwave and RF'97*, London, UK, Sep. 1997.

[Taga90]   Taga,T., "Analysis for Mean Effective Gain of Mobile Antennas in Land Mobile Radio Environments", *IEEE Trans. Veh. Technol.*, Vol. VT-39, No. 2, May 1990, pp. 117–131.

[Tart99]   Tartiere,M., *Three Dimensional Farfield Radiation Measurements of Handheld Phones including the Human Body,* Master's Thesis, Aalborg University, Aalborg, Denmark, Aug. 1999.

[TaSC98]   Tarokh,V., Seshadri,N. and Calderbank,A.R., "Space-time codes for high data rate wireless communication: Performance criterion and code construction", *IEEE Trans. Information Theory*, Vol. 44, No. 2, Mar. 1998, pp. 744–765.

[TBKH99] Tschudin,M., Brunner,C., Kurpjuhn,T., Haardt,M. and Nossek,J.A., "Comparison between Unitary ESPRIT and SAGE for 3-D Channel Sounding", in *Proc. of VTC'99 Spring – 49th IEEE Vehicular Technology Conference*, Houston, Texas, USA, May 1999 [also available as TD(99)062].

[TCJF72]   Turin,G.L., Clapp,F.D., Johnston,T.L., Fine,S.B. and Lavry,D., "A Statistical Model of Urban Multipath Propagation", *IEEE Trans. Veh. Technol.*, Vol. VT-21, No. 1, Feb. 1972, pp. 1–9.

[Tela95]   Telatar,E., *Capacity of Multi-Antenna Gaussian Channels*, AT&T-Bell Technical Memorandum, No. BL011217-950615-07TM, USA, June 1995.

[THST98]   Trautwein,U., Hampicke,D., Sommerkorn,G. and Thomä,R., "Array Based Measurements of the Time-Variant Directive Radio Channel", *COST 259*, TD(98)072, Bradford, UK, Apr. 1998.

[TLDV99]   Torres,R.P., Loredo,S., Domingo,M. and Valle L., "An accurate and efficient method to estimate the local fading statistics from ray tracing", in *Proc. of VTC'99 Fall – 50th IEEE Vehicular Technology Conference*, Amsterdam, The Netherlands, Sep. 1999.

[ToBL98]   Torrico,S.A., Bertoni,H. and Lang,R.H., "Theoretical modelling of foliage effect on path loss for residential environments", *COST 259*, TD(98)111, Duisburg, Germany, Sep. 1998.

[Toel00]   Toeltsch,M., Private Communication, TU Vienna, Vienna, Austria, Apr. 2000.

[ToHA93]   Toftgård,J., Hornsleth,S.N. and Andersen,J.B., "Effects on Portable Antennas of the Presence of a Person", *IEEE Trans. Antennas Propagat.*, Vol. 41, No. 6, June 1993, pp. 739–746.

[ToLa99]   Torrico,S.A. and Lang,R.H., "Simple engineering expressions to predict the specific attenuation of a tree", *COST 259*, TD(99)005, Thessaloniki, Greece, Jan. 1999.

[TrSi97]   Truffer,P. and Sibilia,R., "Wideband Channel Sounder with Optical Antenna Feeding", *COST 259*, TD(97)088, Lisbon, Portugal, 1997.

[TsHT98]   Tschudin,M., Heddergott,R. and Truffer,P., "Validation of a high resolution measurement technique for estimating the parameters of impinging waves in indoor environments", in *Proc. of PIMRC'98 – 9th IEEE International Symposium on Personal, Indoor and Mobile Radio Communications*, Boston, Mass., USA, Sep. 1998 [also available as TD(98)065].

[TVDL99a]   Torres,R.P., Valle,L., Domingo,M. and Loredo,S., "CINDOOR: An engineering tool for planning and design of wireless systems in enclosed spaces", *IEEE Antennas Propagat. Mag.*, Vol. 41, No. 4, Aug. 1999, pp. 11–22.

[TVDL99b]   Torres,R.P., Valle,L., Domingo,M. and Loredo,S., "An efficient ray-tracing method for radiopropagation based on the modified BSP algorithm", in *Proc. of VTC'99 Fall – 50th IEEE Vehicular Technology Conference*, Amsterdam, The Netherlands, Sep. 1999.

[Utsc99]   Utschick,W., "Downlink Beamforming for FDD Mobile Radio Systems based on Spatial Covariances", *COST 259*, TD(99)056, Vienna, Austria, Apr. 1999.

[VaCo97]   Vasconcelos,P. and Correia,L.M., "Fading characterization of the mobile radio channel at the millimetre waveband", in *Proc. of VTC'97 – 47th IEEE Vehicular Technology Conference*, Phoenix, Ariz., USA May 1997 [also available as TD(97)011].

[VaCo98]   Vasconcelos,P. and Correia,L.M., "Fading Dependence on Scenario and Antennas Characteristics at the 60 GHz Band", in *Proc. of VTC'98 – 48th IEEE Vehicular Technology Conference*, Ottawa, Canada, May 1998 [also available as TD(98)100].

[vaSc99]   van den Homberg,M. and Schmidt,H., "Preliminary planar signal strength measurements on handhelds with helix, dipole or patch antennas", *COST 259*, TD(99)084, Leidschendam, The Netherlands, Sep. 1999.

[VaSu99]   Vainikainen,P. and Sulonen,K., "Evaluation of Handset Antenna Configurations", *COST 259*, TD(99)055, Vienna, Austria, Apr. 1999.

[Vazq00]   Vazquez,M.M., "Integraded Antennas for Mobile Communication Handsets", *COST 259*, TD(00)022, Valencia, Spain, Jan. 2000.

[VeCo97]   Velez,F.J. and Correia,L.M., "Optimization Criteria for Cellular Planning of Mobile Broadband Systems in Linear and Urban Coverages", in *Proc. of ACTS Mobile Communications Summit*, Aalborg, Denmark, Oct. 1997 [also available as TD(97)053].

[Vogl82]   Vogler,L.E., "An attenuation function for multiple knife-edge diffraction", *Radio Science*, Vol. 17, No. 6, Nov./Dec. 1982, pp. 1541–1546.

[VVMO99]   Van Lil,E., Van de Capelle,A., Meuris,P., Ocket,I. and Van Troyen,D., "Applications of plane-polar near-field measurements to handset antennas", *COST 259*, TD(99)125, Vienna, Austria, Apr. 1999.

[WaBe88]   Walfish,J. and Bertoni,H., "A theoretical model of UHF propagation in urban environments", *IEEE Trans. Antennas Propagat.*, Vol. 36, No 12, Dec. 1988, pp. 1788–1796.

[WaCh94]   Wagner,R. and Chew,W.C., "A Ray-Propagation Fast Multipole Method", *Micro. Opt. Tech. Lett.*, Vol. 7, No. 10, July 1994, pp. 435–438.

[WaCh95]   Wagner,R.L. and Chew,W.C., "A Study of Wavelets for the Solution of Electromagnetic Integral Equations", *IEEE Trans. Antennas Propagat.*, Vol. 43, No. 8, Aug. 1995, pp. 802–810.

[WaLL97]   Wagen,J-F., Lachat,E. and Li,J., "Performance Evaluation of a Ray Tracing Based Microcellular Coverage Prediction Tool", in *Proc. of PIMRC'97 – 8th IEEE International Symposium on Personal, Indoor, and Mobile Radio Communications*, Helsinki, Finland, Sep. 1997 [also available as TD(97)019].

[WePa99a] Weckerle,M. and Papathanassiou,A., "Performance Analysis of Multi-Antenna TD-CDMA Receivers with Estimation and Consideration of the Interference Covariance Matrix", *International Journal of Wireless Information Networks*, Vol. 6, No. 3, Dec. 1999, pp. 157-170.

[WePa99b] Weckerle,M. and Papathanassiou,A., "A Novel Multi-Antenna TD-CDMA Receiver Concept incorporating the Estimation and Utilisation of Spatial Interference Covariance Matrices", *COST 259*, TD(99)059, Vienna, Austria, Apr. 1999.

[WePH99]   Weckerle,M., Papathanassiou,A. and Haardt,M., "Estimation and Utilisation of Spatial Interference Covariance Matrices in Multi-Antenna TD-CDMA Systems", in *Proc. of PIMRC'99 – 10th International Symposium on Personal, Indoor and Mobile Radio Communications*, Osaka, Japan, Sep. 1999.

[WePS99]   Weckerle,M., Papathanassiou,A. and Schmalenberger,R., "Consideration of the Spatial Covariance Matrix in Multi-Antenna TD-CDMA Systems", *COST 259*, TD(99)018, Thessaloniki, Greece, Jan. 1999.

[Whal71]   Whalen,A., *Detection of Signals in Noise*, Academic Press, London, UK, 1971.

[WMMK99]Winter,J., Martin,U., Molisch,A.F., Kattenbach,R., Gaspard,I., Steinbauer,M. and Grigat,M., *Description of the modeling method*, METAMORP Public Report C-2/1, Deutsche Telekom, Darmstadt, Germany, Feb. 1999.

[WoLa99]   Wölfle,G. and Landstorfer,F.M., "Prediction of the Field Strength Inside Buildings with Empirical, Neural, and Ray-Optical Prediction Models", *COST 259*, TD(99)008, Thessaloniki, Greece, Jan. 1999.

[WoWL99]   Wölfle,G., Wertz,P. and Landstorfer,F.M., "Performance, Accuracy, and Generalization Capability of Indoor Propagation Models in Different Types of Buildings", in *Proc. of PIMRC'99 – 10th IEEE International Symposium on Personal, Indoor and Mobile Radio Communications*, Osaka, Japan, Sep. 1999.

[XiaH96]   Xia,H., "An analytical model for predicting path loss in urban and suburban environments", in *Proc. of PIMRC'96 – 7th IEEE International Symposium on Personal, Indoor and Mobile Radio Communications*, Taipei, Taiwan, Oct. 1996.

[XiaH97] Xia,H., "A simplified analytical model for predicting path loss in urban and suburban environments", *IEEE Trans. Veh. Technol.*, Vol. 46, No. 4, Nov. 1997, pp. 1040–1046.

[XiBe92] Xia,H.H. and Bertoni,H.L., "Diffraction of Cylindrical and Plane Waves by an Array of Absorbing Half-Screens", *IEEE Trans. Antennas Propagat.*, Vol. 40, No. 2, Feb. 1992, pp. 170–177.

[ZDDW99] Zwick,T., Didascalou,D., Döttling,M. and Wiesbeck,W., "Spatial radio channel modeling for the design of new generation mobile communication systems", in *Proc. of VTC'99 Fall – 49th IEEE Vehicular Technology Conference*, Amsterdam, The Netherlands, Sep. 1999.

[ZFDW00] Zwick,T., Fischer,C, Didascalou,D., Wiesbeck,W., "A stochastic spatial channel model based on wave-propagation modeling", *IEEE J. Select. Areas Commun.*, Vol. 18, No. 1, Jan. 2000, pp. 6–15.

[ZhKV00] Zhao,X., Kivinen,J. and Vainikainen,P., "Tapped Delay Line Channel Models at 5.3 GHz in Indoor Environments", *COST 259*, TD(00)025, Valencia, Spain, Jan. 2000.

[ZHMR00] Zwick,T., Hampicke,D., Maurer,J., Richter,A., Sommerkorn,G., Thomä,R., Wiesbeck,W., "Results of double-directional channel sounding measurements", in *Proc. of VTC'2000 Spring – 51st IEEE Vehicular Technology Conferencee*, Tokyo, Japan, May 2000.

[ZwDF98] Zwick,T., Didascalou,D. and Fischer,C., "A wave propagation based stochastic channel model", *COST 259*, TD(98)043, Bern, Switzerland, Feb. 1998.

[ZwFW98] Zwick,T., Fischer,C. and Wiesbeck,W., " A Statistical Channel Model for Indoor Environments Including Angle of Arrival", in *Proc. of VTC'98 – 48th IEEE Vehicular Technology Conference*, Ottawa, Canada, May 1998.

[Zwic00] Zwick,T., "Comparison of Ray-Tracing Results with Indoor Measurements", *COST 259*, TD(00)031, Valencia, Spain, Jan. 2000.

[Zwic99] Zwick,T., *Modeling of Directional Multipath Indoor Radio Channels Using Marked Poisson Processes* (in German), Ph.D. Thesis, Univ. Karlsruhe, Karlsruhe, Germany, 1999.

4

Network Aspects

Thomas Kürner

Both in the evolving second generation (2G) systems and the new third generation (3G) systems, radio network aspects are becoming more important in the planning process.

Apart from the tasks coming from the introduction of new radio interfaces in 3G systems, additional challenges are introduced even to the existing 2G systems. Capacity enhancement techniques like Frequency Hopping (FH), Discontinuous Transmission (DTX), adaptive antennas or even sophisticated hierarchical cellular architectures like Intelligent-Underlay-Overlay (IUO) techniques are introduced in 2G systems [FRGF96]. Today's mobile networks predominantly carry voice traffic. In the future mobile networks will be operated in a multi-service environment carrying both circuit-switched and packet-switched data with different transmission rates [Jabb96]. The traffic characteristics of these services will be both inhomogeneous and asymmetric. The operators will only be able to successfully exploit the full potential inherent to the new network features and fulfil the new service demands if a clear understanding of their implications on the planning process is reached and adequate planning and optimisation strategies are designed [BGMM98]. Working Group 3 'Network Aspects' of COST 259 has been focusing on the research and development of the corresponding algorithms, methods, and strategies:

- spectrum efficiency;
- compatibility (including frequency sharing with fixed services);
- channel allocation strategies (fixed and/or dynamic), best suited to the network and the operating environment;
- studies on efficient protocols for high data rates including voice and video integration (multimedia) with specific reference to the opportunities offered by ATM access;
- assessment of tools for cellular, micro- and pico-cellular coverage, taking into account that for future systems, non-uniformity of traffic distributions and traffic capacity considerations in most cases will impose restrictions on cell planning more than propagation constraints themselves;
- network optimisation algorithms by minimisation of mutual interference;

- development of advanced systematic planning methods, capable of coping with the various different network and environmental situations.

The whole lifetime of the COST259 project has been characterised by the initial standardisation of UMTS/IMT2000 and the improvement of the existing 2G systems. These two main streams in the evolution of mobile cellular radio are reflected in this chapter. Some more mature techniques for GSM systems are presented, together with a couple of basic techniques required for planning of UMTS/IMT2000.

4.1. Compatibility and Spectrum Efficiency

Luis M. Correia

4.1.1. Spectral compatibility

Peter Seidenberg

4.1.1.1. Introduction

Efficient allocation of frequency spectrum to mobile communication providers requires that the characteristics of the planned or existing mobile communication systems are taken into account, to ensure the simultaneous and undisturbed service of mobile radio systems operating in adjacent frequency bands. Mutual interference between different operator's networks can occur if the transmit band of a station in one network is adjacent to the receive band of another station in another system; this interference is caused by both side-band emissions and receiver imperfections. Figure 4.1.1 shows the general interference scenario: two operators with different network deployments operate in the same area using adjacent frequency bands.

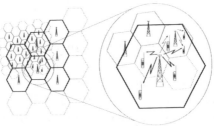

Figure 4.1.1: General interference scenario.

Since a transmit band is always the receive band of a corresponding station, an interfering system also suffers from interference originating in the perturbed system, i.e. if a MS(BS) of network A perturbs a BS(MS) of network B, then the MS of network B also perturbs the BS(MS) of network A. Particularly in networks with small cluster sizes, such as the CDMA-based UMTS/IMT-2000, mutual interference between adjacent channels of different operators is likely to occur. As long as such interferences cannot be bypassed, due to a lack of alternative frequencies within the same system or to an interference perturbing all timeslots of a frame, the Adjacent Channel Interference (ACI) directly affects the capacity of all interference-limited systems.

4.1.1.2. Calculation of minimum frequency separation

The unused frequency band between two different radio systems intended to decrease the possibility of mutual interference is referred to as Minimum Frequency Separation (MFS), Figure 4.1.2. Thus, the MFS can be derived from $MFS = (Tr_2 - B_2/2) - (Tr_1 + B_1/2)$, where Tr_x stands for the carrier frequency of system x and B_x for the bandwidth requirement of a carrier in system x.

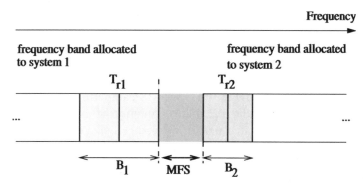

Figure 4.1.2: Minimum Frequency Separation.

The spectral characteristics of interferences are described with the help of masks. The mask for the interfering TX represents the maximum permissible unwanted emission level as a function of the frequency. To define a mask for the emission, the different sources of interferences, like effects of the modulation process, rise and fall times of the transmitted signal, intermodulation products, and wideband noise, are combined in one mask. The RX characteristic is also represented by a mask that can be found by transferring the interference rejection mechanisms defined in the standards to an equivalent Carrier-to-Interference Ratio (CIR), C/I. Depending on a number of parameters, like user's densities, antenna heights, coverage radii, transmitter (TX) power, etc., which determine an interference situation as the one depicted in Fig 4.1.1, the interference at a perturbed RX station is evaluated with the help of the masks using a Monte Carlo (MC) simulation technique.

Each MC cycle starts with the positioning of the RX station by means of an appropriate distribution function for the user path. This RX station becomes the victim RX due to interferences of the interfering stations. After all participating stations have been located, the attenuation between the victim RX and the interferers will be calculated, using the appropriate propagation model.

To determine the interference power at the victim RX, the unwanted emissions of each interferer are calculated with the help of the mask and the loss on the interference path will be subtracted from it. With the information of the signal carrier power on the user link, C, and of the interference power, the C/I present at the victim RX can be calculated as a function of frequency. All C/I values, whether they are measured inside the RX band or outside it, are calculated using the TX masks and are compared with the respective required C/I, C/I_{req}, defined by the RX masks at the respective frequency f_i. The minimum difference between the present and the required C/I is chosen as the value for statistical evaluation, $\min_{f_i}\{C/I(f_i.)-C/I_{req}(f_i.)\}$. The probability that this difference is less or equal to zero is the probability that the C/I value at the perturbed receiver, RX, falls below the required ratio.

4.1.1.3. GSM/TETRA compatibility

Since the GSM uplink band (890–915 MHz) is adjacent to the TETRA downlink band (915–921MHz), the GSM MS can perturb the TETRA hand-held. Figure 4.1.3 depicts the distribution function of the difference between the required and the generated C/I for a fixed cell size of the victim system. The 0 on the x-axis applies exactly at that situation at which the required C/I is still achieved. The probability of an inadequate coverage corresponds to the probability value along the curve at C/I-difference equal to 0. The simulation was based on a density of 20 interferers/km^2; in all cases the MFS was 600 kHz.

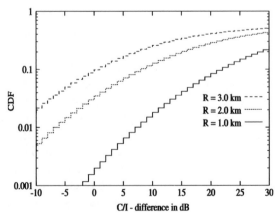

Figure 4.1.3: *C/I*-difference, GSM MS perturbs TETRA hand-held.

BSs with large TX radius serve many mobile users in reception areas with a low signal level. For this reason, mobile users in such areas are much more susceptible to interference from other systems. From this, it follows that hot spots should not be located in poorly served areas of a cell to ensure that a satisfactory receiving level is available to the victim RX.

4.1.1.4. UTRA/DECT compatibility

Figure 4.1.4 represents the spectral power density of UTRA BS and MS TXs, respectively, i.e. the power measured in one DECT carrier bandwidth; the masks are based on values given in [ETSI97] and [ERC97a]. The required performances of DECT in the presence of an interferer are determined in [ETSI92]. The C/I RX mask considers the values given in the standard for interference performance, blocking, out-band and in-band spurious emissions. The maximum allowed interference power due to one of these issues is given for a specified signal power. These values define equivalent C/I. The maximum required C/I values are taken to construct the mask depicted in Figure 4.1.5. This mask represents the ratio between the received signal power and the interference power measured in a 1 728 kHz bandwidth, respectively.

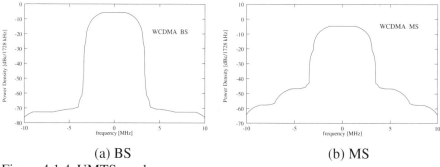

| (a) BS | (b) MS |

Figure 4.1.4: UMTS masks.

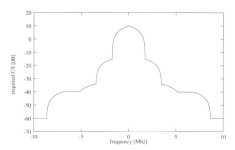

Figure 4.1.5: DECT C/I mask.

The DECT system is located in the band 1 880 to 1 900 MHz, the UMTS in the band 1 900 MHz and beyond that. The maximum distance between a DECT TX and RX is 100 m, whereas the cell radius of the UMTS is chosen to be 1 km. For an outdoor scenario, Figure 4.1.6 shows the distribution function of the difference between the required *C/I* and the *C/I* for the DECT channel adjacent to the UMTS band.

It should be noted that the simulation results for the interfering MSs are valid for a worst-case scenario with 100 UMTS MSs per km^2. Furthermore, no power control is taken into account. The four interfering stations nearest to the perturbed DECT station are considered.

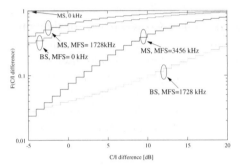

Figure 4.1.6: *C/I*-difference, UMTS perturbs DECT.

If the UMTS uplink band is adjacent to the DECT band, the probability that the required *C/I* cannot be reached is nearly 100 %; for a MFS of 1 DECT channel width, i.e. 1 728 kHz, this probability decreases to about 50 %. An acceptable interference probability is not reached until a MFS of two DECT channels. If the UMTS downlink band is adjacent to the DECT band, one can achieve an adequate interference probability of about 3 % by claiming a MFS of 1 728 kHz.

4.1.1.5. UTRA/UTRA compatibility

The UTRA-to-UTRA compatibility is determined in terms of ACI power leaking from the transmit band of one UTRA system into the receive band of another UTRA system. The interference power can be expressed as

$$I_{adj} = \frac{P_{TX}}{L_c} \int_{-\infty}^{\infty} \Phi(f - \Delta f, P)\left|H(f)\right|^2 df \qquad (4.1.1)$$

where P_{TX} is the interferer's TX power, L_c is the path loss between the perturbed and the interfering station (which is often called the coupling loss), $\Phi(f,P)$ represents the normalised spectral power density of the modulated signal incorporating the transfer function of the power amplifier which depends on the actual transmission power, $|H(f)|$ describes the transfer function of the RX filter, and Δf is the carrier spacing. Since $\Phi(f,P)$ depends on the transmit power, the portion of interference power leaking into the adjacent band is not constant; therefore the ACI ratio is not constant.

In the simulations [SASH99], it has been assumed that the maximum ACI between a UTRA MS and a UTRA BS is –37 dB. The ACI can decrease down to –130 dB due to power control at minimum transmit power. This is caused by a 3 dB decrease of side-band emissions for each dB of total transmitted power decrease. If this decrease factor is chosen to be 1 dB, then the ACI remains constant at its maximum value.

The UTRA-to-UTRA compatibility has been studied in a macro-/micro-cellular mixed-mode environment as depicted in Figure 4.1.7 using a simple signal-strength based power control scheme.

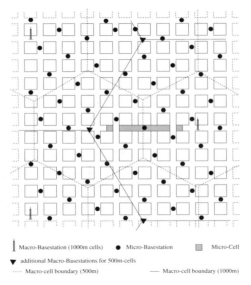

Figure 4.1.7: UTRA network deployment.

For macro-cells a radius of 500 m or 1 km has been used, micro-cellular BSs being placed 360 m apart in a Manhattan like grid. Simulation results are depicted in Figure 4.1.8, for a variable (3 dB) and a constant (1 dB) ACI. It can be seen that for a variable ACI the ACI power is fairly low. If the ACI is constant and never better than the specified value of –37 dB then the interference power leaking into the perturbed band can exceed the reference sensitivity level of the receiving station by more than 20 dB. This indicates a severe interference situation the TDD station cannot bypass without changing the frequency channel.

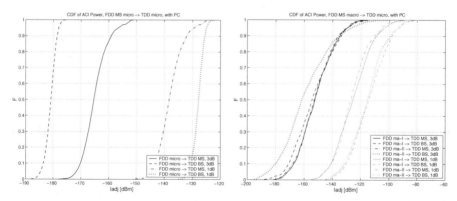

(a) macro FDD perturbs micro TDD   (b) micro FDD perturbs micro TDD

Figure 4.1.8: CDF of interference power between FDD and TDD cells.

4.1.2. Spectrum Efficiency

Luis M. Correia

The worldwide success of mobile communications has put a lot of pressure on the efficient use of the spectrum. In countries where user's penetration is very high, even higher than the fixed network, cellular operators often face very harsh limitations on network expansion, precisely due to the lack of spectrum available. Moreover, spectrum scarcity is imposing tighter schedules for the development of UMTS/IMT-2000, due to the pressures of some countries in getting the new standard as soon as possible. Therefore, it is clear that an efficient use of the spectrum is of key importance today. In what follows, some contributions are given to this area in a quite wide range of systems, from UMTS at the 2 GHz band up to Mobile Broadband Systems (MBSs) at 60 GHz.

4.1.2.1. Optimising spectrum efficiency in UMTS

UMTS is going to be a multi-service system, to be deployed in environments with inhomogeneous traffic distributions, which means that radio network planning will be much more complex than the GSM planning. These inhomogeneous traffic characteristics require an appropriate allocation of resources per cell, i.e. frequency/channel planning needs to be done with adequate care: in GSM, where each radio channel has a bandwidth of 200 kHz, approximately 40 to up 150 channels are available to an operator, hence the increase of traffic can be solved by adding a new TRX to the required BSs; on the contrary, for UMTS, each radio channel has a bandwidth of 5 MHz, and only 2–5 channels will be available to an operator, which may impose adding a new cell to the network in order to cope with traffic increase. Since, for the latter, cell size is very dependent on the capacity required, an integrated approach for coverage and capacity planning is necessary.

Spectrum allocation will affect deployment scenarios and the use of hierarchical cell structures. Some issues are to be considered as far as operators are concerned: on the one hand, they should be able to make optimum use of the scarce resource of the spectrum; on the other, they should also be given the maximum flexibility to cater to the different customer/market requirements. Since it is expected that UMTS will make use of advanced cellular structures (hierarchical ones) aimed at maximising network capacity, these problems may be diminished. Assuming that the spectrum available will be that decided by ERC [ERC97b], i.e. 2×60 MHz for the paired band and 35 MHz for the unpaired band, that traffic estimates follow the values given by the UMTS Forum [UMTS98b], and that the network will be planned using a hierarchical structure (macro-, micro- and pico-cells), some scenarios can be drawn and an evaluation of spectrum allocation can be made [Kürn99c].

A three-layer network is considered: FDD/macro-cells, providing wide area coverage for high-speed mobiles; FDD/micro-cells, to be deployed at street level (for outdoor coverage), ensuring extra capacity; TDD/FDD/pico-cells, mainly for indoors, providing high data rate services, the range of terminals being the limiting factor. As a basis for a network operator, an allocation of paired frequencies for FDD operation is necessary: 2×5 MHz allows a single layer only, hence a hierarchical cell structure is not feasible; 2×10 MHz gives room for a two-layer structure, e.g. a macro-cell layer together with either a micro- or a pico-cell one; 2×15 MHz allows the deployment of a complete hierarchical cell structure where traffic demand is high or a mix of layers such as one of macro-cells and two of micro-cells; 2×20 MHz allows increased flexibility and additional capacity. In addition to the allocation of paired frequencies, an operator may need an allocation of unpaired frequencies for TDD operation, in particular for low mobility indoor applications with asymmetric traffic: 5 MHz may be required in order to give satisfactory capacity; 10 MHz would give more flexibility and additional capacity. The feasibility of minimum spectrum allocation per operator is described in Table 4.1.1, according to different scenarios.

From the scenarios of Table 4.1.1, one can estimate service capabilities, assuming some hypotheses [UMTS98a] on traffic forecast for the various services (ranging from speech up to high interactive multimedia) as well as their asymmetry and spectral efficiencies, maximum available data rates and mobility per cell type (and respective coverage areas), among others.

Table 4.1.1: Spectrum allocation scenarios.

Scenario	Frequencies allocated to one operator [MHz]		Maximum number of operators	Spectrum not allocated [MHz]
	Paired	Unpaired		
1	2 × 5	-	12	35
2	2 × 5	5	7	50
3	2 × 10	-	6	35
4	2 × 10	5	6	5
5	2 × 15	-	4	35
6	2 × 15	5	4	15
7	2 × 20	-	3	35
8	2 × 20	5	3	20

Besides these, one can add some more assumptions on traffic distribution: all UMTS operators will share traffic in an equal way; services will be mainly focused on multimedia; most (90 %) of the speech and low data rate traffics will be carried within GSM networks; 60 % of the traffic in central business districts will be carried in license exempt networks; high multimedia traffic is distributed micro- and pico-cells. These assumptions enable the calculation of the net cell loading for the different scenarios without Quality of Service (QoS) considerations [Kürn99c], as well as the effective cell loading according to some optimisation criteria on traffic distribution among network layers. As a result, one can see that the scenarios with a small bandwidth do not allow a full service capability, Table 4.1.2, which can only be achieved if a minimum of 20 MHz is available, scenario 6 (2 × 15 MHz paired and 5 MHz unpaired) being the recommended one: it allows for one macro-cell, two micro-cells and one pico-cell layers.

Table 4.1.2: Service capabilities for different scenarios.

Scenario	Frequencies allocated to one operator [MHz]		Service capability
	Paired	Unpaired	
1	2 × 5		Limited
2	2 × 5	5	
3	2 × 10	–	
4	2 × 10	5	Some possible Restrictions
5	2 × 15	–	
6	2 × 15	5	Full
7	2 × 20	–	
8	2 × 20	5	

The technical parameters still to be defined in UTRA will have an effect in on the feasibility of the previously mentioned spectrum allocation scenarios. Current uncertainties include maximum cell ranges, spectral efficiency figures, QoS service definitions, and traffic asymmetry characterisation. These issues need to be dealt with in the very near future, if UMTS is to be deployed within the foreseen schedule.

4.1.2.2. Indoor scenarios

RF penetration into buildings and interference in indoor scenarios is an important area of research in order to evaluate the performance of wireless systems, as well as its implications in frequency re-use, hence on spectrum efficiency. Typical channel characteristics are known, but more accurate capacity and coverage predictions are necessary, for which a site-specific propagation model based on ray tracing can be used to provide accurate results, e.g. interference inside a building for both co-channel and adjacent channel studies.

A version of the EPICS ray tracing tool [DMSV97] is used for a study at 2.4 GHz [GDVP99], which is appropriate for indoor environments with obstacles comparable in dimension with the wavelength. The phenomena of reflection, diffraction and penetration were implemented by using Geometrical Optics, Uniform Theory of Diffraction [Pars92] and COST231 [COST99] penetration formulas, second and third order combinations of these phenomena also being included (penetration through walls is the dominant effect). The case of an 8-office building (a $2 \times 2 \times 2$ cube) was considered, with offices of $3 \times 3 \times 3$ m; the material chosen for outside walls was stone with a thickness of 10 cm, while ceilings, floors, and the remaining walls were considered to be made of wood, with the same thickness. The TX is located in the upper floor, in the centre of a room (R8), transmitting with vertical polarisation, while the RX is considered to sweep the rest of the building, with a resolution of half wavelength.

The Grade of Service (GoS) expressed as the outage probability (probability of unsatisfactory reception), P_{out}, was considered as measurement parameter. There are two criteria to specify what satisfactory reception means: for a coverage criterion one calculates the probability that the minimum carrier power, C_0, determined by the RX threshold (from noise), is exceeded; for a co-channel interference, CCI, criterion, one calculates the probability that the wanted received signal, C_w, exceeds the one of interfering sources, I, by a factor known as the protection ratio, R. (the minimum value for the CCI ratio). In cellular systems, it is likely that interference will pose more stringent constraints than noise, hence system performance evaluation is based on this criterion, and an interference margin is defined, $\tau = (C_w - I) - R$ [dB] (S_w, and I are taken here as mean values). Figure 4.1.9

presents P_{out} against R and τ, for the different rooms; one can see the influence of the various penetration losses on the results.

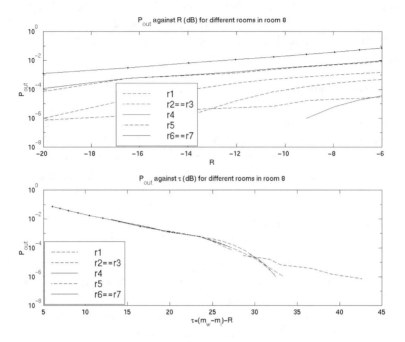

Figure 4.1.9: Outage probability due to interference.

According to results of this particular simulation, preliminary conclusions about interference levels in any particular indoor propagation environment can be extrapolated. At 2.4 GHz, in this eight-office building, the worst case of interference occurs in adjacent rooms, as expected, the CIR being approximately 7 dB, while the maximum value of this ratio is obtained at the other floor, about 19 dB.

4.1.2.3. Microwave video distribution systems

Microwave Multichannel or Multipoint Video Distribution System (MMDS or MVDS) is a local cellular point-to-multipoint radio system, delivering broadcast services from a central TX (hub) to individual houses or blocks of apartments within its cell size, offering rapid infrastructure deployment and the ability to provide local content. It can be significantly cheaper to install than a cable system, since only homes requesting the MMDS services are provided with RXs (extension on demand). Beyond one-way broadcasting distribution, this wireless technology has the potential for providing expanded TV services (like pay per view and video on demand), as well as to become a low-cost integrated medium (perhaps based on ATM) for providing voice and data services, as well, having the potential to integrate

Cable TV (CATV) and ISDN networks. Since this wireless technology can be a quick, simple and cost-effective solution, it might be a possible broadband access to subscribers for the new competitors of the old established operators after the European liberalisation of the telecommunications market.

In Europe, the [40.5, 42.5] GHz band has been harmonised within the CEPT for MVDS [CEPT91]. The allocated 2 GHz spectrum for digital MVDS is divided into a *downstream* for broadcast (including 96 RF channels, with a bandwidth of 33 MHz and a channel spacing of 39 MHz) and an *upstream* for return paths (having a bandwidth of 50 MHz at both ends of the 2 GHz band). To double the capacity in cellular networks, use is made of vertical and horizontal polarisations; hence, the available spectrum may be separated into four groups of 24 RF channels for cellular MVDS networks.

In order to illustrate trends in service areas, typical system parameters were assumed in the link budget of a digital MVDS [HaJa99]. For Line-of-Sight (LoS) situations, the maximum service distance, or the system margin at a given distance, can be obtained, either related to the noise threshold or to the TX power; Figure 4.1.10, exhibits the maximum service distance vs. TX power per RF channel for a required CNR, C/N, of 6.8 dB (a low-power MVDS with 10 to 20 mW meets the 99.9% criterion at a maximum LoS range of 2.5 to 3 km, while a high-power one with 0.5 to 1 W does it at 5.5 to 6 km). Irrespective of the frequency used, all microwave broadband wireless technologies suffer from the same fundamental limitation: they require clear LoS between the hub antenna and the receiving antenna. Hence, in urban areas, only low LoS penetration rates are achievable; estimations of the penetration rate are difficult to achieve, so that they range from 30 to 70 %, depending on parameters of the hub and its environment. This relatively low penetration rate in urban areas is caused by the obstruction due to buildings, vegetation, etc; repeaters and master antennas can be used to achieve higher penetration rates, but an inexpensive option to increase penetration might be the usage of reflected and diffracted waves in the Non-LoS (NLoS) areas, if attenuation due to reflection or diffraction is smaller than a certain available system margin, especially in the vicinity of the hub, below 2.5 km, where the system margin can be greater than 20 dB.

MVDS, especially in industrialised countries, will be competitive to CATV and DBS (Digital Broadcast System) only if it can provide service-on-demand, i.e. if it can become a low-cost integrated medium with high-speed access to subscribers for interactive multimedia applications. For these interactive services, an *upstream* or return path is needed from subscribers to the hub, hence, interactive MVDS includes a unidirectional Forward Broadcast Path (FBP) (for video, audio and data distribution), and a bi-directional interaction system, composed of a Forward Interaction Path (FIP) and a Return Interaction Path (RIP) [ETSI98a].

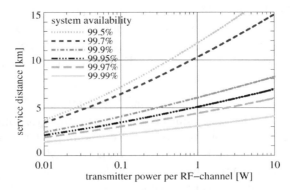

Figure 4.1.10: Service distance vs. TX power, for Climate Zone *H* and horizontal polarization.

The required data rates for interactive services are very dependent on the kind of services being requested: for example, video-on-demand, tele-shopping, or remote learning needs high bit rates of 2–6 Mbit/s in the FIP, whereas 20 kbit/s are sufficient in the RIP; in contrast, a video conference or an exchange of multimedia data needs several Mbit/s in both interaction paths. There are two different strategies to meet these different data rate requirements and future service requests of customers: RIP and/or FIP are implemented out of band, offering low data rates via an external medium; RIP and/or FIP are embedded in the 2 GHz MVDS spectrum (in-band), able to provide higher bite rates up to several Mbit/s. The former causes problems in combining the connections of FBP with external FIP and/or RIP, whereas transporting inter-active broadband services over a single wireless link seems to be more attractive from a customer's point of view (single connection), and it can be the only or at least a cost-effective solution for the competitors of the established operators in some areas. Therefore, only the latter is being considered here. Figure 4.1.11 shows the maximum distance or coverage between customer stations and hub as a function of the RF-bandwidth of the RIP for an TX power of 10 mW, where the 99.9 % availability criterion meets the maximum LoS distance at 3.5 and 2.5 km for a bandwidth of 1 and 8 MHz; the distance can be increased up to 4 and 6 km if the TX power is raise to 100 mW [JaGr96].

For a bi-directional MVDS, the only limiting factor in rural areas is rain attenuation, conversely in urban areas, LoS obstructions and interactive capacity are more likely to limit the maximum cell size than rain attenuation characteristics. Moreover, as a kind of 'public radio network', MVDS can also be easily extended for in-house (office and industrial) radio networks, offering the possibility for restricted mobile and portable communications, e.g. for video and data transmission of moveable robots in industrial plants or radio connections to ISDN or Internet via portable Laptops or PCs.

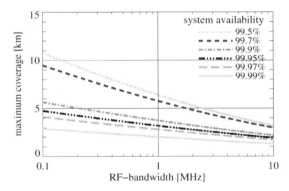

Figure 4.1.11: Maximum coverage distance vs. RF bandwidth.

4.1.2.4. Comparison between cellular systems at the 40 and 60 GHz bands

Mobile Broadband Systems (MBS) will allow extending high data rates provided by the fixed Broadband-ISDN to the cellular communications market, supporting high-speed communications in high mobility outdoor scenarios [Fern95], leading to the use of millimetre wavebands. The following specific bands are being considered for the implementation of MBS: [39.5, 43.5] GHz and [62, 66] GHz, with an interval of 2 GHz in between 1 GHz bands. For such bands, shadowing from buildings is important, the propagation mainly being in LoS; as a consequence, for urban scenarios the system will be based on a micro-cellular structure with cells confined to streets, having dimensions of the order of a few hundreds of metres (the use micro-cells for main roads and highways scenarios is also foreseen). This usage of the millimetre wavebands imposes that the attenuation of atmospheric elements, namely rain and oxygen, has to be taken into account in cellular design, which is not necessary at the UHF band; moreover, propagation characteristics are not equal in these two bands, with oxygen and rain presenting different values for their attenuation coefficients. Oxygen absorption is the major difference between the two bands: at 40 GHz the attenuation is negligible, while at 60 GHz it has to be considered, decreasing from 14 dB/km (at 62 GHz) down to approximately 1 dB/km (at 66 GHz). Since a larger attenuation leads to the possibility of re-using frequencies at a closer distance for approximately the same coverage (the attenuation is not substantial for short distances like those involved in cell coverage), the usage of one or the other frequency bands has significant consequences on system capacity and on its cost/revenue performance. Thus, it is important to establish the correspondence between the maximum coverage and re-use distances, R and D, and the CIR, C/I, for both bands, and to analyse the resulting consequences, in order to decide under which conditions it is preferable to use one band or another.

In the following, a comparison of characteristics between the two aforementioned allocated bands for MBS is carried out. It is assumed that propagation occurs essentially in LoS, and a simple propagation model for the average decay of power with distance is taken [CoFr94]. Only coverage scenarios with regular structure are considered, such as the linear and 'Manhattan grid' geometries, in order to enable an easy analytical treatment.

The C/I ratio has a direct influence on the co-channel re-use factor, and on system capacity. Considering two co-cells, the minimum value for C/I is given by [VeCo97]

$$C/I_{[dB]} = \gamma_{[dB/km]} \, (r_{cc} - 2) \, R_{[km]} + 10 \, \alpha \log(r_{cc} - 1) , \qquad (4.1.2)$$

where the usual assumptions for C/I analysis have been considered, and: γ represents the attenuation by atmospheric elements, α is the average power decay with distance coefficient; $r_{cc} = D/R$ is the co-channel re-use factor. In the absence of rain, Figure 4.1.12, one concludes that, at the 40 GHz band (where the oxygen absorption is negligible), C/I does not depend on the value of R when r_{cc} varies, and a value of the order of 10 dB for $r_{cc} = 4$ is obtained, which is a typical value for the UHF band (where a similar behaviour is found). However, at 60 GHz the behaviour is much different, because the values for the oxygen attenuation are not negligible; for the same value of r_{cc} and different values of R, different values exist for C/I, and the larger R is the larger C/I one gets, with values ranging from 12 dB up to 25 dB at $r_{cc} = 4$, when R varies one order of magnitude from 50 to 500 m. In the presence of rain, a larger value for the attenuation coefficient is obtained, and basically the previous behaviour at 60 GHz, without rain, is observed for the two bands. At the 60 GHz band, the main difference consists in a larger value for C/I; at 40 GHz the previous behaviour changes, and different curves exist for different coverage distances, since the attenuation coefficient is not negligible.

The linear coverage geometry corresponds to the coverage of an indefinitely long street or highway, for which values approximately 3 dB below those of the previous case (only two cells) are obtained; in this situation, there are CCI sources at both sides of the cell, which explains the degradation. A similar difference exists for the 'Manhattan grid' geometry (a regular urban structure with streets perpendicular to each other: cells, which form re-use patterns, have their BSs antennas placed in the middle of the crossroads, with four main lobes in the directions of orthogonal streets, allowing a perfect tessellation within the urban grid. Due to the obstruction by blocks of buildings, interference occurs only in LoS, coming from the directions associated with each street (the West/East and South/North directions); as a consequence, C/I is 3 dB lower than the value corresponding to the linear coverage case (and 6 dB below the value obtained for the pair of interfering cells).

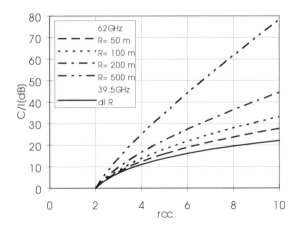

Figure 4.1.12: CIR in terms of the co-channel re-use factor with *R* as a parameter, in the absence of rain, for two cells.

Depending on *R* and r_{cc}, the system operates limited either by interference or by interference-plus-noise. For a proper system operation, *C/N* should exceed its minimum value, which depends on various factors, BER being among them. Using the parameterisation of constant BER contours [BrVe96], for a situation where the same equipment and data rates are used, an equation for the minimum carrier-to-noise-plus-interference ratio can be obtained in terms of $\alpha_c = (C/I)_o/(C/N)_o$ which is specific to each modulation, $(C/N)_o$, $(C/I)_o$ being the minimum values for the respective ratios. Using a typical configuration envisaged for MBS, for each value of *R* the minimum value of *D* that can be obtained, as well as the corresponding value of *M=I/N* in the presence and absence of rain [VeCo97] (*M>>0* dB corresponds to a system operation limited by interference, while the noise limited operation corresponds to *M<<0* dB). r_{cc} is then easily obtained by dividing *D* by *R* for the worst case situation, i.e. for which *D* takes larger values.

Figure 4.1.13 shows an example for the 'Manhattan grid' geometry, comparing the two bands (one should note that the feasible values for r_{cc} need to be even). It is clear that there are in fact different capacities associate to the two bands, since in this case one gets $r_{cc} = 6$ at 40 GHz and 4 at 60 GHz. However, it should be stressed that other simulation conditions lead to different values of r_{cc}; but more important than that is the fact that oxygen absorption is not uniform within the 60 GHz band (with lower values at the higher end, 66 GHz), which may decrease the capacity at this band and make it equal to the 40 GHz capacity. In conclusion, the fundamental difference between the two bands is the oxygen absorption. At 40 GHz, *C/I* does not depend on the value of *R*, but only on r_{cc}, but at 60 GHz the behaviour is much different. There is no significant difference between the values of the co-channel re-use factor for the linear and 'Manhattan grid' geometries, because the presence of obstructions decrease considerably the degree of interference between cells. The use of the 60 GHz band may lead to a higher system capacity, but it depends on the exact implementation to be done.

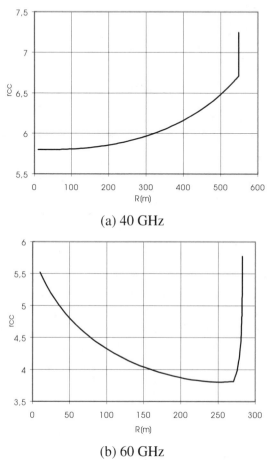

(a) 40 GHz

(b) 60 GHz

Figure 4.1.13: Co-channel re-use factor as a function of the maximum coverage distance for a 'Manhattan grid' geometry.

4.1.2.5. Roadside-to-mobile communications at 60 GHz

Not only have cellular systems like MBS been considered as potential users of the 60 GHz band, but beacon-to-vehicle communication systems have also been explored [ADFV94], with the purpose of allowing significant exchange of information between mobile users in vehicles and the fixed infrastructure. Different environments can be considered with reference to both highway and urban applications; the latter is taken here with the purpose of defining basic design criteria of the mobile and infrastructure systems. An urban 'Manhattan grid' scenario is considered, where beacons can be exploited in order to collect data from vehicles and transmit information about traffic conditions. Taking a mono-dimensional architecture, characterised by a spacing of $2R$ between adjacent beacons, each beacon covers a cell of radius R; links between vehicles and infrastructure are bi-directional, the analysis done here being restricted to the downlink. In [AnDV95] the performance of the uplink is also evaluated. With the aim of considering a simplified and worst-case system, both vehicles and beacons are assumed to be equipped with omni-directional antennas. A multi-user environment is considered, hence a multiple access method has to be taken into account: some kind of Time Division Multiple Access (TDMA) scheme is assumed. As a worst-case situation, every channel is considered to be busy, i.e. the number of vehicles in a cell is equal to the number of channels assigned to the corresponding beacon.

The channel model is a two-ray Rice model, based on the coherent accumulation of the direct path and a road-reflected path, and the incoherent addition of a multipath power component [ACTV92]. System performance is characterised by the outage probability, P_{out}, given a fixed desired link quality in terms of bit error probability. The situation of both absence and presence of interferers is considered for establishing an analytical model, results being assessed by comparing with MC simulations [Verd97]; in order to improve system performance, antenna height diversity is also considered, together with selection combining.

Among the parameters taken for model assessment, one should note the following: carrier frequency of 60 GHz; oxygen attenuation of 15 dB/km; heights of 4.5 and 1.45 m, respectively, for beacon and vehicle antennas; spacing between antennas for diversity of 10 cm; Rice factor of 0 dB. Figure 4.1.14 shows a performance comparison between the results obtained by the analytical model (lines) and a MC simulation approach (dots): the outage probability is computed as a function of the useful user distance both in the absence (P_{out},) and in the presence ($P^{(d)}_{out}$) of diversity with selection combining. In both cases, interference is considered with $R = 100$ m and re-use factor equal to 2. As can be seen in the figure, either in the presence or in the absence of diversity, the agreement is quite good.

The coincidence between the analytical and simulation results in the absence of diversity (curve (a) and dots) points out the validity of the respective model. Moreover, the agreement between curve (b) and dots also confirms the validity of the method proposed to estimate outage probability when interference is present and antenna diversity with selection combining is exploited. It is worth mentioning that good agreement has also been found between MC simulation and the analytical method proposed to estimate outage probability with selection combining for values of the Rice factor up to a few dB. As an example of application, a MSK modulation scheme with non-coherent Limiter-Discriminator (LD) demodulation and a 6-pole Butterworth receiving filter with equivalent noise bandwidth equal to the symbol rate has been considered. System performance can be improved by increasing the cluster size, N_C, since it is interference limited; also, height diversity gives a significant improvement in performance. Due to high oxygen absorption at 60 GHz, a cell radius optimisation problem arises when the re-use factor is fixed. One notes that for small values of R, P_{out} does not depend on the reference signal-to-noise ratio, S/N_{ref}, but only on N_C (in this case, the system is interference limited); however, when increasing the cell radius, P_{out} does depends only S/N_{ref} and not on N_C (in this case, one has a noise limited system). In order to have sufficient levels of availability in the whole cell, a value of $N_C = 3$ is needed.

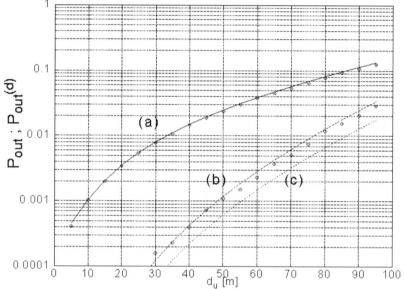

Figure 4.1.14: Comparison of the outage probability from analytical models with MC simulations (dots): (a) without diversity and with interference; (b) with diversity (2 antennas) and interference; (c) as in (b) but with a simplified analytical method [Verd97] (© 1997 IEEE).

From the spectral point of view, one can note that, by increasing N_C, the total bandwidth required increases as well; on the other hand, from Figure 4.1.15 it can be noted that, when fixed, the desired link availability, the minimum value of the product $R \cdot N_C$ is found by choosing $N_C = 3$. One should also note that, when diversity is exploited, the best value of R, from the point of view of system availability, is in the range between 100 and 200 m, depending on the link budget.

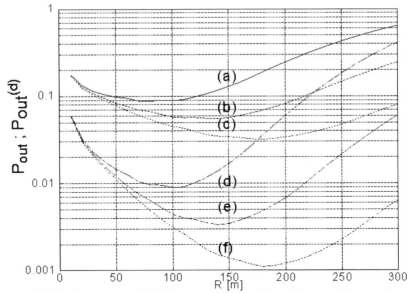

Figure 4.1.15: Outage probability against cell radius without diversity ((a) S/N_{ref} = 24 dB; (b) S/N_{ref} = 29 dB; (c) S/N_{ref} = 34 dB), and with diversity ((d) S/N_{ref} = 24 dB; (e) S/N_{ref} = 29 dB; (f) S/N_{ref} = 34 dB) (© 1997 IEEE).

As final remarks, one should notice that, due to the movement of vehicles, the road-reflected wave can be neglected in numerical computations. The analytical approximated procedure to determine outage probability in the presence of noise, CCI, Rice fading and antenna diversity with selection combining has been checked against MC simulations, and good agreement has been observed. It is possible to establish communication between vehicles and beacons up to 100–200 m, with suitable values of outage probability. Additional rain absorption limits CCI from adjacent clusters.

4.2. Channel Allocation Strategies

Andreas Eisenblätter

4.2.1. Introduction

Andreas Eisenblätter

This section reports on recent developments in automated frequency planning. Frequency planning is a key issue in fully utilising the available radio spectrum in many radio systems. Although the specification of the next generation UMTS is ongoing, it is commonly expected that GSM systems will remain in service for at least the next 15 years. Within this period, an increasing demand for voice traffic has to be covered within a radio spectrum that is already perceived as very limited nowadays. Moreover, enhancements of GSM systems like HSCSD, GPRS, and EDGE enable the introduction of data services with data rates up to 384 kbit/s. Those enhancements come along with a higher sensitivity to interference. All of this will increase spectrum scarcity and add to quality problems.

The automatic generation of a good frequency plan for a GSM network is a delicate task. At the starting point of COST259, commercially available algorithms were already close to their limits despite the fact that only a few TRXs per cell had to be planned. New system features like Slow Frequency Hopping (SFH) or IUO could hardly be taken into account, or not at all.

Three major prerequisites have to be fulfilled in order to generate good frequency plans automatically:

1. A concise formal model of which restrictions apply.
2. All necessary data must be available.
3. An efficient solver has to be used

The importance of each prerequisite is explained in the following.

First, the automatic generation of a frequency assignment by a computer relies on the representation of all relevant aspects in the computer. Hence, a concise format model of the frequency planning problem is necessary. On one hand, this model should be simple for the sake of easy handling. On the other hand, all information has to be captured which is necessary to accurately estimate the quality that a frequency plan would have if it were installed into the real network. This is, for example, a point where the traditional model with hexagonal cell shapes fails (the model still receives attention in literature, however, because all relevant data is easily generated and planning based on this model is more easily accessible). The spectrum of other models currently in use is wide. It ranges from simplistic graph colouring models over graph-based models dealing with the maximisation of satisfied demand or minimisation of interference to models, building directly on pixel-based signal predictions and looking at frame erasure rates.

Secondly, the concise formal model is futile unless the corresponding data is provided. Most problems arise with data related to radio signal levels. This data is needed in numerous ways, for example, in order to estimate how much interference can occur between TXs or between which cells a handover (HO) is possible. Signal levels are provided through measurements in few cases only. For the most part, the signal level is predicted using radio propagation models. The work of the Working Groups on *Radio System Aspects* and *Antennas and Propagation* has therefore to be exploited to a great extent.

Thirdly, with a concise formal model and reliable data in hand, the task of producing a good frequency plan can be reduced to the problem of finding a solution to a hard mathematical optimisation problem. Special software for this purpose is in demand. Operations Research picked up this problem in the early 1960s, but most progress has been made in recent years, accompanying the deployment and extension of GSM networks and stimulated by the co-operation between network operator and research facilities.

The main goal of Sub-Working Group 3.1 on 'Standard Scenarios for Frequency Planning' was to push this development further by establishing a collection of planning scenarios [FAPL00]. These scenarios, provided by operators, are extracted from GSM networks or forecasted planning situations. The scenarios serve as public benchmarks for frequency planning algorithms. From experience with other optimisation problems, it is evident that powerful and efficient solvers for hard optimisation problems are best developed in a community with co-operation and competition, and with access to a common set of test problems. The library of pure and mixed integer programming problems [MIPL96] and the library of travelling salesman problems [TSPL95] are two prominent examples of this.

The section is organised as follows. In Section 4.2.2, an artificial but realistic example of a GSM network is introduced and many parameters relevant to frequency planning are explained. In Sections 4.2.3 and 4.2.4, two traditional models for frequency planning are considered. Interference data is not taken directly into account in either of them. The problem of finding a frequency plan using a narrow range of frequencies as possible is addressed in Section 4.2.3. In Section 4.2.4, the total number of TRXs assigned to cells is maximised under the condition of using a fixed set of available frequencies. In Section 4.2.5, the objective is to minimise interference given a fixed set of available frequencies. The solution methods described comprise simple and fast heuristics, randomised local search approaches like Simulated Annealing, and other methods from Operations Research. Two methods for supplying quality guarantees along with a frequency plan are also discussed.

So far, frequency planning is considered for a static demand profile. There is also research into methods for dynamic frequency assignment. A frequency plan is no longer computed for a time horizon of several months, but is, instead, updated with the changes in traffic during the day. Issues of dynamic channel assignment are addressed in Section 4.2.6. Finally, the benchmark scenarios are the topic of Section 4.2.7. Several frequency plans for these scenarios from different sources are compared.

4.2.2. Concise data for automated frequency planning

Andreas Eisenblätter

A small artificial, but realistic, example of a cellular network called TINY is presented together with some remarks on interference prediction. The example is used to explain the most important parameters taken into account for the 'standard scenarios for frequency planning', see Section 4.2.7. A sketch of the TINY scenario is given in Figure 4.2.1.

TINY comprises three sites, named A, B, and C. Site A has three sectors with sector numbers 1, 2, and 3. Sites B and C have only two sectors, numbered 1 and 2. Each sector of a site defines a *cell*. The numbers of elementary transceivers (TRXs) installed per cell are given in Table 4.2.1.

Table 4.2.1: Number of TRXs installed per cell.

Cell	A1	A2	A3	B1	B2	C1	C2
TRXs	1	3	2	2	1	1	2

TINY is a GSM900 network with the radio frequency band 891.0–893.4 MHz available for the uplink and band 936.0–938.4 MHz available for the downlink. This corresponds to 13 uplink and 13 downlink frequency slots of 200 kHz each. In GSM900 networks, the radio frequency bands for downlink and uplink always differ by 45 MHz (the difference for GSM1800 is 75 MHz). A paired frequency slot is called a *channel*. The Absolute Radio Frequency Channel Numbers (ARFCNs) of those channels are 5–17. The ARFCNs of the channels are called the *spectrum* here.

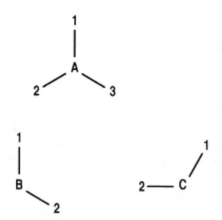

Figure 4.2.1: The TINY network.

Due to technical and regulatory restrictions, some channels in the spectrum may not be available in every cell. Such channels are called *locally blocked*. Local blocking can be specified for every cell. We assume that channels 5 and 6 are blocked in cell B2, and that channel 13 is blocked in cell C1.

Each cell operates at least one broadcast control channel (BCCH) and some dedicated traffic channels (TCHs). Nowadays, one BCCH per cell is usually sufficient, but a number of 3 or 4 TCHs per cell is common.

The difference of the ARFCNs of two channels is a measure for their proximity. For a pair of TRXs, a restriction sometimes applies on how close their channels may be. This is called a *separation requirement*. Its purpose is to ensure that the TRXs can transmit and receive properly, to support the preparation of call HO between cells, or to avoid strong interference. Separation requirements and locally blocked channels give rise to so-called hard constraints, i.e. none of them is allowed to be violated by an assignment.

There are several sources of separation requirements. If two or more TRXs are installed at the same site, *co-site separation* requirements have to be met. A co-site separation of 2 is assumed to be necessary at all sites of TINY. Furthermore, if two TRXs serve the same cell, a *co-cell separation* requirement has to be met. The minimum co-cell separation is 3 in each cell of TINY. In practice, this value may vary from cell to cell due to different technologies in use, but the values given for TINY are typical.

It may be necessary to have TRXs from different cells serving an on-going call at different times when a cellular phone is moved. Technically speaking, the cellular phone switches from using a channel operated in the passing-on cell to a channel used by some TRX in the receiving cell. The HO relation is defined between all ordered pairs of cells and tells from which cell to which other cell a HO is possible. The HO relation for TINY is given in Table 4.2.2.

Table 4.2.2: HO relation for TINY. An 'x' at the intersection of a row and a column indicates that a HO from the cell listed in the row to the cell listed in the column is possible.

->	A1	A2	A3	B1	B2	C1	C2
A1		x	x				
A2	x		x	x			
A3	x	x				x	x
B1		x			x		x
B2				x			x
C1			x				x
C2			x			x	

The HO operation is a sensitive process. Some separation between the channels in the two involved cells is required. Table 4.2.3 lists minimum separations required to support HO for TINY. The BCCH and all TCHs in the source cell have to be separated by at least 2 from the BCCH in the target cell. Whereas, the BCCH and all TCHs in the source call have to be separated by only 1 from the TCHs in the target cell. These values are also typical.

Table 4.2.3: Minimum separation to support HO.

->	BCCH	TCH
BCCH	2	1
TCH	2	1

Interference can occur if two TRXs use channels with little or no separation. For TRXs installed at the same site, interference is ruled out by appropriate separation requirements, as seen above. Between TRXs installed at different sites, only CCI and ACI are relevant. Interference relations do not have to be symmetric, i.e. if cell B1 interferes with cell A1, cell A1 does not necessarily also interferes with cell B1. In case two cells mutually interfere, the ratings of the interference can be different. The rating is normalised such that all interference values lie between 0.0 and 1.0. In Table 4.2.4, the CCI and ACI for TINY is specified in terms of affected cell area. Interference relations are also called soft constraints in literature.

Table 4.2.4: Interference between TRXs for pairs of cells.

Interfering cell	Interfered cell	CCI	ACI
A2	B1	0.30	0.10
A2	B2	0.10	0.02
A3	C1	0.05	0.00
A3	C2	0.20	0.06
B1	A1	0.01	0.00
B1	A2	0.25	0.09
B1	C2	0.25	0.08
B2	C2	0.15	0.04
C1	A3	0.01	0.00
C2	A2	0.06	0.01
C2	A3	0.12	0.03
C2	B2	0.25	0.08

The specification of interference for pairs of cells rather than for pairs of TRXs presupposes that all TRXs in a cell use the same technology, the same transmission power, and emit their signals via the same antenna. If this assumption does not hold, then a sector of a BS can be treated as the host for several cells within which the assumption holds. This is relevant if, for example, IUO is applied, see Section 4.4.

In case interference is very strong, it may not be possible to process calls. Interference should then be ruled out by means of separation requirements with minimum separation of one or two. A minimum separation of one excludes CCI, because the involved pairs of TRXs may not use the same channel. A minimum separation of two excludes CCI and ACI.

The total interference caused (or induced) by a frequency assignment serves as a quality measure for assignments. Although there are other reasonable ways to measure the quality of an assignment, this measure is widely accepted in practice. It is also the basis for comparing assignments in Section 4.2.7. Several ratings of interference are conceivable. Area-based and traffic-based ratings are most often used in practice. The occurrence of interference is either measured or predicted. The interference prediction relies on signal predictions done for points on a regular grid. Typical mesh sizes are 5×5 m^2 (metropolitan), 50×50 m^2 (urban), and 200×200 m^2 (suburban & rural). Each grid point is assumed to be a representative of its surrounding. Recently, there have been attempts to use the signal predictions for the grid points directly as input for frequency planning, e.g. see [Jimé97], [Cami99]. The standard, however, is still to aggregate the grid-based signal predictions into interference predictions at a cell-to-cell level. For an area-based rating this is typically done using the 'best-server' model, where every grid point is associated to the cell from which the strongest signal is received, see Section 4.3.1. Signals from cells are neglected if they are more than t dB below the strongest signal, and all signals of intermediate strength are considered as potential interference. The GSM specifications request that a signal 9 dB above noise (including interfering signals) has to be decoded properly by a RX. As a consequence, a threshold value t equal to 9 is often used in practice. An investigation carried out in [EiKF98] reveals, however, that a threshold value t of 15 or even 20 usually results in frequency plans where interference is more evenly distributed and at a lower overall level.

The accuracy of the interference predictions is a cornerstone for automated frequency planning. An analysis of how accurate interference predictions affect the quality of a resulting frequency plan is given in [EFK98]. For the same planning region three interference predictions are computed on the basis of the 'best-server' model and three different signal propagation prediction models. The three models are a 'free space' model, the Modified Okumura-Hata model, and the 'E-Plus' model. In the 'free space' model the propagation conditions of free space are assumed, but a decay factor of 1.5 rather than 1 is used. The increase of the decay factor from 1 to 1.5 (or similar) is taken as an empirical value between the decay when only the direct ray is taken into account (resulting in a decay factor of 1) and the decay of a two ray model, e.g. see [KüFa94]. In the two-ray model, the interaction between the direct ray and a reflected ray results in a decay factor of 2 for distances larger than the so-called breakpoint. The Modified Okumura-Hata predictor is based on the 1 800 MHz extension COST231-Hata basic path loss equation [COST99]. Land use information is used by empirical correction factors for each land use class [OOKK68], [RACE91]. Terrain variations are taken into account by using an effective antenna height and a knife-edge diffraction loss for topographical obstacles [EpPe53]. The 'E-Plus' propagation prediction model [KüFW96] is the most sophisticated approach used in this comparison. The model consists of a combination of several propagation models (e.g. COST231-Walfisch-Ikegami, Maciel-Xia-Bertoni and Okumura-Hata), which is developed for GSM1800 and calibrated with numerous measurements from the network of the German operator E-Plus Mobilfunk GmbH.

Ranking the propagation prediction models has its difficulties. The crucial question is how to compare assignments computed on the basis of different predictions without implementing the assignments into the live network and performing measurements. In the approach taken in [EiKF98], each assignment's interference is determined according to all three interference predictions. The findings are as follows. The assignments computed using the predictions from the 'E-Plus' model have relatively little interference according to all three predictions. The assignments computed using the interference predictions based on the 'free space' model also have decent interference ratings according to the Modified Okumura-Hata predictions, but are mediocre according to the 'E-Plus' predictions. The worst picture is obtained for the assignments computed using the Modified Okumura-Hata predictions. They are mediocre to bad according to the two other predictions. This observation is the basis for ranking the 'E-Plus' model above the 'free space' model, which, in turn, is ranked above the Modified Okumura-Hata model (in this particular context).

Finally, the frequency assignment shown in Table 4.2.5 is assumed to be in use in the network TINY. If a new frequency assignment for TINY is to be computed, some of the existing assignments might want to be kept fixed. The TRXs whose channel shall not be changed are called unchangeable. Otherwise, the TRX is called changeable.

All this data has to be adequately represented in a computationally tractable fashion as a basis for automated frequency planning. Further requirements arise with the introduction of features into GSM like DTX and SFH or new cell properties as they occur in the context of micro- and pico-cells and IUO. These aspects are discussed in Section 4.4.

Table 4.2.5: Feasible channel assignment for TINY, incurring only ACI.

Cell	A1	A2			A3		B1		B2	C1	C2	
TRX	0	0	1	2	0	1	0	1	0	0	0	1
Channel	9	11	7	16	5	14	14	5	9	8	17	12

4.2.3. Fixed channel allocation using graph colouring

Dirk Beckmann

The model of the frequency assignment problem presented in [KaMo98] is slightly different in comparison to most related work performed within this COST project. The objective considered there is to find a valid frequency assignment with a minimised spectrum span, i.e. the difference of the channel numbers corresponding to the smallest and the largest used frequency channel is being minimised. This problem of minimising the number of frequencies needed has drawn a lot of attention in the literature, and it can be shown that there are close relations to the problem of interference minimisation, which one considered within this COST action.

Two simple variants of the sequential graph colouring algorithm are applied to the minimum span frequency assignment problem. Given some order of the vertices, i.e. the TRXs, the assignment of colours (frequencies) can be performed by either the frequency exhaustive or the requirement exhaustive assignment strategy. Using the *frequency exhaustive strategy*, every TRX is assigned to the lowest frequency that does not lead to any violations of separation constraints, considering the TRXs that have already been assigned. In that fashion, the TRXs receive frequencies according to their order in the list. With the *requirement exhaustive strategy*, first the lowest frequency is assigned to TRXs, then the second lowest frequency, and so on. When processing one frequency, all unassigned TRXs are still considered for assignment according to the order in the list. In case no separation constraint is violated, the frequency is assigned.

Many heuristics have been proposed based on these two strategies, and it is well known that the number of frequencies required in total strongly depends on the order in which the TRXs are listed.

Though the task of finding an optimum order of TRXs which finally results in an optimum solution with a minimum needed frequency spectrum remains open, it has now been proved [KaMo98] that there is always an order of TRXs which leads to an optimum assignment if the frequency exhaustive strategy is used. Furthermore, it has been proved that in some cases, it is not possible to derive such an optimum assignment using the requirement exhaustive strategy, even if all possible permutations of TRXs are considered. Consequently, it can be concluded that the frequency exhaustive strategy generally outperforms the requirement exhaustive approach, provided that a suitable order of TRXs can be determined.

Based on this fact, it is finally concluded that heuristic frequency assignment algorithms like that proposed by [Gams88] can be further improved by substituting the requirement exhaustive by a frequency exhaustive assignment procedure. This seems to be an important result for the development of future frequency assignment algorithms. The task of finding a good order of TRXs for the frequency exhaustive assignment can then, for instance, be solved using genetic algorithms, as proposed in [BeKi99b].

4.2.4. Fixed channel allocation maximising assigned TRXs

Andreas Eisenblätter

Another version of channel allocation that has received attention in the literature is the following: maximise the total number of TRXs assigned to cells in a cellular network when a fixed set of available channels is given. The problem is specified by means of an edge-labelled graph. The vertices of the graph represent the cells of the network. An edge connects two vertices if the TRXs of the associated cells have separation requirements. The necessary minimum separation is the label at the edge. As interference is not captured explicitly, it is translated into separation requirements by means of some threshold. In case the interference does not exceed the threshold, the interference is neglected. In case CCI exceeds the threshold, a minimum separation of 1 is required. If the ACI also exceeds the threshold, a minimum separation of 2 is required.

Binary linear programs for the maximisation problem and some variants are given in [MaHe98], [HLMN99].

- Maximise the total number of TRXs assigned to cells with or without satisfying a minimum demand per cell.
- Maximise the total number of TRXs assigned to cells without satisfying any minimum demand per cell, but the assignment of the first TRX to a cell is accounted for p times (the parameter p is to be specified).

A variant that respects the different type of BCCHs and TCHs also exists.

- Maximise the total number of TRXs assigned to cells while meeting the demand in BCCHs and satisfying a minimum demand in TCHs for every cell.

For realistic problem sizes, none of these variations is easily solved by commercial state-of-the-art Mixed Integer Program Solvers. Hence, a standard constraint strengthening on the basis of (large) cliques in the graph is performed, and a geometry-based partition of the total problem into smaller sub-problems is suggested. These sub-problems are solved in turns while keeping the remaining parts fixed. The geometric partition is based on an embedding of the BSs into the Euclidean plane in such a way that the distances reflect the separation requirements, i.e. a short distance corresponds to a high separation requirement. The embedding is obtained via an appropriate Simulated Annealing procedure.

4.2.5. Fixed channel allocation minimising interference

Andreas Eisenblätter and Hans Heller

Five heuristic procedures are described for finding frequency assignments that satisfy all separation constraints and incur as little CCI and ACI as possible. A frequency plan is called *feasible* if it satisfies all hard constraints. The interference of a frequency plan is sometimes also referred to as the cost of all violations of soft constraints.

The first procedure is a combination of an adapted graph colouring heuristic and local optimisation [BEGM98]. Then, two Simulated Annealing approaches are compared [BeKi99a], [deUr99], and an implementation of a deterministic variant of Simulated Annealing, namely, Threshold Accepting, is discussed [HeHe00]. Finally, a heuristic is presented which, basically, dives down several times in a tree associated to a branch & bound scheme, using a different branching rule every time [NiWi00]. Computational results are compared in Section 4.2.6.

4.2.5.1. Fast construction and improvement

Many heuristics are designed to be run in combination with others. Two such heuristics are described here. The run-time efficiency of these algorithms allows us to compute frequency assignments within a few minutes even on large scenarios.

DSATUR With Costs [BEGM98]: this starting heuristic builds on the graph colouring algorithms DSATUR [Brél79] and uses ideas from [Costa93]. It works best on scenarios with a low traffic load, i.e. 1–2 TRXs per cell. The heuristic differs from other descendants of graph colouring heuristics used for frequency assignment most prominently by taking CCI and ACI explicitly into account. The basic idea is to recursively assign the TRX that seems 'hardest' to assign first: while there are unassigned TRXs, the TRX with the least number of remaining available channels is picked. If there is more than one such TRX, an interference-based tie-breaking rule is applied. The selected TRX is then assigned to a channel that presently incurs the least interference.   Which TRX is assigned at the very beginning has a considerable impact on the quality of the assignment computed in the end. No rule of sufficient generality could be identified concerning how to pick the first TRX, and some percentage of the TRXs, 5 % say, are selected randomly as starting points.   The best assignment generated is returned. Once *DSATUR With Costs* is finished, the following improving heuristic is run.

Iterated 1-Opt [BEGM98]:   an assignment can only be changed in one particular way by this procedure. The channel of one TRX can be replaced by the channel presently incurring the least amount of interference (among all those causing the least number of hard constraint violations, if any). This 1-opt step is repeated over and over again for varying TRXs until no more TRX has a favourable alternative to its present channel.   The resulting assignment is most likely not optimal. The algorithms might instead be trapped in a local minimum. Other heuristics may be called in turns with Iterated 1-Opt, provided those are capable of leaving such a local minimum and creating room for improving 1-opt steps.

4.2.5.2. Simulated Annealing

Simulated Annealing is a well known method used in many areas of discrete optimisation [AaKo89], [Cerný85], [KiGV83].   The solution space is searched by following a random path of 'adjacent' solutions. Basically, it tries not to get trapped at a local minimum in the solution space by allowing random moves that deteriorate the current solution.   The probability of accepting deterioration is decreased with the progress of the algorithm until it is (almost) zero in the end. This probability is controlled by a parameter T called temperature [MRRT53].   Figure 4.2.2 sketches the prototypical Simulated Annealing procedure.   The cost of a move M is the difference between the cost of the assignment before and after the move.

```
initialise temperature T
while (stopping criterion not met) do
    for N steps do
        select random move M
        if (cost of M > 0)
            execute M with probability e^{cost of M / T}
        else
            execute M
    decrease temperature T
```

Figure 4.2.2: The skeleton of the Simulated Annealing procedure.

The only step that is explicitly related to channel allocation is the random move. For all other steps many realisations are proposed in literature. The simple choices in [BeKi99a] and [deUr99] (see also [dUDB00]) are similar enough to allow a joint presentation.

- *Start Solution:* in [BeKi99a], an initial feasible solution is generated using the technique described in [BeKi99b]. This techniques falls into the realm of Section 4.2.2. The initial solution used in [deUr99] is generated at random and may be infeasible.

- *Initial Temperature:* a binary search is performed in [BeKi99a] to identify a temperature value T_0 for which the following holds. Approximately 90 % of the proposed random moves are accepted on the basis on the criterion used in the inner loop of Figure 4.2.2. In [deUr99], the initial temperature is set to 10 % of the cost of the initial assignment.

- *Stopping Criterion:* the implementations in [BeKi99a], [deUr99] do not trigger the termination by a particular temperature value. The percentage of accepted deteriorating moves is monitored instead in [BeKi99a], and once this drops below 5 % the process is stopped. In [deUr99], the difference of the maximum and the minimum cost observed at a temperature is determined. If this value is below a threshold value of 0.001, say, then the process is halted (in both implementations the search is stopped in case a feasible assignment without any interference is found).

- *Inner Loop:* each time, the inner loop is simply performed for 50 times the number of TRXs in [BeKi99a]. A fix value of $N = 600\,000$ is used in [deUr99], with the inner loop being skipped if, at the previous temperature, an acceptance rate of more than 2/3 is observed.

- *Temperature Reduction:* there are many publications on this topic. Depending on the rule used here, Simulated Annealing is capable of converging to an optimal solution with probability 1 (under mild conditions). Such rules, however, lead to run-times growing exponentially in the problem size, see [AaKo89], for example. A rapid decrease of the temperature is used in practise instead. The rule $T_{new} := (3\ \sigma)\ /\ (1 + \ln(1.1) * T_{old}) * T_{old}$ is used in [BeKi99a]; see [AaKo89] for its origin and motivation. The variance σ is taken from the cost values of the assignments accepted within the last inner loop. A simpler rule is used in [deUr99], namely, $T_{new} := 0.85 * T_{old}$.

- *Random Moves:* a significant difference among move types is whether the moves are allowed to introduce hard constraint violations. The moves used in [BeKi99a] do not. Hence, once a feasible assignment is obtained all subsequent assignments are feasible, too. A cell is selected at random under uniform distribution. Within this cell, the TRX incurring the highest cost is determined, and this TRX is assigned the best alternative to the channel it currently uses. If no alternative is available, nothing is done. The moves used in [deUr99] are allowed to introduce hard constraint violations. First, a TRX is chosen, then a new frequency among the available ones is selected. Both choices are made at random under uniform distribution.

Simulated Annealing is not expected to terminate quickly. The choices described above typically lead to run-times on a modern PC of several hours for scenarios with a few hundred cells and one to two thousand TRXs.

4.2.5.3. Threshold Accepting

Threshold Accepting is a deterministic variant of Simulated Annealing. A deteriorating move is accepted if the deterioration is below some threshold. This threshold declines with the progress of the algorithm [DuSc90]. The algorithm [HeHe00] described next is a variation of the basic Threshold Accepting procedure. Random steps are alternated with local optimisation steps. The procedure is developed on the basis of experience made while investigating local optimisation for solving integer linear programs [MaHe98]. The scheme is given in Figure 4.2.3. The cost of a move is again the difference between the cost of the assignment after and before the move.

```
initialise threshold
while (stopping criterion not met) do
    for time T do
        select random move M
        if (cost of M < threshold)
            execute M
        select random cell and optimise it
    decrease threshold
```

Figure 4.2.3: Adapted Threshold Accepting procedure.

For the most part, Threshold Accepting inherits from Simulated Annealing the variety of proposed implementations for each general step. The following choices are made in [HeHe00]:

- *Start Solution:* the random moves described below are not well suited for finding a feasible assignment if many hard constraints apply. In such a case, a feasible assignment should be provided at the start. Otherwise, just any assignment, feasible or not, can be taken, e.g. a randomly generated one.
- *Initial Threshold:* quite some deterioration must be allowed initially so that the optimisation is not trapped in a local minimum right from the

start. A binary search is performed to identify a threshold where 80–90 % of the proposed random moves are accepted.

- *Stopping Criterion:* the termination is triggered if the acceptance rate for random moves sinks below 5 %, compare [KiGV83].
- *Inner Loop/Threshold Reduction:* the run-time of the procedure depends on the initial threshold, the factor by which the threshold is reduced, the length of the inner loop, and the stopping criterion. The length of the inner loop is limited by a time bound of 10 s, say. Taking this, and a maximum desired run-time into account, a factor for reducing the threshold at the end of each outer loop is computed. Notice that the longer the algorithm runs the higher is the probability for a better result, within reasonable limits. Typically, run-times from some 10 minutes to several hours on a modern PC are used for scenarios with a few hundred cells and one to two thousand TRXs.
- *Random Moves:* first, a TRX is chosen at random. Then, a (non-blocked) channel is chosen for this TRX that does not cause any separation constraints. This constitutes the random change. Both choices are done according to a uniform distribution. In case no channel is available for the TRX, the choice of the TRX is repeated.
- *Cell Optimisation:* the optimisation of a cell is done by Dynamic Programming. This is the topic of the following paragraphs.

The essential feature of this variant of the Threshold Accepting algorithm is the cell optimisation step. Cell optimisation is a counterweight to the perturbations and deterioration of random moves. The special structure of the frequency assignment problem allows a complete and yet efficient optimisation of a cell, given the constraints imposed by the current assignment in all other cells that are kept fixed. Prerequisite here is that co-cell separation is at least 2 (in fact, it is typically 3). Therefore, no interference can arise among TRXs within the same cell. The optimisation of a cell is first explained under the provision that a broadcast control channel (BCCH) and a traffic channel (TCH) show no difference with respect to frequency planning. How this restriction can be removed is discussed later. Two observations can be made under this provision:

1. The cost of channel assignments to TRXs in one cell is invariant under redistributing the channels among the TRXs. Consequently, the search for an optimal cell assignment can be restricted by assuming the channels of a cell to be in increasing order.
2. The current channels' costs and the currently forbidden channels have to be computed only once. They are the same for all TRXs in the cell.

Hence, finding an optimal assignment for the cell can be reduced to the following problem. Identify an increasing list of frequencies, whose lengths matches the number of TRXs in the cell, such that successive frequencies are at least the necessary co-cell separation apart, and such that the list incurs minimal cost (compare Observation 2). This optimisation problem can be solved efficiently using a standard method called Dynamic Programming with Memorisation [CoLR90], for example.

The assumption that BCCHs and TCHs have equivalent needs with respect to frequency planning is unrealistic. In general, TRXs may be of different types, which often leads to different separation constraints and an increased or decreased sensitivity to interference. For a BCCH, for example, there are usually more restrictive separation constraints than for a TCH. Moreover, one may prefer less interfered channels for a BCCH. This can be expressed by rating interference for BCCHs more heavily than for TCHs. As a consequence, the costs and separation requirements for TRXs within a cell are no longer uniform. A possible solution is to split the TRXs of a cell into equivalence classes and to perform the optimisation steps on equivalence classes instead of cells.

The perturbations due to random moves used here may not be sufficiently powerful for problem instances which have many and strong separation constraints, because these moves never violate hard constraints. Optimising greater local areas than cells thus appears desirable. However, we know of no efficient way to accomplish this.

4.2.5.4. Run-time bounded enumeration

The following procedure [WNMM98], [NWMM99], [NiWi00] is designed to produce frequency plans for GSM networks in which SFH is used. The effects of SFH, DTX, and partial (or fractional) traffic load are studied on the basis of measurements and simulations. Two parameters, the 'hop gain' and the 'fractional load gain' (see Section 4.4.2) are identified. These parameters are employed to modify the cost function coming from interference predictions. The algorithm uses the modified costs while performing ordinary frequency planning with the objective of finding a minimum interference/cost assignment. Hence, the contribution of [WNMM98], [NWMM99], [NiWi00] is twofold. On the one hand, a frequency planning method is proposed which uses the same objective as the other methods described in this subsection. On the other hand, a systematic way to modify the outcome of the interference predictions is given which allows us to take the effects of SFH, DTX, and partial traffic load into account at the planning stage. With the proposed cost modifications, any other method from this subsection can also be used to plan networks that use SFH.

The procedure builds upon (partially) enumerating all possible assignments. The basic scheme of implicit enumeration is as follows. According to a given order, the TRXs are assigned one after another. For the TRX in turn, an eligible channel is chosen, where a channel is *eligible* if it is neither locally blocked nor incompatible with any of the assignments for preceding TRXs. The choice of the eligible channel is done at random according to a uniform distribution. Two cases may occur.

1. No channel is eligible. Then a backtracking step is taken: the assignment of the preceding TRX is undone, and an alternative channel is looked for (not taking into account the channels that have already been checked out). In case of an alternative, this channel is tentatively assigned and the process continues with the next TRX in the ordering. Otherwise, another backtracking step is performed.

2. All TRXs have received a channel. By construction, the frequency plan is feasible. The cost of the plan is determined. If the plan is better than the best known, it is stored away. Finally, a backtrack step is performed, as described in the previous case.

Once all eligible channels for the first TRX are checked out, the enumeration is complete. All feasible assignments are then enumerated, and the best possible assignment is memorised. The run-time of this enumeration scheme, however, grows roughly as much as the spectrum size to the power of the number of TRXs to assign. An exhaustive enumeration is thus out of the question for real-world planning for more than a few TRXs.

If only a single cell is to be planned while the remaining cells are kept fixed, then this exhaustive search is indeed possible. A local optimisation method may, for example, optimise each cell's assignment in turn using the enumeration scheme until no more improvement is obtained. In case all cells have only one TRX, this method coincides with Iterated 1-Opt, see above, but it is more powerful in the general case.

Figure 4.2.4 depicts a situation where all TRXs have a channel assigned. The order is reflected by the levels in the figure, with the first TRX being at the top level, the second on the next level, and so on. The selected channels are highlighted. The cost contribution of each assignment is listed to the right of each level. Making use of these cost values, it is sometimes possible to see whether the completion of a partial assignment can lead to a better assignment than the current best. If this is not the case, trying out all those completions is useless, and therefore skipped (in the literature on branch & bound algorithms this is also called a 'bounding step').

Instead of the impracticable exhaustive enumeration of all feasible assignments, a partial enumeration is done as follows in [NiWi00]. Initially, the cells are arranged randomly. An order of the TRXs is obtained from the order of the cells. The TRXs from different cells are listed block-wise, where the TRXs within the same cell are listed consecutively. The enumeration process, according to the initial order, is aborted once the first feasible assignment is found. For the subsequent phase, the cells are reordered such that the cost they incur in the feasible assignment is increasing with the order. Another enumeration is launched on the basis of the new order, but this enumeration does not start in the beginning, it 'resumes' at the feasible assignment computed before, as if this assignment had just then been enumerated according to the new order. Hence, the first step to come tracks back from the bottom most level, and the subsequently enumerated assignments have alternative, and hopefully better, choices for the cells which presently incur the highest cost. The second enumeration process is aborted if for some time span, 10 s, say, no new best assignment is produced.

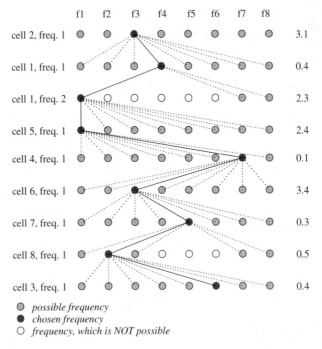

Figure 4.2.4: Snapshot of the enumeration process [NiWi00].

The two enumeration processes, starting with a random order until a feasible assignment is found and then intensifying the search for alternative assignments for the most costly cells, can be repeated any given number of times or until a given time span has elapsed. Another turn is, however, taken in [NiWi00]. Prior to repeating the process, extra separation constraints are added. These constraints are selected in order to rule out severe occurrences of interference observed in the present best assignment. The selection of constraints to add is significantly refined over what is traditionally done in the context of graph colouring-oriented methods, see Section 4.2.2, where interference is either converted into a separation constraint or dropped by means of a threshold, [NiWi00]. If the added constraints are too strong in the sense that no feasible assignment is found within some specified time span, then weaker constraints are added for the next trial.

4.2.5.5. Quality guarantees

Five heuristics for frequency planning have just been described. Their common objective is to produce frequency plans incurring as little interference as possible. Several more have been published, and yet others are used in commercial tools. However, given a feasible frequency assignment, how may one decide whether it is good?

If an assignment is better than every other in a set of alternatives, is it already the best possible? Of course, that is not the case. It is observed in experiments that a given (randomised) heuristic produces assignments incurring a particular amount of interference for a fixed scenario, with some spread. This amount varies significantly depending on the heuristic analysed. See [JAMS91] for a documentation of such behaviour on the related problem of graph colouring. Hence, even if a heuristic consistently outperforms some set of competitors there is no guarantee at all that the produced assignments are anywhere near optimal. If, on the other hand, the interference of an optimal assignment were known, then the question concerning the quality is easily answered. Even without knowing the exact optimal value, quality guarantees can be given by means of lower bounds. A *lower bound* is a numerical value that bounds the value of the optimal solution from below. The closer a lower bound is to the optimal value, the better it is. Several ways in which such bounds can be obtained are described in the Operations Research literature, see [NeWo88], for example.

A simplified version of the frequency planning problem is considered in [Eise98b]. The optimal solution to this problem yields a lower bound for the optimal problem. There are three simplifications. ACI as well as locally blocked channels are dropped, and the separation is limited to at most 1. The resulting problem is (almost) the so-called *min k-partition problem* that is studied in the Operations Research literature. The k stands for the number of available channels in the spectrum.

The simplified problem is still too hard to be solved exactly in general. In [Eise98b] the optimal solution is computed only for subsets of TRXs, for which each pair of TRXs has a mutual relation (this is a clique in an associated graph). The subsets leading to non-trivial bounds have to have a size larger than the number of available frequencies. The first non-trivial lower bounds are computed in this way for several scenarios, but the obtained bounds are often too weak to give good quality estimates.

Starting from the same simplified problem as above, significantly better bounds are obtained in a different fashion in [Eise00]. From the literature, a relaxation of the min k-partition problem is known, where a linear function is minimised over the set of positive semi-definite square matrices subject to linear constraints. For scenarios with up to one thousand TRXs, this relaxation can be solved by sophisticated mathematical software [HeRe97], [HeKi99], [BuMZ99]. The result often supplies a significantly better lower bound than the first approach.

4.2.6. Distributed dynamic channel allocation

Andreas Eisenblätter

An intrinsic feature of a Fixed Channel Allocation (FCA) is that no adaptation to varying traffic profiles during a day, for example, is performed. For a GSM network, the FCA plan is typically designed to accommodate an expected peak traffic with a blocking probability of at most 1 %. In contrast, Dynamic Channel Allocation (DCA) deals with the problem of assigning frequencies to a cell on-line as the need arises.

A general distinction is made between centralised and distributed dynamic strategies to accomplish this. In the former case, there is one supervision unit that collects traffic data from all cells and tries to assign the frequencies as well as possible. This approach is very similar to FCA. In a central unit the same problem is solved using the same information, apart from the traffic profile. Only in the dynamic case is the problem solved again as soon as the demand profile changes. In the distributed case, there is only co-ordination among a few neighbouring cells whose frequency use will affect each other.

The performance of two algorithms for distributed DCA from the literature, namely Timid DCA and Persistent Polite Aggressive DCA, is assessed in [KeDK98]. Their behaviour under two different traffic profiles is studied. A static homogeneous traffic is assumed on one hand, and a time-dependent inhomogeneous traffic on the other hand. In the latter case, the traffic data is observed in a live GSM network in Amsterdam, The Netherlands. Both algorithms are simulated for both traffic conditions on sets of 49 and 144 cells arranged on a torus (in order to eliminated boundary effects).

On call arrival or HO request, the *Timid DCA* algorithm scans all channels from bottom to top in search of a channel not used by any of its neighbours. The first available channel is assigned. In case no channel is available, the call is either blocked or dropped. The *Persistent Polite Aggressive DCA* algorithm initially behaves like the Timid DCA algorithm. However, if no channel is available, then it checks for the first one that is used in only one of its neighbouring cells. If this search is successful, the neighbouring cell is asked to clear the frequency by trying an intra-cell HO using the Timid DCA scheme. In case the HO is not possible, a persistence parameter controls for how many times the search is resumed.

The findings in [KeDK98] are as follows. The traffic capacity gain of DCA is greater for time-dependent inhomogeneous traffic than for static homogeneous traffic. Moreover, the more persistent the Persistent Polite Aggressive DCA scheme, the greater is the capacity gain, and also the more extra HOs are required. The second effect declines with the traffic becoming more inhomogeneous.

In [GrBT97], [GrBT98] (see also [Grace99]), several schemes of (unsupervised) distributed dynamic channel assignment in a point-to-point communications scenario are compared on the basis of simulations. The link lengths are 500 to 5 000 m. All the schemes obey a 'coexistence etiquette' aimed at minimising the amount of interference inflicted on the other users in the same radio spectrum. The decision as to whether or not to use any given channel is based on interference measurements at the TX prior to transmission and on carrier-to-interference (C/I) ratio measurements at the RX during link establishment. A new call is either set up correctly, possibly causing other calls using the same channel to be dropped or reassigned, or it is blocked.

Four policies are investigated concerning their effect on offered system capacity in terms of call blocking, call dropping, and successful call completion: 1. *First Available* assigns a call to the first available channel (in the scanning order); 2. *Least Interfered Channel* orders the available channels according to increasing interference and uses this order to search for the first channel that allows call establishment from the point of view of the RX also; 3. *Least Interfered Channel with Reassignment* uses the previous policy for call assignment, but also when checking for alternative channels for calls that would otherwise be dropped; and 4. *Least Interfered Channel with Hysteresis in C/I* uses policy (2) for channel selection, but the C/I ratio at the transceiver must exceed the usual threshold value *t* by a given hysteresis value *h*.

The following effects are observed for varying threshold ratio *t* and hysteresis value *h*. An overall observation is that the *Least Interfered Channel*-policies provide an increased capacity for a specific grade of service. There is a trade-off of call dropping vs. call blocking. Tight interference thresholds *t* can be used to reduce the level of call dropping at the expense of increasing call blocking. In case the probability of successful call completion is taken as a measure, using a low interference threshold *t* provides the highest capacity for all the *Least Interfered Channel*-policies (2, 3, and 4) with the converse being true for the *First Available*-policy (1). When a positive hysteresis value *h* is used, then the call dropping is only reduced if there is a high level of call dropping. At modest levels of call dropping the use of hysteresis does not provide improvements with respect to call dropping, but decreases overall capacity due to increased call blocking.

4.2.7. Benchmarking frequency allocation strategies

Andreas Eisenblätter and Thomas Kürner

4.2.7.1. Initial considerations

Many of the algorithms, commercially available when Subgroup 3.1 was initiated, were already close to their limits when planning a GSM system with few TRXs per cell. From the operators' point of view, it was therefore pressing to support continued research and development in the area of automated frequency planning. For researchers and developers, on the other hand, it is important to have realistic test problems available upon which to try out new ideas.

The goal of Subgroup 3.1 was to establish a collection of planning scenarios, extracted from real world or forecast situations by network operators. The opportunity to compare different algorithms on the same set of data stimulates competition among developers of planning algorithms and leads to significant improvements of the planning algorithms. The CNET of France Telecom pursues a similar idea, and has also made one benchmark problem publicly available [Cami99].

The focus of Subgroup 3.1 was predominantly on fixed (or static) frequency allocation for GSM networks. Two file formats, the *scenario format* and the *assignment format,* are used to uniformly represent planning problems and frequency plans [Eise98a]. The *scenario format* is used to represent the frequency assignment scenarios, whereas the *assignment format* is used to represent (partial) frequency. All parameters given for the TINY network in Section 4.2.1 can be specified in the scenario format. The assignment, including an indication of whether a TRX's channel is changeable, can be specified in the assignment format. A scenario either alone or together with a (partial) assignment constitutes a planning problem. By adding an objective, like minimising the overall interference, the planning problem turns into an optimisation problem.

Alternatively, the formats of a commercial tool could have been used. There are two major reasons against this. First, restrictions as to which properties of a scenario may be expressed are imposed. Even if some of the commercial formats would be fine for the time being, it was questionable whether more complex future scenarios can also be adequately represented.

The formats used here can easily be extended. Secondly, no proprietary rights have to be dealt with.

4.2.7.2. Available scenarios

At the time of writing, there are 32 scenarios available altogether. The scenarios have been contributed from three sources, E-Plus Mobilfunk GmbH, Siemens AG, and Swisscom Ltd., in alphabetical order.

- **bradford_nt-*t-p*** (E-Plus Mobilfunk GmbH): data for a GSM1800 network with 649 active sites and 1 886 cells. The wildcards *t* and *p* stand for five different traffic loads and three interference predictions on the basis of different signal propagation prediction models. The basic traffic load is drawn at random according to an empirically observed distribution [GoGR97], [Kürn99a], see Section 4.3.2.1. This traffic is then scaled with the factors *t* equal to 0, 1, 2, 4, and 10 prior to applying the Erlang-B formula in order to obtain the required number of TRXs per cell. The resulting average numbers of TRXs per cell are 1.00, 1.05, 1.17, 1.47, and 2.20, respectively. The signal predictions (*p*) are done assuming free space propagation with a decay factor of 1.5 (free), using a Modified Okumura-Hata model (race), and using the model developed by E-Plus (eplus), see [EiKF98] for details. The available spectrum consists of 75 contiguous frequencies. There are also variants called **bradford-*t-p***, which are used in [EiKF98]. Due to an inaccuracy in the Monte-Carlo simulation used to derive the random traffic distribution, the TRX demand per cell in these variants is overestimated. The average numbers of TRXs per cell are 1.00, 1.57, 1.82, 2.12, and 2.60, respectively. Hence, **bradford_nt-*t-p*** is the more realistic scenario, whereas **bradford-*t-p*** provides a more complex frequency planning problem and can therefore be used as a worst-case scenario.

- **siemens1** (Siemens AG): data for a GSM900 network with 179 active sites, 506 cells, and an average of 1.84 TRXs per cell. The available spectrum consists of two blocks containing 20 and 23 frequencies, respectively.

- **siemens2** (Siemens AG): data for a GSM900 network with 86 active sites, 254 cells, and an average of 3.85 TRXs per cell. The available spectrum consists of two blocks containing 4 and 72 frequencies, respectively.

- **siemens3** (Siemens AG): data for a GSM900 network with 366 active sites, 894 cells, and an average of 1.82 TRXs per cell. The available spectrum comprises 55 contiguous frequencies.

- **siemens4** (Siemens AG): data for a GSM900 network with 276 active sites, 760 cells, and an average of 3.66 TRXs per cell. The available spectrum comprises 39 contiguous frequencies.

- **Swisscom** (Swisscom Ltd.): data for a GSM900 network in a city with many locally blocked channels and a partial assignment that has to be completed. There are 148 cells with 1 to 4 TRXs, and 707 neighbour relations. In general, 47 frequencies are available, but 136 cells suffer from restrictions. There are only 10 available frequencies in the worst case; the median of available frequencies per cell is 19.

4.2.7.3. Contributed assignments

At the time of writing, there are 115 assignments available, with more than one assignment for every scenario. Thus, there is the opportunity to compare several planning methods on the basis of their results. Most of the methods used are described in Section 4.2.4. Table 4.2.7 gives details. No assignments are available that are computed using run-time bounded enumeration, which is also described in Section 4.2.4.

A number of characteristics are determined for each assignment. These characteristics taken together should allow a fair comparison of assignments for the same scenario. A sound comparison of the run-times is unfortunately not possible. In most cases the run-time is, in fact, not known. The following is a rough estimate: *DC5_IM(ZIB)* and *U(Siemens)* typically compute an assignment within several minutes, whereas all other methods from Table 4.2.6 typically have a run-time of a few to several hours on a modern PC. Notice also that results of particular implementations of heuristics are compared. Especially for the run-time intensive two local search heuristics considered here, namely, Simulated Annealing and Threshold Accepting, the frequency plans produced (and their quality), as well as the run-time (where not set explicitly), depend significantly on the implementation, and on how much effort is spent in tuning the program.

Table 4.2.6: Sources of contributed assignments.

Acronym	Method	Supplied by	Reference
DC5_IM(ZIB)	'DSATUR with Costs' (5 %) followed by 'Iterated 1-Opt'	Andreas Eisenblätter, ZIB, Germany	[BEGM98]
SA(Telefonica)	Simulated Annealing	Luis de Urries, Telefonica, Spain	[deUr99] [dUDB00]
SA(TUHH)	Simulated Annealing	Dirk Beckmann, TU Hamburg-Harburg, Germany	[BeKi99a]
TA(RWTH)	Threshold Accepting	Martin Hellebrandt, RWTH Aachen, Germany	[HeHe00]
TA(Siemens)	Threshold Accepting	Hans Heller, Siemens AG, Germany	[HeHe00]
U(Siemens)	Unpublished	Reinhard Enders, Siemens AG, Germany	

The assignment characteristics presented for comparison are the following:

- The *total interference* is the sum over all CCI and ACIs occurring between pairs of cells.
- The *CCI and ACIs* are given in terms of the maximum, average (among the occurrences), and standard deviation of interference of each type. These figures are computed on the basis of the sum of the (two) directed interference predictions for pairs of cells.
- The *interference for TRXs* is given in terms of the maximum, average (among the occurrences), and standard deviation. The basis is here the amount of interference a TRX causes in, as well as what it suffers from, other cells.
- The *histogram of interference* displays how many times the interference between two carriers exceeds the value of 1, 2, 3, 4, 5, 10, 15, 20, and 50 %, respectively.

The complete evaluation of these characteristics for all available combinations of scenarios and frequency assignment algorithms can be found in [EiKü00]. Figures 4.2.5–4.2.8 show a selection of these characteristics for the assignments corresponding to the five **bradford_nt-*t*-eplus** scenarios with traffic factor $t = 0, 1, 2, 4, 10$. The assignments computed using *SA(TUHH)*, *TA(RWTH)*, and *TA(Siemens)* give comparable results in terms of all the evaluation criteria stated above. The only noteworthy difference can be observed in Figure 4.2.8, where the number of carrier pairs with interference larger than 15 % is displayed. *TA(RWTH)* shows worse results there. In comparison, the assignments computed by *DC5_IM(ZIB)* have up to 25 % more interference in terms of the total as well as in terms of the maximum interference between two carriers. The picture is, however, reversed for the number of carrier pairs with interference more than 15 %. *DC5_IM(ZIB)* produces fewer occurrences of severe interference than the other three algorithms.

Notice that only three assignments produced by *SA(Telefonica)* are available. None of these assignments is feasible. This seems to be caused by an error in the data conversion. The effect is that in case of a HO relation between two cells, a reduced minimum separation of 1 rather than 2 is imposed. Disregarding the infeasibility, the amount of incurred interference is consistent with results obtained by a SA implementation of ZIB using the same type of moves. The assignments are slightly inferior to those produced by *SA(TUHH)*.

In Figure 4.2.9, the total interference for the scenarios **siemens1**, **siemens2**, **siemens3**, and **siemens4** as well as **Swisscom** is shown. Again, the assignments produced by the faster *U(Siemens)* method show a 20–25 % higher total interference than the assignments produced by *SA(TUHH)*, *TA(RWTH)*, and *TA(Siemens)* implementations.

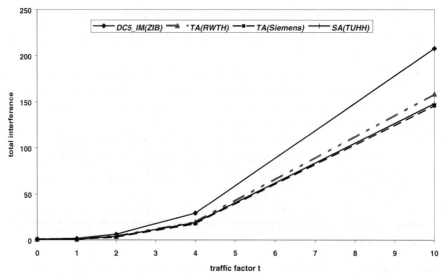

Figure 4.2.5: Total interference of various assignments for the **bradford_nt-
t-eplus** scenarios.

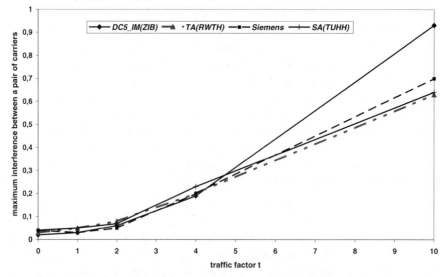

Figure 4.2.6: Maximum interference between carrier pairs (TRXs) of various
assignments for the **bradford_nt-*t*-eplus** scenarios.

In summary, the following conclusion can be drawn. The implementations
of frequency planning algorithms referenced in Table 4.2.6 with the
objective of minimising the overall interference in a cellular network
produce assignments with a recognisable spread in terms of the
characteristics analysed here. It is, nonetheless, fair to say that all of them
are reasonable choices from a practical point of view.

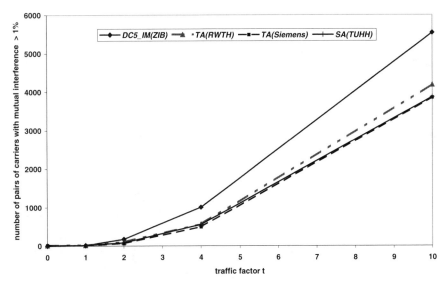

Figure 4.2.7: Number of carrier pairs with a mutual interference of more than 1 % for various assignments for the **bradford-nt-*t*-eplus** scenarios.

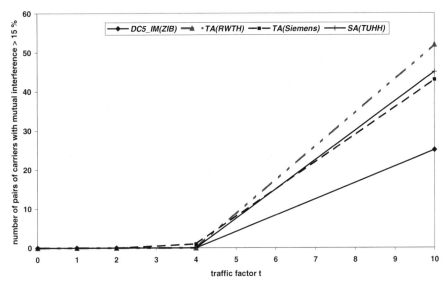

Figure 4.2.8: Number of carrier pairs with a mutual interference of more than 15 % for various assignments for the **bradford-nt-*t*-eplus** scenarios.

Within the actual planning cycle, one may prefer a method like *DC5_IM(ZIB)* or *U(Siemens)* with shorter run-time for the frequency planning at intermediate stages, and produce the final assignment using something like *TA(Siemens)*, spending several hours of computing time.

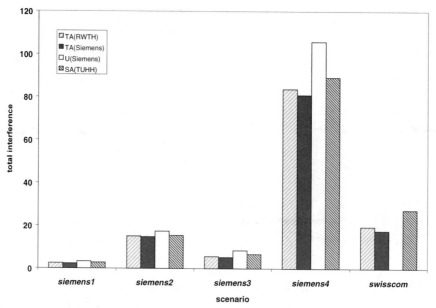

Figure 4.2.9: Total interference for Siemens' and Swisscom's scenarios. No assignment by U(Siemens) is available for the Swisscom scenario.

4.2.7.4. Related observations

The outcome of the frequency planning process is clearly not solely determined by the assignment algorithm in use. Two other aspects, considerably affecting the outcome, are briefly addressed, too. The first topic is the impact of the propagation model used for the interference predictions. The second topic is the choice of the threshold value to be applied when interference is measured in terms of affected cell area. The threshold value discriminates between pixels (grid points) that are affected by interference and those that are not in the best server model.

Influence of the propagation model on channel allocation

Three different interference predictions are available for the same basic planning scenarios **bradford_nt-*t*** for traffic factors *t* = 1, 2, 4, 10. These predictions stem from the use of three different propagation models, namely, *eplus*, *free*, and *race*, see the introduction of the scenarios earlier in this subsection. Among others, Table 4.2.7 shows some statistical parameters of the propagation models that are determined from measurements of five randomly selected BSs, with 15 cells in total, in the planning area. These values indicate that the model *eplus* is more accurate than the model *race*, which in turn is more accurate than the model *free*. An evaluation of the corresponding interference matrices reveals that the model *eplus* predicts the most interference in terms of the overall value, as well as in terms of the number of interfering carrier pairs. The model *free* is second according to that criterion, the model *race* last.

An analysis of the simultaneous influence of the propagation model and the frequency assignment algorithm on the resulting assignment can be performed, see also Section 4.2.1. A detailed such analysis is already presented in [EiKF98] for the basic scenarios **bradford-*t*** using the planning algorithms *DC5_IM(ZIB)* and a Simulated Annealing implementation that relies on the same neighbourhood as *SA(Telefonica),* see Section 4.2.4. The earlier results are confirmed here using the slightly different set of **bradford_nt-*t*** scenarios and more planning algorithms. A subtle question is how to compare the frequency assignments in a fair way.

Table 4.2.7: Performance measures (ΔE: mean value of prediction error [prediction minus measurement]; σ: standard deviation of prediction error) for three propagation models and total interference for different assignment algorithms; scenario **bradford_nt-10**.

Propagation model	Performance measure of propagation model			Frequency assignment method: incurred total interference			
	ΔE/dB	σ/dB	RMS/dB	ZIB	RWTH	Siemens	TUHH
Eplus	1,2	8,8	8,9	208	159	146	148
Free	-8,1	8,9	12	15	8	6	9
Race	-4,9	8,6	9,9	4	2	1	2

The implementation of each of the plans into a live network and the subsequent measurement of relevant quality indicators would be fair, but it is practically infeasible, at least within the study. Instead, the interference incurred by an assignment is evaluated according to all three interference predictions. Focussing on the total interference incurred with respect to the same model as is used in the planning, the following is observed. The model *eplus* yields higher numerical values than the model *free*. The model *race* yields the smallest values. This is in accordance with the overall amount of (potential) interference recorded in the corresponding interference matrices. This picture changes in the cross-comparison. The *race*-based assignments incur the most interference when evaluated by the predictions of the other models, the *free*-based assignments yield intermediate values, and the *eplus*-based assignments show the least deterioration when changing the model.

Tables 4.2.8 and 4.2.9 display the total interference and the number of carrier pairs with a mutual interference of more than 20 %, respectively, for the frequency plans generated by *DC5_IM(ZIB)*, *SA(TUHH)*, and

TA(Siemens) accounted for on the basis of the model *eplus*, which is taken as the most accurate propagation model among the three.

The assignments based on the *eplus*-predictions are clearly dominating. The results for models *free* and *race* show little dependency on the traffic factor or on the assignment algorithm. This can be explained by the fact that both propagation models underestimate the interference. The difference in total interference between the worst and the best solution at a traffic factor $t = 10$ is 300–400 %, if the propagation model is varied. Contrast this with a difference of approximately 25 % when the assignment algorithm is varied, compare Figures 4.2.5 and 4.2.9. This underlines the importance of selecting a proper propagation prediction model for automatic frequency assignment. The observation is more remarkable taking into account that many cellular network operators use the model *race* for interference prediction!

Table 4.2.8: Total interference for scenario **bradford_nt-10**; assignment with *DC5_IM(ZIB)*, *SA(TUHH)*, and *TA(RWTH)* based on different propagation models; interference analysed according to the model *eplus*.

traffic factor t	frequency assignment algorithm / propagation model used for the assignment								
	SA(TUHH)			TA(RWTH)			DC5_IM(ZIB)		
	eplus	*free*	*Race*	*eplus*	*free*	*race*	*eplus*	*free*	*race*
1	1	224	312	1	112	125	2	211	301
2	4	232	336	4	137	159	6	238	344
4	19	234	248	20	231	248	29	304	409
10	148	617	617	160	617	627	208	617	610

Table 4.2.9: Number of pairs of carriers with interference larger than 20 % for scenario **bradford_nt-10**; assignment with DC5_IM(ZIB), SA(TUHH), and TA(RWTH) based on different propagation models; interference analysed according to the model *eplus*.

traffic factor t	frequency assignment algorithm / propagation model used for the assignment								
	SA(TUHH)			TA(RWTH)			DC5_IM(ZIB)		
	eplus	*free*	*Race*	*eplus*	*free*	*race*	*eplus*	*free*	*Race*
1	0	239	307	0	104	125	0	216	322
2	0	221	345	0	119	149	0	222	366
4	0	229	227	0	216	222	0	287	406
10	27	616	602	24	659	628	9	600	588

Influence of the C/I-threshold on channel allocation

The interference is most often predicted in terms of affected cell area nowadays. The best server model is typically applied to cluster pixels (grid points) into cells. A pixel assigned to one cell is considered to be interfered with by another cell if the quotient of the field strength of that other cell's signal is not at least x dB below the field strength of the serving (strongest) signal. The impact of choosing the threshold value x is investigated in [EiKF99]. This study relies only partially on the scenarios listed at the beginning of this subsection. The scenarios **bradford_nt-*t-p*** are obtained for a threshold value of 20 dB. Two more sets of scenarios are used, obtained for the threshold values 15 and 9 dB.

Two heuristics are employed, one after the other, to compute frequency assignments. First, a Simulated Annealing implementation is applied, which uses the same neighbourhood as *SA (Telefonica)*. The second heuristic improves on the assignment and is a similar fashion as 'Iterated 1-Opt', see Section 4.2.4. The basic difference between the two methods is that the one used here may aggregate several 1-Opt steps into a single one in order to achieve an improvement (at the expense of a moderately higher run-time, the method is typically capable of finding significantly larger improvement).

The assignments are evaluated according to the interference predictions obtained for the tightest threshold value of 20 dB using similar characteristics as before, e.g. the histogram of interference. For low traffic factors, the threshold of 20 dB turns out to be best. For the high traffic factors of 4 and 10, in particular, the plans obtained for the 15 dB threshold have a more favourable distribution of the interference over the network. The value of 9 dB generally results in inferior plans. Figure 4.2.10 shows interference area plots of two plans for the scenario **bradford_nt-10-eplus**.

Obviously, using a threshold of 9 dB rather than 20 dB results in drastically more interference. The range of reasonable threshold values depends on the particular scenario at hand, and 20 dB will be inappropriate for networks with a high traffic load. Nonetheless, 9 dB as a threshold value seems to be an inappropriately small choice.

a)

b)

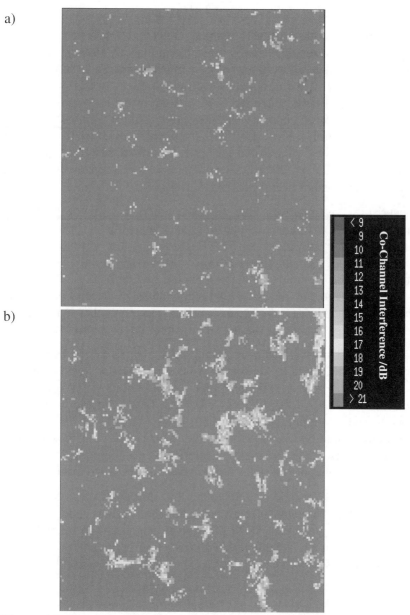

Figure 4.2.10:   Area interference plots for a part of the network of scenario
bradford_nt-10-eplus.   The same assignment method is
applied using different *C/I*-thresholds: a) *C/I* = 20 dB;
b) *C/I* = 9dB.

4.3. Cellular Aspects

Narcís Cardona

For future systems, the inhomogeneous traffic distributions and traffic capacity considerations, especially in multiple service environments, will in many cases, impose restrictions to cell planning more than propagation constraints themselves. Apart from prediction data, future planning tools and algorithms will be based on proper traffic and mobility databases. Hence, it is necessary to develop methods that from are derived these databases. These tasks include the development of traffic and mobility models taking into account heterogeneous sources of multimedia traffic.

4.3.1. Cell modelling

Thomas Kürner

In the radio network planning process a proper description of cell shapes is important for many tasks, e.g. interference analysis, HO planning or traffic calculations. In Figure 4.3.1 a typical radio planning process for GSM is displayed. Based on site data of the BSs, a path loss prediction is done for every site.

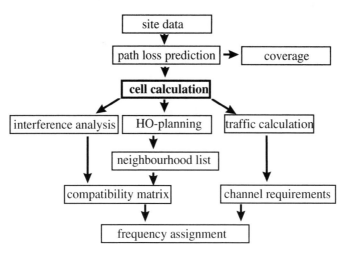

Figure 4.3.1: Radio network planning process.

The first output of this planning process is the coverage map. The cell shapes are determined in the following step, called cell calculation. The results of the cell calculation have a direct impact on interference analysis, HO planning and traffic calculation. In these three steps, the neighbourhood list, the compatibility matrix and the channel requirements are derived. The latter two are used as input data for frequency planning. In this way cell calculation has an indirect impact on frequency assignment. Therefore, the focus also has to be on cell calculation methods when automatic frequency planning methods are considered.

Traditional cell descriptions assume hexagonal cell shapes, see Figure 4.3.2 (a). However the hexagons are based on the assumption of isotropic propagation conditions and a regular distribution of BSs. Furthermore, it is assumed that for every mobile location a dedicated assignment to a certain cell exists and that the influence of the HO process is neglected. These assumptions are generally not fulfilled in real radio networks. An improved method is based on the best server model. The best server model is based on real BS locations and evaluates the path loss using a propagation prediction model. This model produces a more realistic cell map with arbitrary cell shapes, see Figure 4.3.2 (b). The drawback of this model is that a dedicated assignment of each mobile location to a single cell is still assumed, i.e. overlapping of cells is neglected. A more advanced method is a cell description based on Cell Assignment Probabilities (CAPs) [GBSZ97], [SBGZ87], see Figure 4.3.2 (c). CAP is computed by means of a Markov process, where the transition (or HO) probabilities are derived from path loss predictions and a statistical description of the HO process. In order to enable a reasonable calculation time, a number of simplifications and approximations are necessary. It is therefore essential that results achieved with this model are compared with measurements.

In [Kürn98] a quantitative analysis of the CAP model is presented, where the predicted CAP is compared to a measured assignment frequency based on test-mobile measurements in a GSM1800 network. From such measurements, the cell identifier of the serving cell and the RxLev value, which corresponds to the received power, can be retrieved for every position of the mobile. If the same test route is measured n times a relative *Measured Assignment Frequency (MAF)* for cell i at the pixel with the centre co-ordinates $\vec{x}$ can be computed by

$$MAF(i,\vec{x}) = \frac{n_A(i,\vec{x})}{n_v(\vec{x})} \ , \tag{4.3.1}$$

where $n_A(i,\vec{x})$ is the number of assignments to cell i at $\vec{x}$ and $n_v(\vec{x})$ is the number of visits at $\vec{x}$.

In principle, the CAP can be computed either from predicted power levels or from the measured RxLev values. When the CAP calculation is based on measurements an average over all measured RxLev values of cell i at $\vec{x}$ is used. Furthermore, it is possible to analyse the performance of the best-server model based on predicted or measured power levels. The following notation is used in the comparison for the results of the different cell models:

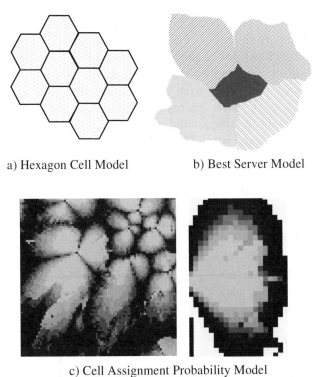

a) Hexagon Cell Model   b) Best Server Model

c) Cell Assignment Probability Model

Figure 4.3.2: Cell maps: a) Hexagon model: cell maps consist of hexagons of the same size. b) Best server model: the cell map consists of arbitrary areas that are formed by the propagation conditions (cells are distinguished by different colours). c) Cell assignment probability model: for each location there is a probability that the mobile is connected to a certain cell. On the left the maximum CAP for every location is displayed for a network, whereas the right map shows the CAP of a single cell.

- *CAPp:* predicted cell assignment probabilities based on propagation predictions,
- *CAPm:* predicted cell assignment probabilities based on measured RxLev values,
- *BSPp:* predicted best server probabilities based on propagation predictions,
- *BSPm:* predicted best server probabilities based on measured RxLev values.

In the best server model BSP is 100 % for the cell with the maximum received power level and 0 % for all other cells.

Figure 4.3.3 shows one example of the analysed measurement runs, where the influence of the prediction model on the accuracy of the cell model is demonstrated. For the prediction of CAPp or BSPp two subsequent modelling steps are required. The first step includes the prediction of the received power level by a propagation model. Although considerable progress in developing sophisticated propagation models have been made, these predictions still suffer from inaccuracies when they are compared to measurements. With macro-cellular propagation models in the 1 800 MHz band, typically a mean error around 0 dB and a standard deviation of about 7 to 9 dB can be achieved [COST99], [KüFW96]. On the other hand, the HO process is modelled using a couple of assumptions. Therefore, the comparison of CAPp or BSPp with MAF includes the inaccuracies from the propagation model and the cell assignment model. In order to eliminate the influence of the error caused by the propagation model, the CAP model and the best server model are applied to the measured power levels yielding CAPm and BSPm.

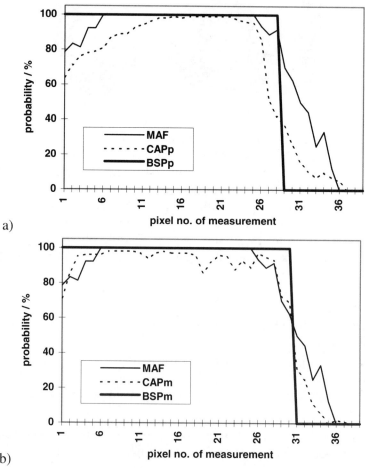

Figure 4.3.3: Comparison of MAF with CAPp, BSPp (a) and CAPm, BSPm (b) for one cell along a measurement run of about 2 km length.

Figure 4.3.3 shows the corresponding values for one cell along a measurement run of about 12 km length. It can be observed that the agreement between CAPm and MAF (Figure 4.3.3 (a)) is better than between CAPp and MAF (Figure 4.3.3 (b)). The difference between CAPm and MAF can be interpreted as the inaccuracies caused by the approximation in the CAP model. The inaccuracies due to the propagation model are described by the difference between CAPm and CAPp. Although a couple of deficiencies of CAPp can be observed, it seems to be even more realistic than BSPm, i.e. the inaccuracies introduced by the best server method are higher than those caused by the propagation model.

Additional investigations in [Kürn98] include the impact of the fading parameter as well as mobility aspects. Although a large number of approximations and simplifications have been made, a reasonable agreement with measurements was achieved. It is shown that the cell description by a prediction-based cell assignment probability is more accurate than the traditional best server method. Furthermore, an optimum value of 9 dB for the standard deviation to describe the signal fading at 1 800 MHz has been observed.

4.3.2. Tele-traffic engineering

Luis M. Correia

Modelling and characterisation of tele-traffic has always been a challenge, since its random nature makes it difficult to produce very accurate forecasts. Moreover, for mobile systems this challenge is even greater, not only because there are more random variables at stake, but also because some of these variables are very difficult to model and characterise. This subsection presents some contributions to this area, not only by proposing some models for traffic distribution in both time and spatial domains, but also by introducing algorithms for the optimisation of cellular planning.

4.3.2.1 Temporal and spatial traffic models

The temporal variation of traffic can be analysed using different perspectives: instantaneous variations, studying parameters like holding time and calls arrival rate; average variation over a period of time, looking at rush hour traffic variations, daily variations, and other aspects. In the former, Erlang's model [Yaco93] (applicable to fixed networks) has to be modified mainly due to mobility. One important effect of mobility in a cellular system is that the call holding time is no longer necessarily equal to the channel occupancy time at one particular cell, therefore modifying traffic statistics. In the latter, the whole cell traffic is considered, and previous works (e.g. [YaFo93]) consider long-term rush hour traffic evolution, since cellular planning is made according to these values; studies on daily traffic variations are less common, though its shape is reasonably known [Faru96]. Traffic depends on various factors, such as the number of users within the cell and their activity, the speed of mobiles and mobility levels, and cells size.

The spatial variation of traffic is also of importance. Due to the usual urban geography (in which services and commercial activities are located in the centre, while residential areas have moved to the periphery), the traffic density distribution is highly non-uniform and variable along the day, as people move to and from the city centre. Past works have shown that spatial traffic decay can be approximated by an exponential function [AsFu93], large spatial variations in traffic density leading to the use of cells with different sizes and capacities, in order to provide an efficient QoS.

In order to model both spatial and temporal variations of traffic, data was collected from a GSM network, Telecel's, in the area of greater Lisbon (in a total of 227 km^2) [AlQC99]. It consisted of hourly traffic values for three weekdays from 199 BSs, and of a map with their coverage area (from drive tests), defined on a quadrangular grid of 250 m × 250 m.

Observing traffic variation as a function of time t for each BS, one can see that traffic is concentrated in the hours corresponding to normal work schedules: it rises rapidly in the early morning period (from 8 to 10 h), after which it experiences a small decrease at lunch time (usually around 13 h); by the end of the afternoon (between 18 and 22 h), traffic decreases slowly; the minimum typically occurs in the middle of the night (circa 5 h). In order to enable a better comparison of different traffic shapes, all curves were normalised to its maximum value; curves were also shifted 5 h, to obtain an easier way to model the traffic shape, $ap(t_{sh})$, i.e. $t_{sh\,[h]} = t_{[h]} - 5$.

After testing several functions, and having the MMSE as a criterion, two functions were found to model the traffic shape. The *double-gaussian*, (4.3.2), consists of two gaussian functions, centred at the rush hours (morning and afternoon), with a breakpoint at the lunch hour,

$$ap_{gauss}(t_{sh}) = \begin{cases} p_1 \cdot e^{-\frac{(t_{sh}-h_{1sh})^2}{2d_1^2}} & , t_{sh} < h_{l\,sh} \\ p_2 \cdot e^{-\frac{(t_{sh}-h_{2sh})^2}{2d_2^2}} & , t_{sh} > h_{l\,sh} \end{cases} \qquad (4.3.2)$$

where p_1 and p_2 are the gaussians' amplitudes, h_{1sh}, $h_{l\,sh}$ and h_{2sh} are the shifted morning, lunch and afternoon peak hours, and d_1 and d_2 are the gaussians' deviations. Continuity is ensured at the transition point, $h_{l\,sh}$, where ap_{gauss} takes the minimum value of the two gaussians. The *trapezoidal*, (4.3.3), is an upper limited gaussian,

$$
ap_{trap}(t_{sh}) = \begin{cases} p \cdot e^{-\frac{(t_{sh}-h_{tsh})^2}{2d_t^2}} & , t_{sh} < t_{q1sh} \\ c & , t_{q1sh} \le t_{sh} \le t_{q2sh} \\ p \cdot e^{-\frac{(t_{sh}-h_{tsh})^2}{2d_t^2}} & , t_{sh} > t_{q2sh} \end{cases}
\tag{4.3.3}
$$

where p is the gaussian's amplitude, $h_{t\,sh}$ is the shifted gaussian peak hour, d_t is the gaussian's deviation, t_{q1sh} and t_{q2sh} are the shifted breakpoints, and c is the upper limit, continuity being ensured at $t_{q1,2sh}$.

To improve the accuracy of the model, and to analyse the influence of geography in traffic's daily variation, cells were grouped into classes with similar characteristics, as follows: urban centre (C); urban centre with main roads (CE); residential area (R); residential area with main roads (RE); suburban area (S); suburban area with main roads (SE). The C class is occupied primarily by services, while R presents residencies and some commerce, and S has lower terrain occupancy level; the distinction 'with main roads' is used for areas that have important traffic routes. The analysis of the data reveals that it is reasonable to consider models based on the *trapezoidal* function only for the R and S classes (especially for areas where the evening traffic has a significant weight), while the *double-gaussian* is of general application. Average values for the parameters are presented in Table 4.3.1. The values for the different classes are very similar among each other, which highlights the fact that the traffic shape is not highly affected by geography. Nevertheless, it is possible to draw some trends: as one moves away from the centre ($C \rightarrow R \rightarrow S$), traffic peaks move away from each other, thus widening the curve; the presence of main roads has an effect similar to the previous one, mainly in R and S; in classes with main roads, models are always *double-gaussian*, following the temporal variation of road traffic.

Table 4.3.1: Average parameters for temporal models.

Double-gaussian							Trapezoidal			
h_1 [h]	h_1 [h]	h_2 [h]	p_1	p_2	d_1 [h]	d_2 [h]	h_t [h]	P	d_t [h]	c
11.5	13.2	16.8	0.92	0.94	2.14	4.19	16.0	3.05	3.64	0.85

As one could expect, the differences among classes are revealed when the amplitude of the traffic density is considered. The average traffic density peak per class was obtained by averaging the values from all cells within each class, Table 4.3.2. It is important to stress that relatively high errors are associated to these peak values, due to the simplistic classification that was done (only one peak amplitude for each geographical class). It is possible, using parameters like the number of stories per building and terrain occupancy ratio, to develop models with a better performance; necessarily, these models will be more complex.

Table 4.3.2: Traffic density models amplitude [Erl/km^2].

C	CE	R		RE	S		SE
		DG	T		DG	T	
27.6	33.6	11.0	10.5	2.5	3.0	3.2	3.2

In order to have a more comprehensive representation of the spatial traffic distribution in the service area, traffic data was mapped into the cells on the coverage map, leading to a map with a traffic density value for each area element, corresponding to a specific hour, but being variable in time. In order to model spatial traffic variations, traffic density profiles departing from the centre to the periphery were studied; the centre is located in the city's business centre. Traffic profiles were outlined with a 10° interval between them, and were grouped in tree sectors with distinct geographical characteristics: the North sector is characterised by having the most recent and fast growing area of the city, presenting an important traffic concentration; the West sector is dominated by a natural park area, which presents a high traffic density decay, after which traffic raises again, mainly due to the existence of important peripheral areas; the Southeast sector characterised by the absence of suburban areas, since Lisbon is limited in South and East directions by the Tagus river.

To model the traffic density variation with distance, three functions were tested: *exponential*, *exponential/linear* and *piecewise linear*. The first one does not perform very well, in part because it assumes that traffic density is not constant in peripheral areas, which is not close to reality. The other two can be described by (4.3.4) and (4.3.5):

$$mod_{el}(d) = \begin{cases} e^{-d/Del} & ,d \leq dq \\ Cel & ,d > dq \end{cases} \tag{4.3.4}$$

$$mod_{pl}(d) = \begin{cases} 1 - Apl \cdot d & ,d \leq dq_1 \\ Bpl - Cpl \cdot d & ,dq_1 < d \leq dq_2 \\ Dpl & ,d > dq_2 \end{cases} \tag{4.3.5}$$

where *Del* is the decay factor, *dq*, *dq$_1$* and *dq$_2$* are breakpoints, *Apl* and *Cpl* are slopes, and *Cel*, *Bpl* and *Dpl* are constant factors. These parameters are functions of time, as well as of the sector in which they are applied, and their specific values can be found in [AlQC99].

Working on the establishment of standard scenarios for frequency planning requires simulating the influence of inhomogeneous spatial traffic distributions, which should be as realistic as possible. Therefore, traffic data for each cell is required, which can be very sensitive for network operators, and that is why this type of data is not usually available for the public, as it is required for investigations. Recent investigations on spatial traffic distributions in cellular networks [GoGR97] have revealed that the spatial cellular traffic follows a log-normal distribution, based on measurements in a real GSM1800 network. In order to create a realistic traffic distribution for an entire network, a random number generator can use these statistical parameters (mean value and standard deviation). The derived traffic distribution and the resulting channel demand profile are a reasonable basis to investigate frequency planning scenarios without giving away the detailed data of the network.

Considering the monthly average of busy hour cell traffic, $T_{[Erl]}$, its distribution is given by (4.3.6):

$$f(T) = \frac{1}{\sigma T \sqrt{2\pi}} \exp\left[-\frac{1}{2\sigma^2} (\ln T - \mu)^2 \right], \quad \text{for} \quad T > 0 \qquad (4.3.6)$$

where μ and σ are the mean value and the standard deviation of the stochastic variable. Once the traffic vector $T = (T_1, T_2, ..., T_N)$ has been created, the scaled traffic vector, $T_s = sT$, can be derived, N being the total number of cells in the network and s the factor that describes the increase of the traffic load. The latter is assumed to be uniform for all cells, hence for each T_s the corresponding channel demand vector d_s can be calculated according to Erlang-B. A relationship between traffic and channel demand, assuming a GSM network with a specified blocking probability, P_b, and a traffic/signalling channel configuration, can then be established. One should note that this configuration varies from operator to operator, i.e. there are many possible parameterisations of the network.

As an example, a traffic distribution has been generated using the procedure described above [Kürn99a], and the demand for TRX has been evaluated. This scenario consists of 1 886 cells, having $P_b = 2\ \%$ as reference and parameters $\mu_{dB} = -3.5$ dBErl and $\sigma_{dB} = 4$ dBErl have been selected according to measurements (note that $\mu_{dB} = (10/\ln 10)\ \mu$, and similarly to σ_{dB}): for $s = 1$, which corresponds to the original traffic distribution, a maximum demand of 3 TRX per cell is obtained; for $s = 1, 2$ or 4 the majority of the cells requires only 1 TRX; the maximum number of TRX is 5 for $s = 2$ and 9 for $s = 4$. When traffic increases by a factor of 10, roughly half of the cells require 2 TRXs and a few cells even need up to 12 TRXs making frequency planning more difficult. Based on this procedure, each user of the frequency planning scenarios should be able to generate realistic channel demand profiles in a well-defined and understandable way.

The previous models enable a fair description of space and time variation of traffic in large metropolitan areas. Although the specific values of the parameters do depend on the specific geographical characteristics of area that originated them, as well as on the habits of its population, it is the authors' belief that these models can be used for other situations.

4.3.2.2 Influence of mobility

Traffic models derived from the Erlang-B formulation can be used in cellular systems only as a first step towards cellular design. One characteristic of cellular systems against the use of these models is mobility, yielding a continuous changing number of users in a cell; moreover, users with an ongoing call may perform a HO to an adjacent cell, implying that a call may not be associated to only one cell. Besides this, due to the large number of subscribers, very small size cells are required in order to have a good QoS, hence the mobility effect has an even greater importance. As mobility affects the offered traffic in a cell, and consequently P_b, mobility models are very important in the performance analysis of cellular networks. A significant step in cellular planning is the evaluation of traffic caused by users' mobility and its contribution to the offered traffic in a cell.

Usually, the initial cellular planning is based on population density, and traffic (new calls) is estimated from these values by establishing Traffic Density Functions (TDFs) [Erl/km^2]. When such procedure is used, i.e. when traffic is estimated from Originated TDFs (OTDFs), corresponding to a certain hour (usually the busy one), all cells satisfy the maximum P_b criterion, max$\{P_b\}$. However, users' mobility causes an increase of the offered traffic, thus the Total TDF (the total traffic offered to the cell), TTDF, is different from OTDF. The way that mobility influences TTDF depends on the hour of the day, since, as mentioned before, substantial movement of users occurs during working days in urban areas. In the morning there is a concentration of people towards the city centre, while in the evening there is a dispersion away from the centre, towards the suburbs.

During the total call duration, τ, a user can either be served by only one cell, the call being originated and terminated inside the same cell, or perform a number of HOs before the call is finished, with probabiliies P_h. Therefore, the channel occupancy time, τ_c, is either the total call duration for a mobile that completes its call within the same cell, or the time spent in the cell before HO, τ_h, if the call continues, i.e. $\tau_c = \min\{\tau, \tau_h\}$. As a consequence, the mean channel occupancy time for a call originated in a cell is [Jabb96] $\bar{\tau}_c = 1/\mu_c = 1/(\mu+\eta)$, where μ_c is the channel service rate corresponding to channel occupancy in a cell, μ is the service rate associated to call duration, and η is the HO rate, which can be calculated via [ThMG88]. The total traffic can be decomposed into two components, the new calls one and the one coming from HO, i.e. $T = T_n + T_h$; besides their dependence on μ_c, the former depends on the new calls arrival rate, λ_n, while the latter depends on the incoming HO calls arrival rate, λ_h^{in}. λ_n does not depend on the network, but λ_h^{in} is proportional to the outgoing rate from a neighbouring cell, λ_h^{out}, which is associated to the network structure. On the other hand, λ_h^{out} depends on the total offered traffic in the cell and on the HO probability within it, and as only non-blocked calls participate in outgoing HO, one gets $\lambda_h^{out} = P_h \cdot (1 - P_b) \cdot (\lambda_n + \lambda_h^{in})$.

In order to enable a simple approach, it is considered that an urban area is covered by a regular pattern with cells of the same size, organised in rings: the central cell is surrounded by six cells of the first ring, followed by $6 \cdot 2$ cells of the second one, and so on. The TDF is the same for all directions (from the central point), therefore all cells of a ring will process the same traffic, as they have the same distance to the centre. The mobility of users is considered preferentially on a radial direction (*periphery→centre* or *centre→periphery*), corresponding to the aforementioned pendulum movement of people. In these circumstances, it is easy to obtain analytical expressions for η and λ_h^{in} [KrCo99], from which the two traffic components can be calculated. An analysis of the influence of mobility is done for a circular service area, in terms of the parameters that have the largest influence on traffic generated by users' mobility: users' mean velocity, varying in [0, 70] km/h; cell radius, ranging between 0.6 and 3 km; and mean total call duration, ranging in [30, 90] s. Three different OTDFs have been considered: *uniform* (i.e. the traffic density is constant all over the service area), *exponential* and *linear*. As an example, Figure 4.3.4 presents the influence of velocity on TTDF for the exponential case, where one can see an increase of the total traffic of 84 % for 70 km/h, which corresponds to $P_b = 43$ %; this value of P_b is clearly much higher than the usual one (around 1 %), which can pose real problems in a network.

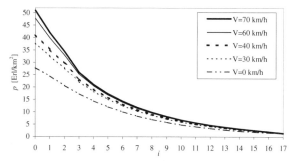

Figure 4.3.4: Effect of velocity on the TTDF for an exponential OTDF (the distance to the centre is expressed in terms of the index of the cells ring) (© 1999 IEEE).

The general conclusion is that the OTDF cannot be used instead of the TTDF in the cell planning process, because mobility may cause a large offered traffic increase in a cell, thus deteriorating the QoS, by increasing P_b.

4.3.2.3 Planning optimisation

Usually the deployment of cellular systems starts with cells providing the coverage of large areas, the so-called macro-cells, but the constant increase of the number of subscribers, especially in urban areas, leads to the degradation of the grade of service in these macro-cells, since they cannot cope with the offered traffic with the required quality. As a consequence, a system expansion has to be done by dividing macro-cells into smaller ones, micro-cells, only where necessary, since traffic increase is not uniform in the whole service area (the opposite would lead to a very high cost of the network). A lower cost system can be achieved if the service area is constituted as a mixture of micro- and macro-cells: the central region of it (where high traffic density occurs) is covered by micro-cells, and the peripheral region (with low traffic density) is covered by macro-cells.

However, there is no unique way to implement this mixture, since many parameters are at stake: traffic distribution, micro- and macro-cells radii, $r_{c\,i}$, and delimitations of the various cell size regions, $R_{r\,i}$, among others. A design method has been proposed [AsFu93], which determines the optimum mixed cell structure under a specified spectrum efficiency condition, but it is difficult to implement in a real system, since it does not account for parameters as the number of available traffic channels or P_b. Hence, optimisation algorithms have been developed, by minimising the total number of BSs, taking into consideration a specified max$\{P_b\}$, which enables their application to real networks.

In a theoretical approach, [KrCo98], it is assumed that the service area is circular and that the cell shape is hexagonal; a ring structure is considered for micro- and macro-cells, i.e. in each ring all cells have the same size, enabling a simple approach to the traffic density distribution. The total number of BSs in the service area, N_{BS}, can then be calculated as the sum of cells inside each region. Assuming a radial TDF (e.g. as the previously mentioned exponential or linear ones), the traffic in each cell can be calculated by integrating over the cell's surface, from which the associated P_b can be determined (using Erlang-B). Since all cells from the same ring have the same distance to the centre of the area, they will have the same traffic as well, hence P_b will be the same for all cells of the ring; the average blocking probability, $\overline{P_b}$, on the whole network is then easy to calculate, which will constitute the design parameter.

The relationship among N_{BS}, R_{ri} and r_{ci} is analysed for a given $\overline{P_b}$, r_{c1} being the parameter under variation (it is considered that the radius of each increasing cell size doubles the previous one, i.e. $r_{ci} = 2r_{ci-1}$). This procedure is repeated for increasing values of r_{c1} and then, by analysing N_{BS}, the optimum value is chosen. However, a problem occurs only if the criterion of $\overline{P_b}$ is used: although it may satisfy the design criterion, the central micro-cell and cells from the first ring usually present very high values of P_b (in some cases 40 %), due to the fact that the TDF is very high at the centre and decreases rapidly to the periphery; as a consequence, there are many cells with $P_b \approx 0$, which compensate a high P_b of the central cell. Therefore, a $\max\{P_b\}$ criterion is used as well, the solution (N_{BS}, R_{ri}, r_{ci}) being accepted only if $\max\{P_b\}$ for any cell is lower than a given value. Figure 4.3.5 shows an example of the existence of a minimum for N_{BS}.

In a more realistic approach for network optimisation, [SFGC98] presents an algorithm that can be applied to a service area of any shape, although hexagonal cells are still considered. The algorithm calculates the optimum dimensions of the micro- and macro-cells and estimates their location in the service area, again ensuring that $\max\{P_b\}$ is not exceeded at any cell. Its input parameters are the TDF (now a realistic one, distributed over the service area without any restrictions), $\max\{P_b\}$, number of cells per cluster, and number of available frequencies for the system. This algorithm has two possible applications: one is to propose an optimum cellular planning to cover a certain service area that still has no system deployed; another is to solve traffic congestion problems in a certain region of an existing system (e.g. the centre of a town), proposing a micro-cellular structure to cover the high traffic density region. Grids of micro -and macro-cells are defined over the service area, and their position adjusted in order to minimise N_{BS}. Optimum radii of the cells are obtained in an iterative way, the variation obtained for N_{BS} in terms of r_{c1} being similar to the one of Figure 4.3.5.

Also addressed in [SFGC98] is the issue of obtaining the TDF from real network data, coming from a 24 h operation. Results are compared on the basis of the following parameters: N_{BS} (coming from the optimisation algorithm), P_{over} (percentage of BS working hours in a day that exceed the supported traffic per cell) measuring the over-design of the network; P_{under} (percentage of BS working hours in a day that do not sufficiently exploit the cell capacity) related to under-designed networks. An example is shown in Figure 4.3.6, considering three types of criteria: traffic values at either 9, 11 or 17 h; 60, 80 or 100 % of the worst traffic case; average value in [11, 17] h, [9, 18] h or [9, 23] h. The optimum cellular planning is the one that corresponds to the lowest values of the parameters; the criteria that seem to lead to the best compromise among the three parameters are the ones corresponding to average values in [11, 17] h or 80 % of the highest value.

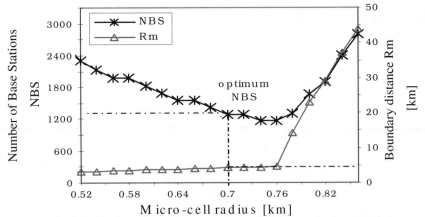

Figure 4.3.5: Variation of N_{BS} and $R_m = R_{r\,1}$ (radius of the first micro-cell region) vs. $r_{c\,1}$ (cell radius of the first micro-cell region) (© 1998 IEEE).

Although still in an initial form, the algorithms referred to here show that it is possible to optimise network planning, not only to reduce its expansion cost for a given QoS, but also by taking into account adequate design criteria.

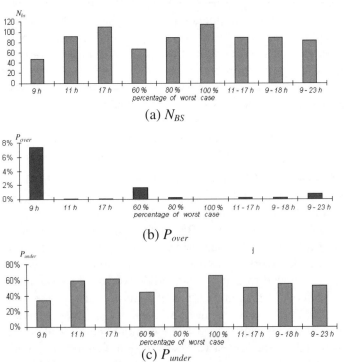

Figure 4.3.6: Comparison of planning efficiency for cellular structures resulting from various traffic criteria (© 1998 IEEE).

4.3.3. Hot spot location

José Jiménez

4.3.3.1 Initial considerations

COST 259 has devoted some effort to locating the hot spots for traffic management purposes. In a GSM network, the presence of high traffic locations is an important planning issue. Ideally, BSs should be located as close as possible to those high traffic spots. In that way, interference to other BSs is reduced and the global capacity and network efficiency is increased. The problem is related to the more classical problem of mobile location, but presents some special features.

In order to find those high traffic spots, several approaches have been suggested. The first approach uses a 'dummy' BS, i.e. a BS that does not carry channels, but which identifies itself to the mobile. The procedure is simple. A dummy BS is located in a candidate area. That station transmits a BCCH, but it is not allowed to take users. Then the mobiles in the area report to the network whether they would connect to the candidate BS. A software analysis of the reported results permits us to decide whether the candidate area is a good one. Unfortunately, that approach is expensive, since a number of 'dummy' BSs have to be deployed. Those BSs consume radio resources (BCCH), and not many can be used at a time.

Another possibility is to use mobile measurement reports. As is well known, during normal GSM operation, the mobile measures and transmits to the BS the received BCCH levels of a neighbouring BS. The use of that information does not require extra equipment. However, it is subject to important errors due to the erratic nature of 900 / 1 800 MHz propagation. Besides, some operation and maintenance systems provide measurement reports only for a reduced period of the call. This means the location has to be found with only 10–15 reported measurements.

Following the same research line, the problem of traffic segregation has also been considered. The objective now is to try to analyse the traffic within a cell and to discriminate the proportion from the total traffic of that coming from inside a given area, normally a building.

4.3.3.2 Algorithms for mobile location using power level reports

The objective is to find a candidate position $\mathbf{r}$ (x, y) for the mobile station (MS), taking into account the collection of reported measurement levels $\mathbf{v_m}$ to a number of bases (normally around 6). The algorithm compares those reported levels to a collection of planned or previously measured value $\mathbf{v_p}$ for every location of the area under study, usually of some square kilometres.

If there is only one set of measured $\mathbf{v_m}$ (corresponding to one location), the usual approach [HeMS97] is to give as candidate position with the minimum square value of the distance between the reported and planned values:

$$f(r) = \sum_{i=1}^{M} \left(v_{im} - v_{ip} \right)^2 \tag{4.3.7}$$

(M being the number of BSs with reported and planned measurements).

If the number of reported measurements of a mobile is greater than one, there are several possibilities.

1) The simpler approach simply averages the results of [HeMS97].
2) The optimum approach would be to use a new expression for $f(r)$ that would optimise in the whole set of combinations for the components of $\mathbf{r}$ [JiGA98], [JBGA98]

$$f_1(r) = \sum_{\substack{all\ possible\ paths}} \sum_{i=1}^{M} \left(v_{im} - v_{ip} \right)^2 \tag{4.3.8}$$

This approach is too complex for a practical implementation. For that reason, a simpler possibility suggested by the Viterbi algorithm for code detection has also been tried. First a candidate starting point $\mathbf{r}$ is determined using (4.3.7). The following elements $\mathbf{r}_j$, $j = 2 \ldots k$, can be determined taking into account they cannot be separated more than the maximum mobile velocity (**vel**) multiplied by the measurement interval. This implies that (4.3.7) will be evaluated only for those values v_p inside a circle of radius **vel**· Δt.

The algorithm can be rather time consuming for large l. One possible way of reducing computing time is only to pursue the analysis of those trajectories with a small accumulated error, in a manner similar to Viterbi's Algorithm.

Location algorithm results

The comparison of different algorithm performed during COST 259 shows better performance for the more complex algorithm. However, the differences are very small (of the order of 10 %) and it was considered not relevant when compared to the experiment variability. In summary, the simpler approach, followed by an averaging and smoothing process leads to reasonably good results.

The results of using algorithms in a simulated area of 4×6 km^2 with seven trisect BSs are provided. The coverage is simulated using a distance model. Measured values are evaluated from the actual values adding a log-normal random variable. BS locations are indicated in the map by a 'o' symbol. Those BSs are tri-sectorial ($120°$).

In Figure 4.3.7 [JAGR99], the average location error for a log-normal error in reported measurements of 7 and 9 dB is given. The location is performed using 10 reports from the mobile, and the simpler averaging algorithm is used. The main conclusion seems to be that for those locations close to a BS, errors would be in the region of 200 m. As can be expected, location error grows with measurement error, but it does not seem to be very sensitive to constant errors such as those likely to appear in indoor locations.

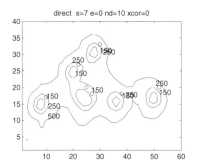 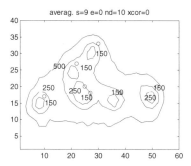

Figure 4.3.7: Average location error.

4.3.3.3. Traffic segregation

A similar problem would be to determine what proportion of traffic is coming from an area within a cell, as compared to the total cell traffic [JAGB99]. The objective is to decide whether a new BS should be located near the area where the traffic is generated. Normally, this problem arises when an operator suspects that most of the traffic of a cell is generated inside one or two big buildings inside the cell. The objective is to determine, by using sample measurement reports inside and outside the candidate buildings, the proportion of the total cell traffic corresponding to that generated inside.

Two approaches have been analysed to solve the problem. In the first, an algorithm based on distance concepts was used. A second approach has been based on the use of neural networks. Both algorithms use a collection of samples for training or adjusting purposes, as it will be explained in the algorithm description. As can be expected, the larger the training set the better performance is obtained. The experiments reported here correspond to a sample size of 800 measurements.

Algorithm based on distance estimates

Each report received can be assigned to a point in the *N-dimensional* space, where *N* is the number of receiving stations. The first step is to find the region in the space that corresponds to the building under consideration. To do so, a number of *accumulation areas* are defined. An accumulation point is defined as an **N** *dimensional* location containing (within a radius *d)* a number of reports generated inside the building.

The following step is to determine the proportion of **T** (total traffic) included in the previously calculated *accumulation areas*. That proportion is considered representative of the proportion of the total traffic generated within the cell under study originated in the building.

The process described above implies the definition of a *metric* or distance between two reports. That definition is, within itself, a parameter of the algorithm. Several possible definitions of distance have been tried, the most appropriate seems to be:

$$distance(P,Q) = \sum_{i=1}^{N} A_i^2; \begin{cases} A_i = (P_i - Q_i) & \text{if } P_i \neq 0 \text{ and } Q_i \neq 0 \\ A_i = K & \text{if } P_i = 0 \text{ or } Q_i = 0 \end{cases}. \qquad (4.3.9)$$

Algorithm based on neural networks

The main disadvantage of the previous approach is that the radius of the accumulation areas and the number of areas considered are parameters that are difficult to handle because they depend on the building and cell. For that reason, an adaptive approach based on neural networks has also been suggested. Several neural networks architectures have been tried, but the *neuron tansig* followed by a linear have proved to be the most useful. The design of the neural network as described implies the selection of an input and target functions. The input function used has been the measurement report. That is a multidimensional function (the measurement vector contains information of around 15 BSs; however, since the measurement report of the MS only reports about less than five stations, most of the vector contains a null). For the target function, a simpler function can be tried proposing 0 for the outside and 1 for inside. That target function can be complicated if more information is required assigning, e.g. 0 for the outside, 1 for the first floor, 2 for the second, etc. Training is performed using a Levenberg–Marquadart routine.

Results

In Fig 4.3.8, a comparison between the two approaches is presented. In the curves the probability of error p_e (probability of not detecting a call effectively made from inside the building) and probability of false alarm are displayed. p_{fa} (probability of indicating a call made from outside as being made from inside). As a reference, a distance parameter has been used for the heuristic algorithm and the threshold value for the neural network. The first figure corresponds to the heuristic algorithm and the second to the neural network solution.

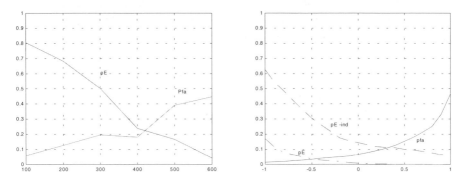

Figure 4.3.8: Probability of error p_e and probability of false alarm p_{fa} for the heuristic algorithm (left) and the neural network solution (right).

The main conclusion of the study is that the neural network approach has better performance and it is also easier to adjust. Besides, it can be easily generalised to more complex scenarios (such as traffic segregation within a building). One disadvantage of neural networks is the relative instability of the training algorithms.

4.3.4. Mobility models

Narcís Cardona

The ever-increasing number of mobile subscribers and the desire to improve their QoS has forced network operators to reduce cell size and to introduce hierarchical micro-/macro-cell systems. A decrease in cell size leads to the reduction of cell residence time, as well as to an increase in cell boundary crossings. Both factors are major contributors to the growth of signalling traffic load [PoGo96].

In two-tier micro-/macro-cell systems, cell selection must be based on the subscriber speed. Micro-cells cover the slow moving units and macro-cells serve the rest. A cell selection approach is based on knowledge of the time spent by the subscriber unit in a call-initiated cell [Benv96], [YeNa96]. Therefore, determination of a suitable model to estimate and analyse movements of subscriber units is of major interest [LaCW97], [Nand93], [ZoDF96]. Actual and selective mobility models, which are applicable to urban as well as suburban traffic situations, are indispensable for the development of a full coverage cellular mobile communication network.

Currently available models tend to be either too simple or too sophisticated. Analytical models using rectangular or hexagonal radio cell bases do not take the actual road system into account, whilst models that are accurately based on the actual traffic system of a certain area require the collection and processing of extensive data [SMHW92].

4.3.4.1. Mobility models for vehicle-borne subscribers

In COST259 a mobility model has been proposed in [BrPB97]. This model considers the heterogeneity of both the traffic systems and the subscriber units, which may move along major roads as well as along crossroads. The model includes the road network pattern, street length between crossroads, street width, traffic regulations and subscriber behaviour. A mobile call can be initiated at any point within the cell along the path of the vehicle as shown in Fig 4.3.9. From these parameters, the Cell Residence Time (CRT) t_s or the remaining residence time t_{rs} are calculated according to the call initiating position of the mobile. The path the mobile follows during the call is modelled by the vectors $\vec{d}_i$, which include the street length value between crossroads and the direction of movement. The average velocities v_i are related to the different sections of the traffic path. The direction of a mobile is uniformly distributed between $[-\pi, \pi[$ and then the PDF of the starting angle is $1/(2\pi)$ for the range $[-\pi, \pi[$.

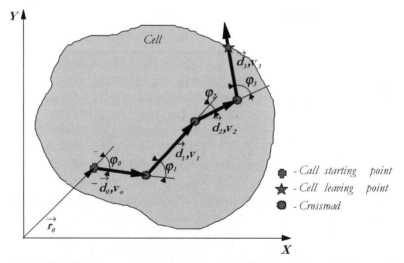

Figure 4.3.9:   Tracing a mobile within the cell (remaining residence time in the call initiated cell).

For the incoming call, the cell due to HO, the starting point corresponds to a point somewhere on their boundaries. The PDF of direction [XiGo93] is given by:

$$
p.d.f.(\varphi_{(\ )}) = \begin{cases} \dfrac{1}{2}\cos\varphi_{(\ )} & \text{for} \quad -\dfrac{\pi}{2} \le \varphi_{(\ )} < \dfrac{\pi}{2} \\ 0 & \text{otherwise} \end{cases} \tag{4.3.10}
$$

The relative direction changes at each crossroad, φ_i, depends on the street network pattern and the traffic situation. The angle is expressed as the realisation of one of four normally distributed variables, with means estimated 90° apart. The PDF of φ_i is then given by

$$
p.d.f.(\varphi_i) = \frac{1}{1 + w_{90°} + w_{-90°} + w_{180°}} \cdot \frac{1}{\sigma_\varphi \sqrt{2\pi}} \left(e^{-\frac{\varphi_i^2}{2\sigma_\varphi^2}} + w_{90°} e^{-\frac{\left(\varphi_i - \frac{\pi}{2}\right)^2}{2\sigma_\varphi^2}} + w_{-90°} e^{-\frac{\left(\varphi_i + \frac{\pi}{2}\right)^2}{2\sigma_\varphi^2}} + w_{180°} e^{-\frac{(\varphi_i - \pi)^2}{2\sigma_\varphi^2}} \right)
$$

$$
\tag{4.3.11}
$$

where $w_{90°}$, $w_{-90°}$ and $w_{180°}$ are the weight factors corresponding to probabilities, and σ_φ is the standard deviation of the direction distributions, assumed to be equal for the four variables.

Since streets take a random course with respect to the axes of the co-ordinate system, their projections $d_{i,X}$ and $d_{i,Y}$ will be regarded as normally distributed random variables. In areas with an irregular street network pattern, the random variables $d_{i,X}$ and $d_{i,Y}$ can be characterised as statistically independent, with zero mean and showing the same variance. Therefore, the street-length between crossroads turns out to be a Rayleigh distribution:

$$
p.d.f.(d_i) = \begin{cases} \dfrac{d_i}{\sigma_d^2} e^{-\frac{d_i^2}{2\sigma_d^2}} & \text{for} \quad d_i > 0 \\ 0 & \text{for} \quad d_i \le 0 \end{cases} \quad \text{where} \quad \sigma_d = \overline{d}\sqrt{\frac{2}{\pi}} \tag{4.3.12}
$$

In [BrPB97] the Rice distribution is proposed as an alternative when the majority of subscriber units use major roads only.

The model considers that the velocity of the subscriber unit does not change while covering the distance d_i, allowing an equation with the average velocities v_i. Then the average velocity can be expressed as a Rayleigh / Rice distributed random variable. Measurements in Vienna and Helsinki [VaNi94] suggested adding a second term to the distribution to account for the users in major roads, whose velocity is better described by a normally distributed random variable.

$$p.d.f.(v_i) = \begin{cases} \dfrac{1}{1+w_{mr}} \left[\dfrac{v_i}{\sigma_v^2} e^{-\frac{v_i^2+\bar{v}^2}{2\sigma_v^2}} I_0\left(\dfrac{v_i\bar{v}}{\sigma_v^2}\right) + w_{mr}\dfrac{1}{\sigma_v\sqrt{2\pi}} e^{-\frac{\left(v_i^2-\bar{v}_{mr}\right)^2}{2\sigma_v^2}} \right], & v_i > 0 \\[4mm] 0, & v_i \le 0 \end{cases} \quad (4.3.13)$$

where w_{mr} is the weight factor for fraction of cars on major roads.

4.3.4.2. Cell residence time distributions

Since new calls can be initiated anywhere in the cell area and HO calls can cross the cell boundaries at any point to its adjacent cells, the residence time t_s and the remaining residence time t_{rs}, are also random variables.

A simulation based on the presented mobility model [BrPB97] enables the calculation of the distributions of these times for cells of arbitrary shape.

Figure4.3.10 shows the PDF of a Monte Carlo simulation with 2 500 runs for a hexagonal cell with the radius $r = 500$ m. Simulation parameters have been set as follows: $w_{-90°} = 0.75$, $w_{90°} = 0.5$, $w_{180°} = 0.065$, $\sigma_\varphi = 0.125\pi$, $\bar{d} = 100$ m, $\sigma_d = 100$ m, Rice distributed, $w_{mr} = 0.5$, $\bar{v} = 10$ km/h, $\bar{v}_{mr} = 35$ km/h, $\sigma_v = 10$ km/h. The results show that the CRT follows the generalised Gamma distribution.

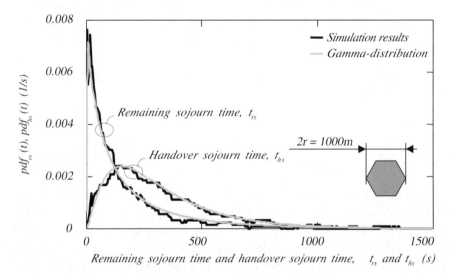

Figure 4.3.10: The PDF of remaining residence time and residence time for a hexagonal cell.

Figure 4.3.11 shows the results for a 120° sectorised umbrella cell. The sector makes up one third of a cluster comprising seven hexagonal cells (each with $r = 500$ m.). Cell enlargements lead to a growth of residence times. Table 4.3.3 compares the calculated mean remaining residence times and mean residence times of hexagonal cells ($r = 500$ m and $r = 250$ m, respectively) with corresponding $120°$ sectorised umbrella cells.

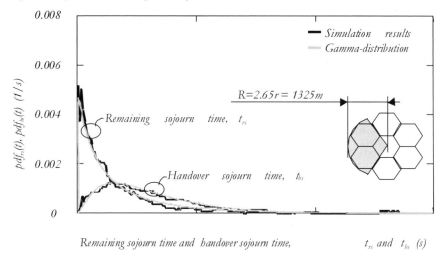

Figure 4.3.11: The PDF of remaining residence time and residence time for a $120°$ sectorised umbrella cell.

Table 4.3.3: Mean remaining residence time and mean residence time for hexagonal and sectorised cells.

Cell	Mean remaining residence time, $\bar{t}_{rs}(s)$	Mean residence time, $\bar{t}_s(s)$
Hexagonal cell $r = 250$m.	65	134
Sectorised cell, $R = 662.5$ m.	135	268
Hexagonal cell $r = 500$m.	171	344
Sectorised cell, $R = 1\ 325$ m.	345	695

Other approaches

The analysis of HO in second generation cellular networks [Rapp93], [Pavl94] has assumed that in most cases the CRT is an exponentially distributed random variable or the sum of exponentially distributed (Erlang) random variables [Rapp93]. However, in practice, the CRT distribution may not be exponential, because of continuous changes in cell sizes and shapes [JoSk95], especially in micro-cellular environments. For example, the constant CRT assumption may be acceptable for Manhattan micro-cells [Ostr96] where users enter one side of the cell, travel through the cell in a fixed amount of time, and leave at the other end of the cell. The CRT is distributed uniformly when mobiles transiting a cell travel a distance uniformly distributed between 0 and $2R$ at a constant speed, where R is the cell radius [dRFG95]. In [ZoDF95] the suggested model for the estimation of the cell residence distribution traces mobiles in an environment where their movement is governed by a set of random variables. In that study, it is shown that the generalised gamma distribution and the negative exponential distribution provide the best approximation for the CRT and the Channel Holding Time (CHT) distributions, respectively. However, in [JeLe96], who uses real cellular data, it is shown that the negative exponential assumption for the CHT is not correct, and that a log-normal model approximation fits much better. Due to the disparity of these findings, it appears important to evaluate the effects of CHT and CRT distributions over the micro-cellular network performance.

In [KhZe97a], the effect of CRT distribution on the performance of cellular mobile networks is studied. It is assumed that

a) The new call arrives in a given cell form a Poisson process. The mean new call arrival rate is λ_o.

b) The CRT is a random variable having a general distribution with mean equal to η_1^{-1}.

c) The unencumbered session duration of a call is the amount of time that the call would remain in progress if it experiences no forced termination due to HO failure. The unencumbered session duration of a call is distributed as negative exponential with the density function $f_c(t) = \mu\, e^{-\mu t}$ and mean equal to μ_1^{-1}.

In [KhZe97a], a comparison for call blocking, HO failure and forced termination probability when the distribution for the CRT is Exponential, Erlang, Gamma, Uniform and Deterministic is presented. The simulation results show that the call blocking probability is insensitive to the CRT distribution while the forced termination probability of a call varies with the CRT distribution. It is also found that the effect of CRT distribution on the forced termination of a call vanishes as the CRT mean decreases. For large average CRT ($\eta/\mu < 1$), the statistical behaviour of the CRT is critical and has non-negligible influence on performance. On the other hand, when average CRT is small compared to the call duration ($\eta/\mu > 1$) the effect becomes negligible. Consequently, the exponential assumption for the CRT can be retained when CRTs are small, and can only be used as a loose upper bound on performance in the case that CRT is large compared to the call duration.

Applications

The capacity of a cellular system can be increased by reducing the coverage area of a cell. A decrease in the coverage area, however, is constrained by the requirement that a subscriber unit remains in the cell for a time interval sufficient to complete the call set up and HO functions. Conversely, in order to achieve smaller cell sizes, it may be necessary to reduce the processing times for these functions. In either case, one needs to know the relationship between the cell size and call set up and HO processing times or, more generally, the call processing times.

Such a relationship is derived by computing the probability that a subscriber unit, initiating a call anywhere in the coverage area of the cell, will be covered by that cell for a time interval equal to the time it takes for call set up and HO combined. By fixing the cell size one can determine the coverage probability for a given call processing time T_{cp}, and *vice versa*. Table 4.3.4 [BrPB97] compares the coverage probabilities of a cellular system that would meet specified call processing requirements for different cell sizes and shapes.

Table 4.3.4: Coverage probability for hexagonal and sectorised cells with different sizes.

Call processing time T_{cp} [s]	Coverage probability, Prob($t_{rs} > T_{cp}$) %			
	Hexagonal cell, $r = 250$ m	Sectorised cell, $r = 662.5$ m	Hexagonal cell, $r = 500$ m	Sectorised cell, $r = 1\,325$ m
5	82.8	90.1	91.8	95.3
10	75.5	85.8	88.4	93
15	68.8	81.3	84.8	91
20	63.1	77.6	82.1	89.5

In [VeCo98] an approximate traffic model which takes into account both HO and new calls traffic was presented for a configuration without using guard channels for HO. Models proposed in [VeCo98] allow study of the influence of coverage distance and velocity on the supported traffic and on the new calls' traffic linear density, and some results were obtained for typical scenarios in a Mobile Broadband System (MBS) with linear coverage geometry. The solutions achieved for new calls and HO traffics for configurations without using guard channels for HO were obtained via the consideration of a Markov chain, which is an approximate model that allows use of the theory from Jackson networks. One shows that for homogeneous traffic in the whole coverage area, although it has been usual to consider the HO traffic as Poissonean, it is not actually necessary.

The independence among the number of calls served in each cell is also presented as a conclusion in [VeCo98], and the need of further research for configurations with guard channels for HO is highlighted. For situations without using guard channels for HO, for a fixed bounding value for the blocking probability, the new calls traffic linear density has been analysed, increasing with the decrease of R, being upper limited by a value which depends on the characteristics of the mobility scenario. However, call-dropping probability requirements also need to be fulfilled, leading to a new calls' traffic density that decreases with the decrease of R, being lower for the scenarios with higher mobility. This situation leads to limitations in system capacity for lower values of the coverage distance, mainly for high mobility scenarios. In order to overcome these limitations, the use of guard channels for HO is studied, mainly for high mobility scenarios. For these scenarios one concludes that there is a degradation in system capacity because, for the typical coverage distances foreseen for MBS, the new calls' traffic linear density is one order of magnitude bellow the values obtained for the pedestrian scenario (where it is approximately 15 Erl/km), decreasing from 3.5 Erl/km, in the urban scenario, down to 1.3 Erl/km, in the highway scenario. The models developed in [VeCo98] for homogeneous traffic can be further generalised in order to take into account more general spatial distributions of traffic, by solving a system of flow equilibrium equations, so that the overall traffic must be locally homogeneous, i.e. approximately equal among neighbour cells.

4.3.4.3 Location area code planning

In current second generation mobile networks accurate planning of Location Areas (LAs) is becoming an essential task due to the growth in customers and the consequent signalling load increase. Such a prerequisite will become even more crucial in future mobile networks, due to the expected large number of users and the provisioning of a variety of services ranging from plain speech telephony to data transmission. Different location management procedures will be available allowing, for instance, a separation between the LAs and the Paging Areas (PAs) that, in current networks, coincide. This novel feature will allow us to achieve an extra gain in signalling load but could increase the delay experienced in procedures like the paging and location updating. It is therefore necessary to study the most effective procedures to manage the user mobility and, in parallel, design some appropriate methodology to support network planners in the dimensioning and optimisation of LAs and PAs.

In COST259 some efforts have been devoted to LA Code (LAC) planning, very closely related to the STORMS project. The STORMS planning tool [MePi99] has been considered in COST259 [Mark98], since it is a suitable tool to jointly optimise LAs and PAs in UMTS networks. In [Mark98], a genetic algorithm is used for the dimensioning and optimising of LAs and PAs. The intelligent paging algorithm assumed is called Recent Interaction Paging: the network tracks the location (cell) where the user last did any interaction with the network, e.g. a call set up or a registration session.

When a mobile-terminating call occurs, the network first pages the user in the indicated cell and, if the paging fails, it extends the paging process to the whole location area. Figure 4.3.12 shows the effect of mobility on the percentage of successful first paging step.

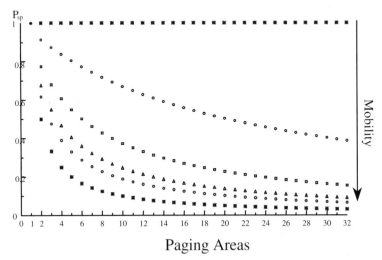

Paging Areas

Figure 4.3.12: Probability of successful first paging step vs. the number of paging areas

The optimisation process in the LAs and PAs optimisation is used to determine a fair trade off between signalling load saving and the extra time delay necessary to safely complete the paging operations. Simulation results suggest that the combined LA/PA planning strategy is the most effective procedure, as shown in Figure 4.3.13.

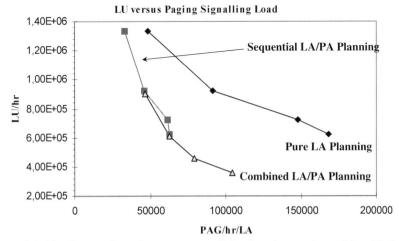

Figure 4.3.13: Comparison between pure LA planning and combined LA/PA planning strategies.

4.4. Network Optimisation

Jean-Frédéric Wagen

This section describes a number of techniques that can be used by cellular network operators to increase or to trade-off the quality and the capacity in their radio network. The previous sections concentrated on modelling techniques that enable better planning and design of a radio network. This section first considers some HO and Power Control (PC) algorithms that insure that sufficient quality is achieved in each cell. Then, very popular and effective capacity enhancement techniques based on frequency hopping are discussed. Other improvements based on adaptive techniques such as link adaptation and smart antennas are finally considered in the third and last sub-section. Planning methods and special protocols that could also be used to optimise the usage of the radio resources are not described here, but are the topics of the following two sections.

4.4.1. Handover, power control and directed retry

Roberto Verdone

Handover is the procedure used to pass an ongoing call from cell to cell (inter-cell HO) or from channel to channel within the same serving cell (intra-cell HO); it can be executed starting from power, mobility or traffic measurement based algorithms, or from link quality estimations, etc. The decision to perform HO depends in any event on the network status and affects the network status.

Power control is the procedure having the aim of reducing the average transmitted power of mobile terminals and BSs, thus minimising the battery consumption (in the uplink) and reducing the impact of interference (in both links); it can be driven by CIR or signal level measurements. In the former case, the transmitted power depends on the network status. In all cases, PC has an impact on the network status.

Therefore, both HO and PC belong to the class of techniques (for Radio Resource Management) that must be investigated by taking complex computation environments into account, that is, both network and link level considerations have to be performed. Two types of investigation can be carried out in this context: based on simulation or analytical approaches.

4.4.1.1. Handover and power control: a simulation approach

The simulation of third generation mobile radio interfaces is a particular challenge. Complex systems are required to back up the demands of different services and high data rates of up to 2 Mbit/s.

The physical influences of mobile radio channels on radio resource management schemes should be considered within the system level (that is, taking the whole access network into consideration). On the other hand, varying BERs, burst error structures and changing delays during the data transmission, at link level, cannot be neglected. They are affected by the current network conditions and their effect on the carrier-to-interference level. Thus the joint evaluation of transmission techniques (at link level) and system level protocols is necessary in many cases. However, the computational environment in a complete link and system simulation based approach is rather complex, and this requires the definition of different approaches.

One solution could be the evaluation during system level simulations of the BER for raw uncoded transmission from a CIR / BER graph using the estimated carrier-to-interference level. This CIR / BER graph could be generated in separate link level evaluations. Still complex from the computational point of view is the need for simulating channel coding/decoding. Another possibility would be to use a CIR / BER graph for encoded data transmission to establish only the residual error ratio. The disadvantage of this method is that the delay of hybrid ARQII schemes cannot be modelled. On the other hand, no calculation of channel coding and decoding is needed and simulation at system level is less complex. The CIR / BER graph can be provided by means of a look-up table.

An additional advantage of having separate link and system level assessment sessions is in the possibility of having only large-scale aspects taken into account at system level; for circuit-switched services this can be easily understood if one considers that network decisions (HOs, etc.) are performed with a clock rate in the order of hundreds of milliseconds, since those services (e.g. voice) are normally sensitive to long-term variations of the perceived quality. The fading rate is usually larger than this clock rate in mobile environments, and this means that fast fading can be considered to be averaged out at system level, whereas it has to be managed at link level. This is no longer true if packet-switched services are investigated, where network decisions (channel assignments, etc.) can be based on a smaller time scale, thus being dependent on channel fluctuations due to fast fading, too.

The approach based on a look-up table providing the link level CIR / coded BER relationship was used in the definition phase of a simulation tool based on SDL (Specification and Description Language, ITU Rec. Z.100) at the University of Hanover [StLJ98], [StGJ99], and of another simulator developed at CSITE-CNR, University of Bologna [VeZa98], [VeZa99]. The former is being completed in order to investigate the performance of some HO algorithms based on the Direction of Arrival (DoA). The latter was used to assess the performance of power and traffic driven HO algorithms and signal-level based PC schemes for a circuit-switched network, and the results obtained are illustrated below.

Power control

Hereafter, the performance of a signal-level-based PC scheme for the downlink of a cellular mobile radio system (whose layout is in Figure 4.4.1) based on Frequency-Time Division Multiple Access is given; the PC strategy considered compensates the estimated path loss only partially, by setting the transmitted power (in dBm) proportionally to a given fraction a of the path loss: $a = 0$ represents the case with no control, $a = 1$ the full compensation PC and $a = 0.5$ (half compensation) is known to be the optimum case when outage probability is evaluated in the absence of HO and mobility [Whit93].

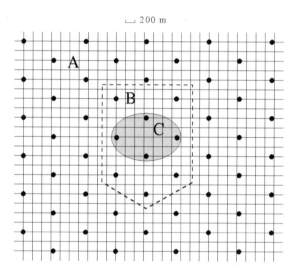

Figure 4.4.1: Scenario for Power Control scheme (© 1999 IEEE).

A Manhattan cellular geographical layout, Figure 4.4.1, is considered, with 54 BS sites; 120° antennas are assumed at each site (three cells per site), with nc channels per cell. The site locations have been chosen according to a hexagonal layout; a re-use factor equal to four has been set (12 cells per cluster). Mobile users are uniformly distributed over the whole area A, travelling at constant speed v and with uniform probability of choice at each crossing. Propagation is modelled through the COST231 Walfisch-Ikegami formula [COST99] to take LoS/NLoS path loss evaluations into account, and shadowing is assumed to be log-normal with standard deviation equal to 5 dB and correlation distance equal to $dc = 20$ m. The total number of users in the area is denoted by Nu. To avoid border effects, the performance figures are evaluated only for mobiles connected to one of the sites in area B (evaluation area). The call inter-arrival time for a user is modelled as a negative exponentially distributed random variable with mean 720 s. Call duration is also negative exponentially distributed, with mean 120 s, and different calls are uncorrelated. Hence, the total offered traffic is given by $E = Nu \ 720/120 = Nu/6$. The offered traffic per cell is $Ec = E/162$. Figures 4.4.2 and 4.4.3 refer to a case with $Ec = 5.6$ Erl/cell and $v = 10$ m/s, $nc = 7$.

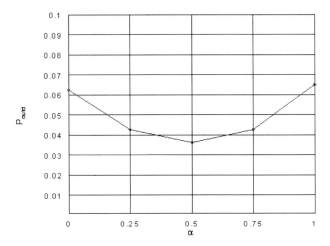

Figure 4.4.2: Downlink outage probability against a.

The outage probability (defined as the probability that the CIR goes below a specified level fixed to 10 dB) is shown in Figure 4.4.2 as a function of a. However, one of the most important performance figures for mobile radio cellular systems is the drop-out probability, which is related to the duration of outage events and hence to the HO strategy; in fact, here it is defined as the probability that CIR falls below the specified threshold of 10 dB for more than 6 s. It is plotted in Figure 4.4.3.

The HO strategy in this case is based on a simple power driven HO algorithm, named ST, which exploits an absolute threshold S (to be compared to the averaged received useful power) and a time out $T = 3$ s (no HOs can be performed before a time-out that started after the last HO expires). It is clear from Figure 4.4.2 that the half compensation scheme gives the best performance in terms of outage probability. This result is identical to that found in [Whit93], where the mobility of users and HO were neglected.

On the other hand, in Figure 4.4.3 the downlink dropout probability is plotted as a function of a; in this case the optimum choice is $a = 1$. That is, outage events are shorter, on the average, if a full compensation scheme is used. This is clearly due to the impact on outage events duration of the HO mechanism.

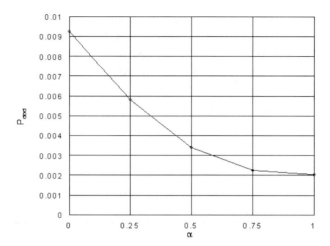

Figure 4.4.3: Downlink dropout probability against *a*.

Handover

In Figure 4.4.4, for the same cellular layout but with nc = 15, uniform (scenario 1, Ec = 7.7 Erl/cell, v = 6.7 m/s) and non-uniform traffic distributions (scenarios 2 and 3 have an additional hot spot of pedestrian users walking at speed of 0.67 m/s and with offered traffic equal to 6.8 and 10.2 Erl/cell, respectively, inside the area C of Figure 4.4.1), the dropout probability is plotted for four different HO algorithms: ST, STG (which includes a control on the gradient of received power to check if sudden degradations of the received power are introduced due to street corner effects), ASTG (which also includes an evaluation on the actual channel occupancy of bases), and tr-ASTG (which adaptively determines the maximum amount of channel occupancy per base to be allowed). The figure shows that tr-ASTG and STG offer the best performance in terms of dropout probability. On the other hand, it was also found that the blocking probability was minimised with tr-ASTG, leading to the definition of tr-ASTG as an interesting (and simple) candidate to optimise the performance of HO algorithms.

In [FeVe99] the impact of HO algorithm design on the topology of connections between BSs and radio network controllers, for a third generation mobile system the aim of which is that of providing seamless service for wireless users, even at high bit rates and for ATM-based services, was investigated. More precisely, an approach was proposed to establish the optimum BSs partitioning method starting from the actual knowledge of HO fluxes. This knowledge can be represented by the HO Matrix, the elements of which represent the amount of HOs performed between two cells during an observation time interval. The HO Matrix was evaluated using the same simulation tool used to provide the results of Figures 4.4.2–4.4.4, by assuming an ST algorithm for different values of T, and other HO algorithms. The most important result found was that the partition which is usually found to be the optimum one for a given HO algorithm still remains a 'good' partition (once a proper metric is defined to measure this 'goodness') if the HO algorithm changes.

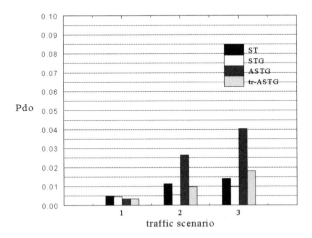

Figure 4.4.4: Dropout probability for the different scenarios (© 1999 IEEE).

4.4.1.2. Handover and power control with frequency hopping: an analytical approach

The same signal-level based PC algorithm described above was investigated in [CCVZ99], in the presence of Frequency Hopping (FH), through the analytical modelling approach discussed in the subsection on FH. Figure 4.4.5 shows the outage probability as a function of collision probability in the case of No PC (NO PC), Half Compensation (HC, $a = 0.5$) and Full Compensation (FC, $a = 1$) PC, with $L = 1$ and 2 branches of diversity and with or without sectorisation.

It is found that the half compensation strategy in many cases provides the best performance. Therefore, the conclusion is that both in the presence and in the absence of FH a half compensation PC scheme provides best performance in terms of outage probability, even if different conclusions could be drawn in terms of dropout probability. Finally, in [CMVZ98] a HO algorithm based on a BER threshold and a hysteresis value is analytically treated, taking FH into account.

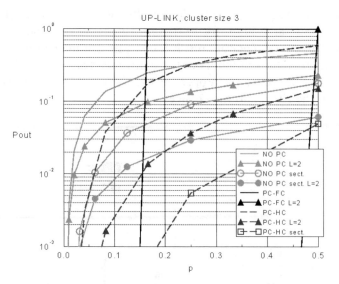

Figure 4.4.5: Outage probability against collision probability (© 1999 IEEE).

4.4.1.3. Statistical models for connectivity between mobiles and base stations

An analytical model that can be used to assess various performance measures for radio networks, such as mobile cellular systems and WLANs has been proposed in [OrPB99], [OrBa99]. It is based on the concept that communication is assumed to be possible if the power loss of the signal transmitted by the mobile terminal does not exceed some specified value by the time it reaches the BS (or vice versa). The propagation model is characterised by an inverse power law, with log-normal shadowing, whereas mobiles and BSs are both assumed to be uniformly and randomly distributed on the infinite plane. Several statistical models are derived. First, a probability distribution can be obtained for the distance between a generic mobile and the BSs, subject to a given maximum loss. Under the previous hypothesis, the power loss in decibel at (random) distance d from a mobile is:

$$L = k_0 + k_1 \ln d + s,\tag{4.4.1}$$

where k_0, k_1 are constants, and s is the log-normal shadowing with variance σ^2. If the mobile is assumed to be in communication with a BS (its loss is less than a given value l_1), the PDF of the distance separating them can be derived:

$$f_d(r) = 2Kr\Phi(a - b\ln r)\tag{4.4.2}$$

where K, a and b depend on the propagation parameters k_0, k_1, and σ, and are given by

$$K = e^{-(2/k_1)\left[l_1 - k_0 - \sigma^2/k_1\right]} \; ; \; a = \left[l_1 - k_0 - 2\sigma^2/k_1\right]/\sigma \; ; \; b = k_1/\sigma \qquad (4.4.3)$$

whereas $\Phi(x)$ is defined as:

$$\Phi(x) = \frac{1}{\sqrt{2\pi}} \int_{-\infty}^{x} \exp(-u^2/2)du. \qquad (4.4.4)$$

Moreover, it is shown that the number of BSs with loss not exceeding a suitable threshold is a random variable with a Poisson distribution, $Q_N(N)$, and mean

$$N_m = \pi \rho e^{\frac{2a}{b} + \frac{2}{b^2}} \qquad (4.4.5)$$

where ρ represents the BS density. This result gives the probability distribution of the number of BSs that can be in communication with a given mobile.

The previous considerations lead to the following distribution probability:

$$P(U_1 \leq u_1) = (1 - e^{-2u_1/k_1})^n \qquad (4.4.6)$$

where U_1 is the difference in loss between the stations with the (m)th and $(m+n)$th smallest losses, for $m = 1$. The equation above gives the probability that the signal received at the BS receiving the strongest signal from the mobile in question is less than $C+u_1$ when the $(n+1)$th strongest is controlled at strength C.

This result can be applied to CDMA-based cellular networks, where signal strength is controlled by the BS receiving the strongest signal from the given mobile, to obtain the probability distribution of the next, next but one, and so on, strongest of the signals received by other BSs, thereby allowing investigation of the overall capacity of the system. These results can also be used when considering protocols [PhOB99b], and in a situation where the signal strength of a mobile is controlled by a BS other than that receiving the strongest signal [KPOB99].

Further results concerning expectations of power ratios can be found in [OrBa00a]. The results lead to a measure of the reduction in the overall capacity of a system brought about by high power interferers affecting one BS, but controlled by a neighbouring one. The model proposed in [OrBa00a] extends the results of [OrBa99], assuming that the signal strength is controlled by the BS receiving the (m)th strongest signal; by defining U

the difference between $L_{(m+n)}$ and $L_{(m)}$, where $L_{(m+n)}$ is the $(m+n)$th smallest loss between the mobile and BSs, the expected value of U is:

$$E(U) = \frac{k_1}{2}\left(\frac{1}{m} + \frac{1}{m+1} + \dots + \frac{1}{m+n-1}\right).$$

(4.4.7)

Moreover, an expression for the expected value of the power ratio between $P_{(m+n)}$, the $(m+n)$th most powerful signal received, and $P_{(m)}$ can be given:

$$E\left(\frac{P_{(m+n)}}{P_{(m)}}\right) = \frac{(m+k_1(\ln 10)/20-1)!(m+n-1)!}{(m+k_1(\ln 10)/20+n-1)!(m-1)!}$$

(4.4.8)

provided that $k_1 \ln 10/20 > 1$.

A similar expression can be derived for the ratio between $P_{(m-n)}$ and $P_{(m)}$:

$$E\left(\frac{P_{(m-n)}}{P_{(m)}}\right) = \frac{m-n}{m-n-k_1(\ln 10)/20}E\left(\frac{P_{(m-n-1)}}{P_{(m)}}\right).$$

(4.4.9)

In this case, $k_1 \ln 10/20$ has to be less than m-n. This result gives the expectation of the power ratio for stations receiving a stronger signal than the (m)th. The expectation is infinite if the above-mentioned condition is broken; this arises as the model is assumed to apply down to zero distance: to face this condition it will be necessary to truncate the distribution at some small but non-zero distance.

The previous results also allow an analytical measure of the impact on capacity of introducing spatial density in a power controlled CDMA system [KeOB99].

4.4.1.4. Directed retry

4.4.1.4.1. Uniform environments

Directed retry is the procedure which allows the routing of a mobile-originated call attempt, during call set up, towards a BS different to the serving one. A call attempt is blocked if all channels are busy in the M closest cells which provide the user with a received power larger than a given threshold, where M is the minimum between the maximum number, R, of BSs considered during call set up and the number, N, of BSs that can be heard at a generic mobile position. This strategy is effective if M is larger than one throughout the service area; however, this is typical of all cellular networks providing a continuous coverage, due to the effects of signal fluctuations. Therefore, the effectiveness of directed retry has to be checked by taking slow signal fluctuations into account, namely, path loss that is distance dependent, and shadowing.

Let us consider a bi-dimensional cellular scenario where both mobile users and BSs are uniformly distributed over an (infinite) plane, as shown above; the cellular scenario illustrated in Figure 4.4.1 can also be considered, under the mentioned conditions. Let us assume that new call attempts (and new call attempts due to retry) follow two independent Poisson processes with rate per cell λ, and λ_r, respectively, and that the call holding time is exponentially distributed with mean $1/\mu$.

For the sake of simplicity, each base is also assumed to have the same number, n_c, of available channels, and mobility is neglected. The offered traffic at each base is given by

$$A = \frac{\lambda + \lambda_r}{\mu} \tag{4.4.10}$$

and the blocking probability, under some assumptions, can be evaluated as

$$P_b = B^R[n_c, A] + P_0 \sum_{N=0}^{R-1} \frac{N_m^N}{N!} \left(B^N[n_c, A] - B^R[n_c, A] \right) \tag{4.4.11}$$

where $B[., .]$ represents the Erlang B formula and $P_0 = e^{-Nm}$.

The rate of new call attempts due to retry can be estimated as

$$\lambda_r = \lambda \left[P_0 - 1 + \sum_{k=1}^{R} B^{k-1}[n_c, A]\overline{P}_k \right] \tag{4.4.12}$$

where

$$\overline{P}_k = 1 - \sum_{N=0}^{k-1} Q_N(N) \quad k > 0, \tag{4.4.13}$$

and allows consideration of the effects of directed retry on the offered traffic increase. Equation (4.4.11) is based on some assumptions concerning the independence of carried traffic in different cell, which is certainly not realistic; on the other hand, this assumption is typical of works investigating directed retry and its impact on the results has to be checked in a realistic environment. Therefore, the same simulation tool used to obtain the results of Figures 4.4.2–4.4.4 is utilised to compare simulation and analytical results obtained by means of the above described methodology.

In Figure 4.4.6 the blocking probability, for the scenario of Figure 4.4.1, is plotted as a function of R, with the offered traffic $As = \lambda/\mu$ as a parameter. The curves show a very good match between simulation and analytical results and quantify the advantage of using directed retry ($R > 1$); it is also clear from the figure that values of R larger than 3–4 do not provide further benefits.

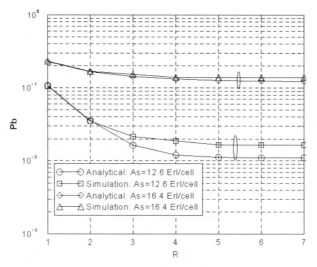

Figure 4.4.6: Blocking probability against R, both simulation and analytical
results are reported ($N_m = 5.66$) (© 2000 IEEE).

4.4.1.4.2. Non-uniform environments

In [ZOBV00] an extension of the previous model for a non-uniform scenario
has been provided, starting from the results of [OrBa00b].

If we assume that a mobile is located in the centre of a circular area H
(having radius r_H) characterised by a large local peak of traffic load (hot
spot), and we denote by n the number of BSs belonging to H that can be
heard by the mobile, and N is the number of BSs heard by the mobile in the
whole area, the conditional probability of n given N can be written as

$$Q_{n|N}(n \mid N) = \frac{N!}{n!(N-n)!} \left(\frac{\mu_1}{\mu_1 + \mu_2} \right)^n \left(\frac{\mu_2}{\mu_1 + \mu_2} \right)^{N-n} \tag{4.4.14}$$

where μ_1 and μ_2 are the mean values of n and $N-n$, respectively, given by

$$\mu_1 = \pi\rho \left[e^{\frac{2a}{b} + \frac{2}{b^2}} + r_H^2 \Phi(a - b \ln r_H) - e^{\frac{2a}{b} + \frac{2}{b^2}} \Phi(a - b \ln r_H + 2/b) \right] \tag{4.4.15}$$

$$\mu_2 = \pi\rho \left[-r_H^2 \Phi(a - b \ln r_H) + e^{\frac{2a}{b} + \frac{2}{b^2}} \Phi(a - b \ln r_H + 2/b) \right] \tag{4.4.16}$$

where a and b have been defined above. So, the non-uniform environment is
modelled assuming the superposition on a uniform scenario of a circular area
H, which can be characterised by a different offered load or radio capacity
(the number of channels per base).

By means of $Q_{n|N}$, the blocking probability of a mobile located in the centre of H can be derived, as a generalisation of the previously presented approach:

$$P_b = \sum_{N=0}^{\infty} Q_N(N) \sum_{n=0}^{N} P_{b|n,N} \cdot Q_{n|N}(n \mid N),$$ (4.4.17)

where $P_{b|n,N}$ can be evaluated following a methodology similar to (4.4.11) and (4.4.12). Unfortunately, in this case the derivation of the offered traffic is much more complex: in fact, it has to be evaluated as a function of the traffic generated by the re-try mechanism, that is, for each cell the probability of a re-try coming from all other cells, which could belong to H or not, should be evaluated, and this would require the consideration of many different combinations of events. For this reason, here we consider two cases:

i) the whole re-try traffic is generated by the cells belonging to H;
ii) the whole re-try traffic is generated by the cells not belonging to H.

This approach provides two different approximations of the blocking probability, namely P_{bH} and P_{bL}, based on the case i and ii, respectively. One of them is an upper bound to the blocking probability, and the other is a lower bound; however, which of the two cases gives the upper bound is not known *a priori*.

Figure 4.4.7 shows the blocking probability as a function of R for three different values of the radius of H.

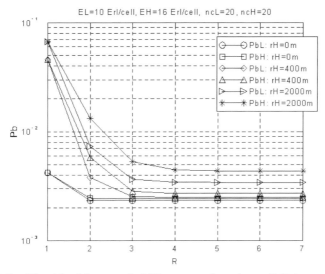

Figure 4.4.7: The blocking probability as a function of R in non-uniform environments.

We have fixed As = 10 Erl/cell outside H and As = 16 Erl/cell inside H, and 20 channels per cell. Let us note that r_H = 0 m represents the case of uniform distribution with offered traffic of 10 Erl/cell whereas the case r_H = 2 000 m is representative of a uniform users distribution with offered traffic of 16 Erl/cell. Figure 4.4.7 shows also that the presence of a hot spot increases the blocking probability; however, the retry mechanism allows the reduction of P_b as R moves from 1 to 4.

4.4.2. Frequency hopping for capacity enhancement

José Jiménez and Marco Chiani

COST 259 has addressed the problem of Slow Frequency Hopping (SFH) planning both from the purely simulation approach and from a semi-analytical perspective. The first approach is very useful and has been frequently used in literature [KHGK96], [MoWF95], [VJMW95]. However, it is difficult to identify the different factors in the solution and it is difficult to see what has to be done to improve the results.

Semi-analytical analysis is a powerful approach, and it allows extra analysis to be performed on aspects such as PC algorithms and HO control (see Section 4.4.1). Moreover, it has allowed us to analyse the throughput of pure Selective Repeat ARQ, and of Hybrid FEC/ARQ, in the presence of fading and SFH.

The semi-analytical approach is considered first. Next, the use of SFH for planning is discussed. In this case, a more conventional simulation is used. To overcome some of the computational burden, a procedure in two steps has been followed: first a link simulation allows the evaluation of the $(FER)_{min}$. Then that requirement is used for network optimisation.

4.4.2.1. Analytical solution of the SFH problem based on the evaluation of codes over block fading channels

The main difficulty in dealing with SFH is that it is absolutely crucial to take coding and modulation into account, otherwise SFH means no network improvement. Moreover, it is not trivial to model the interference, because a description based on the average interfering power is incorrect. To give an example, in GSM for voice transmission a codeword of about 380 bits is obtained by a 1/2 block code (implemented by a convolutional encoder with tail biting). These bits are interleaved and transmitted over eight timeslots, that (especially with SFH) undergo different impairments (fading level, CCI level). It results in the so-called block fading channel or channel with block interference [Chia98]. Now, the same total average interference power can be obtained with few timeslots with high interference or more timeslots with low interference. However, the two situations are very different for FEC. The relation between the error rate and system coding, modulation and collision probability for a channel with block interference has been shown in [CCAA97]. The derivation is provided for block codes; the same analysis has been carried out by simulation for convolutional codes in [ChCA99b]. It was shown then that SFH can be studied as usual cellular systems, but considering a protection ratio (minimum required CIR) as a function of the collision probability (probability of colliding with another interferer in a given hop).

Note that all the results were derived starting from the target FER, whereas conventional simulative approaches, discussed below, must deal with thresholds on the CIRs and on the frame error probability: these thresholds, that have a great impact on the final results, are quite subjective and not clearly related to the perceived QoS.

To clarify the benefits due to frequency hopping, let us consider a pure TDMA system without FH. Assume, for the sake of simplicity, the uplink with the desired user on the cell edge (cell radius R), one interfering mobile in the centre of the re-use cell (re-use distance D), path loss proportional to d^β, d being the TX-RX distance and β the propagation coefficient. If $(C/I)_{min}$ is the minimum required C/I (calculated assuming one interferer always active during the conversation with power I), the re-use distance D must satisfy $(D/R)^\beta > (C/I)_{min}$. Note that we are designing the system by assuming that an interferer is always active. From this point of view, we cannot take advantage of the interferer activity, so that neither DTX nor traffic considerations can be exploited to improve the system capacity.

On the contrary, assuming SFH, we have shown that, given the modulation and coding scheme, the required $(C/I)_{min}$ (defined on the collided hop) is a function of the collision probability: hence, a decrease of the collision probability due to, for example, to DTX or traffic control is translated into an increase in capacity, similar to what happens in CDMA. As an example, we report in Figure 4.4.8 on the protection ratio $\alpha = (C/I)_{min}$ as a function of p N_{IC}, p being the collision probability from each re-use cell and N_{IC} the number of re-use cells. For instance, let us refer to the lower curve (convolutional code, CSI) in Figure 4.4.8 and assume one full loaded interfering cell ($p = 1$, $N_{IC} = 1$). Then, the protection ratio is about 6.5 dB. If SFH and DTX (e.g. VAF = 50 %) are active, the collision probability goes down to 0.5 and the required protection ratio is decreased at about 4.5 dB. This allows for tighter frequency re-use and consequent greater network capacity.

In general, it is possible to increase the system capacity by properly controlling the system load. An example is reported in Table 4.4.1, obtained from Figure4.4.8, with reference to a Reed Solomon code RS(48, 24, 12) on

GF(256). For the sake of simplicity, six interfering cells are assumed, with interferers in the centre, and useful mobile at the cell edge. The propagation coefficient is 3. Similar results can be simply derived from Figure 4.4.8 for other codes.

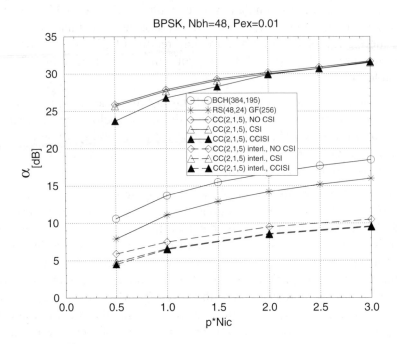

Figure 4.4.8: Protection ratio $\alpha = (C/I)_{min}$ as a function of the collision probability, for BPSK, FER = 10^{-2}, block Rayleigh fading with N_{bh} = 48 bits per hop, rate 1/2 codes, codeword length 384 bits. The codes are: BCH binary; Reed Solomon (RS) on GF(256); convolutional (CC) with 16 states, soft decisions, without channel state information (NO CSI), channel state information (CSI), channel and interference state information (CCISI). Convolutional code with or without intra-codeword interleaving (© 1999 IEEE).

Table 4.4.1: Spectral efficiency, channel spacing 200 kHz, FER < 10^{-2}, N_{TDMA} = 8 time-slots/frame.

	Cluster size K	η [Us./Cell/MHz]
No FH	7	5.7
FH, S_L = 100 %, VAF = 50 %	4	10.0
FH, S_L = 70 %, VAF = 50 %	3	9.3
FH, S_L = 15 %, VAF = 50 %	1	6.0

Moreover, by following this approach we partition the problem of finding the outage in a complex scenario in a two-step methodology:

- The first is the evaluation of the $(C/I)_{min}$ for given required maximum frame error rate, modulation-demodulation strategy, antenna diversity and coding. This can be achieved analytically in some cases (e.g. block codes with hard decision decoding [Chia98], [CCAA97]) or by simulation [ChCo99]). The characterisation must be carried out for the modulation/demodulation system under consideration, but does not depend on the cellular network parameters, such as the re-use, the shadowing characteristic, the propagation loss, etc.
- The second step is the outage evaluation, including shadowing, PC, sectorisation, etc.: it requires a light simulation for the downlink and is completely analytical for the uplink [ChCA99a], [ChCA99b]. Note that the time-consuming simulation at hop level is never required.

An example of results is shown in Figure 4.4.9, where the outage probability is evaluated against the collision probability, p, for the downlink, MSK modulation, Rayleigh block fading, and other conditions.

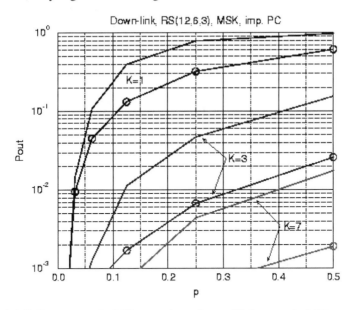

Figure 4.4.9: Outage probability against the collision probability, p, for the downlink, MSK modulation, Rayleigh block fading, RS(12,6,3) code on GF(256), $nsh = 2 = 16$ bits/hop, $120°$ sectorisation, shadowing with $\sigma = 6$ dB, imperfect PC with $\sigma_e = 1$ dB, scenario with a total of 37 cells, propagation coefficient $\beta = 4$. With circles the curves for two branch antenna and Maximal Ratio Combining (© 1999 IEEE).

4.4.2.2. Simulation approach based on thresholds

The second approach, based on thresholds, has been used in [Jime97], [JiGB98]. Its use is suggested in [VeMS84], [DoVe87], [Stee72]. As before, a two-step methodology is applicable:

- Evaluation of a (FER)$_{min}$ requirement (denoted by q).
- Outage evaluation via a simplified simulation.

The (FER)$_{min}$ requirement is obtained using a hop by hop simulation of the link. When the C/I for a given hop is larger than a threshold γ, the hop is deemed correct and the FER obtained. The value of γ is normally 5 dB.

One advantage of the simulation approach is the possibility of considering different statistical behaviour for C and I. Figure 4.4.10, gives the results for a Rayleigh/Rayleigh situation, though other results are also available [Jime97]. Results of Figure 4.4.10 are given as a function of the collision probability.

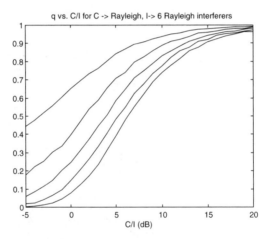

Figure 4.4.10:   FER simulation results for a Rayleigh distributed C, Rayleigh distributed I. The FER normally required in the GSM system is 0.7 (see text).

The collision probability is proportional to the VAF, to the cell traffic (a_k) and to the number of frequencies used in the cell (N_a) as a proportion to the total (N_h).

$$p_k = \frac{N_a(VAF)}{N_h}.a_k \qquad (4.4.18)$$

Once an approximate expression for (FER)$_{min}$ = q as a function of the mean C/I, number and characteristic of interferers is known, it is possible to evaluate q at every location. It can be demonstrated that in the case of a Rayleigh situation q can be exactly evaluated as

$$q = \prod_{i=1}^{M}\left(1 - \frac{p_i \gamma}{C/I_i + \gamma}\right)$$ (4.4.19)

γ being the threshold for correct detection, C/I the CIR from every interferer i at a given location. M the number of interfering cells (in general, the total number of cells) and p_i the probability of a mobile in cell i being active and interfering.

Figure 4.4.11 is an example of a q *map* for an example area, obtained assuming Rayleigh signal and Rayleigh interferers [JiGo98]: in grey the areas under a probability of frame detection 0.7 have been outlined. As can be seen in the comparisons, the sensitivity to the value of p (interferer activity) is very significant.

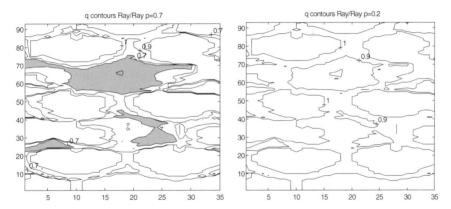

Figure 4.4.11: q contours for a 3/1 plan. Rayleigh statistics are assumed both for the desired and interferer. Different interference probabilities are assumed.

4.4.2.3. Application to planning

The simulative approach has been applied to network planning optimisation. The procedure is directly evaluating a q value at every physical location (q map). The optimisation objective would be the *minimisation of the number of locations* having a q value larger than the detection threshold (normally around 0.7).

That process is normally time consuming, since the number of locations where q has to be evaluated is very large. However, as pointed out in [JiGB98] the behaviour of the optimisation function is much softer than in the conventional non-SFH case, and the difference between local minima is less marked. This implies a simpler and shorter optimisation process.

A different possibility would use the information derived from the *compatibility matrix*. In that case, as pointed out in [JiGo98], the results will not be so good, since the information on the amount of interference C/I is lost and the general procedure of the previous subsection cannot be used. In any case, a very straightforward solution would be evaluating the Kronecker matrix product:

$$\mathbf{R} = \mathbf{CM} \bullet \mathbf{P},\qquad\qquad\qquad\qquad\qquad (4.4.20)$$

CM being the *Compatibility Matrix* and **P** a matrix formed by the probability of interference of cell i over cell j. The objective is to minimise **R** or the sum of its elements.

Unfortunately, and in contrast to the non-SFH case, the results obtained are not too good. Another possibility would be to use a more informative compatibility matrix instead of one just giving 0 when the frequency cannot be re-used or 1 when it can. For instance, a new CM_1 would have a (i, j) element proportional to the interference from cell i to cell j.

Other alternatives try to make direct use of the expression for q (obtained via simulation or directly in the case of Rayleigh interferer) at a location that is considered 'representative' (instead of performing the optimisation over the whole area), which normally gives the best solution.

4.4.2.4. Applications of SFH modelling to system improvement

The analytical methodology permits us to appreciate the role played by system parameters, and also by PC and HO where used jointly with SFH (see Section 4.4.1). An interesting result is that, as observed for pure TDMA system, half compensation PC outperforms full compensation PC even in the presence of SFH. Results on the impact of SFH on the HO algorithms are presented in Section 4.4.1.

The services that can be offered by mobile radio systems are not only concerned with voice, but also with data transmission. Data transmission has inherently different requirements with respect to voice communication. In fact, data transmission is usually of a non-real-time nature and hence can tolerate delays in transmission. Moreover, it should be highly reliable, imposing a low BER (in the order of 10^{-6}). To achieve this quality in a wireless channel environment, ARQ strategies could be applied. For this reason, it is interesting to know if the transmission techniques adopted for voice are also suitable for data transmission. In particular, we have analysed the effect of SFH on ARQ, taking into account thermal noise, fading, non-ideal interleaving and FEC. We first study the errors distribution due to a block fading channel. A typical behaviour is reported in Figure 4.4.12, where the probability $p_X(m)$ of m symbol errors in a packet of $n = 64$ is shown [Chia97], [Chia00].

In the figure the symbols are composed of 8 bit, to allow the analysis of non-binary codes; the number of symbols per independent hop is nsh, so that $nsh = 1$ means perfect symbol interleaving, whereas $nsh = 32$ means two hops per packet. Here the probability of correct reception without FEC is $p_X(0)$. When FEC is applied, with error correction capability up to t errors, the probability of successful reception is given by $p_X(0)+p_X(1)+...+p_X(t)$. Note that $p_X(0)$ increases as the memory of the channel, nsh, increases. The error distribution $p_X(m)$ can be used as shown in [Chia97] to derive the throughput for two important ARQ strategies for data transmission, Selective Repeat (SR) and Hybrid Type I ARQ/FEC. Numerical results show that, for SR, the performance deteriorates when using interleaving and/or SFH.

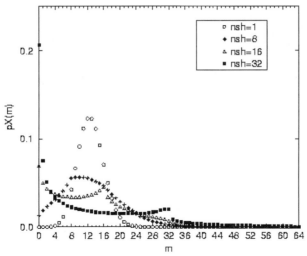

Figure 4.4.12:   Probability $p_X(m)$ of m symbol errors in a packet of $n = 64$
(© 2000 IEEE).

The throughput of hybrid ARQ/FEC is deteriorated by SFH at low signal-to-noise ratio values: the ARQ/FEC technique analysed here gives marginal improvement with respect to pure SR only when perfect interleaving plus SFH are applied.

For the system parameters used here, we have found that the best performance is obtained with pure Selective Repeat ARQ without SFH: in fact, with SFH and/or interleaving the errors are time-dispersed. Since a packet must be retransmitted even if it contains only one error, the best situation is where the errors appear in bursts, corresponding to a non-interleaved and non-SFH case. In this regard, we report in Figure 4.4.13 on a comparison between SR and ARQ/FEC-I with RS(12, 6), RS(12, 8), RS(12, 10) giving code-rates $Rc = 1/2, 2/3, 5/6$, respectively, all defined on GF(256).

In conclusion, real-time and data services should use different strategies for the error control: the former requires memoryless channels (by interleaving and/or FH) and FEC, whereas for the latter it is convenient to leave the channel memory as is (neither interleaving nor FH within a packet), and ARQ protocols should be used.

Two different approaches to the SFH problem have been used in COST 259. In spite of the methodology differences, the conclusions are very similar. SFH is a very useful device for interference reduction in TDMA (GSM) systems, particularly for voice applications. Consideration of the *collision probability* is essential in the process. This requires the knowledge of the cell traffic. The studies have also shown that SFH is more useful when signal statistics are Rayleigh and there are few interferers. That is the reason why tight frequency planning using SFH and fractional loading to achieve re-use of 3 or even 1, need a rather homogeneous layout of cells to decrease the number of potential interferers. Conventional optimisation procedures (e.g. using *simulated annealing*) have been used in SFH. Contrary to the non-SFH case, the differences between different frequency assignments are less marked and, therefore, the optimisation requirements are relaxed. A powerful semi-analytical procedure has been developed. That analysis allows a better understanding of the SFH improvement, and allows for some studies.

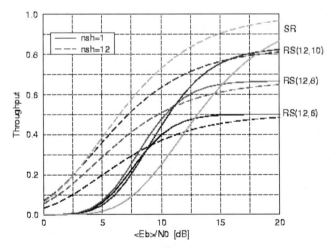

Figure 4.4.13: Throughput vs. the average signal-to-noise ratio for SR-ARQ and FEC/ARQ (© 2000 IEEE).

4.4.3. Quality and capacity enhancement by adaptive techniques

Jean-Frédéric Wagen

Adaptive techniques, such as adaptive antennas [FuKB97], [BaSa97], [StGJ99], [PaHW99] adaptive modulation and coding [PeBT98], and adaptive transmission rate [KhZe97b], [SaAg98], [PeBT98], [RCGO99], have been investigated to improve the quality and capacity of mobile communication networks.

4.4.3.1. Performance measures

Jean-Frédéric Wagen

To compare and quantify the efficiency of various techniques, widely accepted performance measures must be used.

The spectrum efficiency [PaHW99] η, measured in [bit/(s Hz)], is defined as the ratio between the total available information rate R_{tot} [bit/s] in a single cell and the total system bandwidth B_{sys} [Hz]. The spectrum efficiency is used to compare different systems in term of raw bit rate.

To include the trunking efficiency in the performance measure and to allow comparisons in term of traffic instead of bit rate, a useful measure is the spectrum capacity [PaHW99] κ, measured in [Erl/(MHz km^2)]. κ is defined as the total offered traffic T_{tot} [Erl] divided by the product of the system area A_{sys} [km^2] and the system bandwidth B_{sys} [MHz].

System capacity is sometimes simply defined as the average number of users N_{user}. This definition, however, requires a clear definition of a 'user' and of the 'acceptable (perceived) quality' for a fair comparison.

4.4.3.2. Adaptive antennas or smart antennas

Jean-Frédéric Wagen

Smart antennas or adaptive antennas at the BSs have the potential to increase the capacity of mobile radio networks. This capacity increase is achieved since more users could be served by a BS antenna array while keeping the link quality at an acceptable level. An antenna array can maintain the required CIR for each user using either or both of the two main advantages of smart antennas, namely antenna gain and interference rejection for the uplink, and antenna gain and directivity for the downlink. The signal processing and the antenna designs for these antenna arrays have been described in Section 3.3.

Smart or intelligent antenna, switched beams antenna, adaptive array, phased array or simply array antenna, are some of the many ways used to describe antenna systems used to enhance the performance of cellular systems by processing the signals of each individual antenna element of the array using some special techniques.

SDMA (Space Division Multiple Access) describes the generic idea of using the properties of an array antenna to separate users based on their physical location, usually by forming for each user a beam with the array antenna. Ideally, a SDMA system allows us to have several users on the same carrier frequency (if FDMA is used), same time slot (if TDMA is used), and using the same code (if CDMA is used). In SDMA, the DoA (Direction of Arrival) of the user signal plays an important role to steer and shape the beam dedicated to a particular user. Beam switching and beam forming are two self explanatory terms also used in the context of SDMA system.

SFIR (Spatial Filtering for Interference Reduction) systems, unlike SDMA systems, do not support several users per traffic channel, but improve the CIR of each user. This improvement can be traded off for capacity. However, a single SFIR antenna will not lead to any significant capacity gain. Thus, SFIR antennas must be widely deployed to achieve a gain in capacity.

Using antenna arrays in a FDMA/TDMA cellular system has been reported to lead to large capacity increase, e.g. [FuKB97]. Capacity can be calculated in many ways, and the reference scenarios are not always comparable. Thus, care should be taken when comparing the results in Table 4.4.2, concerning the capacity increase.

Table 4.4.2: Capacity increase for various antenna types according to results found in the literature.

Antenna type	Capacity increase	Reference
8 beams adaptive array	3 (vs. omni)	[SBEM90]
M element SFIR	$\sqrt{M}$	[Tang94]
M elements SDMA	up to $1.7\sqrt{M}$, for $M = 8$	[Tang94]
12-beam switched beam antenna	2	[Lope96]
$M = 8$ or 12 for best size-performance trade-off	$0.4\sqrt{M}$ to $2\sqrt{M}$ Performance was found to be sensitive to M, the side lobe level or null depth (ND), the front-to-back ratio (FBR) and the CIR threshold.	[FuKB97]

SFIR systems are quite robust against pointing errors and PC errors [FuKB97]. The performance is not significantly affected as long as the standard deviations of the pointing errors is less than half the main beam width (e.g. 8°) and of the PC errors is less than 6 dB.

SDMA are more sensitive. Pointing errors of the main beam of the antenna array and/or PC errors lead to capacity impairments, for example:

- 20 % less users can be served if the SDMA pointing errors are of the order of half the main beam. A 5 % reduction is only achieved when pointing accuracy of 2° is achieved.
- 17 % less users can be served if PC errors have a standard deviation of 6 dB. Accuracy of 3 dB or less must be achieved to reduce the capacity impairment to less than 5 %.

Using antennas arrays is also a promising technique to increase capacity in CDMA systems [Goda97]. Sample results have been computed and investigated in [PaHW99] for a TD-CDMA system considered in UMTS. Compared to omni-directional antennas, adaptive antennas were found to enable an increase of spectrum efficiency from 0.06 to 0.77 [Erl/(MHz km^2)] in rural areas and from 0.71 to 4.83 [Erl/(MHz km^2)] in dense urban and suburban areas. Antenna arrays are relatively more beneficial in rural environments because the trunking efficiency has diminishing return as the number of traffic channels increases. Performance on the downlink was not considered in this contribution [Goda97].

Antenna array are also beneficial to handle the time-varying or non-uniform distribution of the traffic [BaSa97]. Simulation results presenting blocking rate vs. traffic shows that a switched-beam antenna system leads to double the capacity of a conventional 3-sector system.

4.4.3.3. Link adaptation schemes

Jean-Frédéric Wagen

Adaptive modulation and coding in TDMA cellular communications have been reported to lead to potentially large increases in overall system capacity [PeBT98]. For example, a 42 % increase in capacity can be obtained when 75 % of the users could switch to ½ rate channels under good radio link conditions [KhZe97b]. Another example is given by the GSM EDGE scheme. EDGE has the potential to allow an increase of the average data throughput for a given user by as much 300 or 400 % [FGPR99]. However, it has been shown that in theory, these adaptive data rate schemes are better used without PC in a TDMA system [PeBT98].

4.4.3.4. Proactive resource allocation

Jan Steuer and Nabil Elouardi

The objective of proactive network management is to find a means to predict fault situations and enhance network performance parameters automatically and in advance [ESRF00]. Proactive in this context means that available information (e.g. trends of degradations of QoS or network performance parameters) is evaluated with the goal of achieving a forecast into the future. This forecast concerns the adaptation of parameters and thresholds of the network in order to improve the quality of the network. The proactive behaviour is in contrast to classical reactive network management methods.

The work presented was focused on proactive resource allocation in radio networks with the objective to improve the network performance by decreasing the blocking rate. The idea is to optimise the resource allocation by exploiting knowledge available in the network. Each BS shows, in general, a fairly regular traffic profile. One approach was to extract cycles in this regular traffic pattern by learning. By this, irregularities could also be detected and the network could be trained to react on them. Since BSs are not equally occupied in space and time, free capacities could be used at other places.

The research work presented was mainly based on frequency borrowing methods with varying control mechanisms [CoRe73]. The basic proposed network management architecture is split into two levels: cell agents that determine the resource requirements and a central cluster manager. The agents send their resource requests to the cluster manager that is responsible for distributing the resources. Of course, some aspects have to be considered, such as the number of TXs in an antenna sector, BCCH-carrier frequencies, CIR distances as well as criteria for borrowing and returning resources, especially in overload situations.

Initial results were presented which are based on a fully decentralised mechanism, i.e. autonomous agents in each BS borrow and return resources by exchanging them with their nearest neighbours, based on their local knowledge.

Two basic decentralised algorithms have been investigated. The first one is a threshold-based scheme using cost functions without (1) and with prediction (2), respectively.

$$Costs_{borrow}(t) = \sum_i^C w_{new}{}^i B_{new}{}^i{}_{A_i(t), N_{newi}(t)}(t) - \sum_i^C w^i B^i_{A_i(t), N_i(t)}(t) \qquad (4.4.21)$$

$Costs_{borrow}(t+\Delta t)=$

$$\sum_i^C w_{new}{}^i B_{new}{}^i{}_{A_i(t+\Delta t),N_{newi}(t+\Delta t)}(t+\Delta t) - \sum_i^C w^i B^i_{A_i(t+\Delta t),N_i(t+\Delta t)}(t+\Delta t)$$

$$(4.4.22)$$

If a potential acceptor cell detects a demand for new resources, the cost function is used to calculate the mean traffic loss B^i for the donator and acceptor cell for the actual situation (second sum), and for the situation (first sum with a new loss B_{new}) when allocating new resources (N_{new}). Thus, it is determined whether or not the resource shall be moved. A second method is based on a control engineering algorithm which controls the number of borrowed channels so that the mean loss B for a mean demand A or for a predicted demand $A(t+\Delta T)$ is approximated to an objective value, based on the estimated offered traffic. The control algorithm is able to allocate more than one channel/frequency in one step. The equation that is used for the control element is an approximation of the Erlang distribution. One of the criteria was stability of the resource allocation methods used. Borrowing a frequency in one cell must not lead to significant periodic processes in the radio network.

All methods investigated perform better than the fixed channel allocation. The threshold-based algorithms seem to perform best. Work which focuses on centralised solutions including the dynamic adaptation of the thresholds for HO, PC and link adaptation, as well as antenna beam forming to optimise the global quality of the network, is continuing. For a centralised solution, an approach using optimisation methods proposed for frequency planning seems to be promising for the allocation of resources to a radio cell.

4.4.3.5. Conclusions

Jean-Frédéric Wagen

The introduction of adaptive or smart antenna arrays at the BSs allows us to increase the capacity of a cellular system even by 300 % [FGPR99]. Thus, the adoption of adaptive antennas, although expensive, may constitute a viable alternative to the deployment of new BSs to solve congestion problems. Furthermore, it is worth mentioning that remarkable Shannon capacity gains have recently been pointed out by a number of researchers when using multi-element antennas at both the RX and TX of a wireless system. The capacity gain is theoretically proportional to the number of transmit/receive antennas. A limit is reached, however, in practical situations due to the limited number of significant scatterers [Burr00].

The adoption of link adaptation techniques, such as Adaptive Multi-Rate speech coding in GSM and UMTS, EDGE for GSM or multi-rate transmission in GPRS and future UMTS or IMT2000 systems, allows us to increase the average data throughput for a given user by as much 300 or 400 % [FGPR99]. Combined, these adaptive techniques could triple the number of subscribers, and even increase the global throughput by an order of magnitude [FGPR99].

Other combinations of adaptive techniques with resource allocation techniques such as frequency hopping, power control, advanced planning, dynamic channel allocation, etc. remain for further investigations, but seem to be promising according to recent results [ESRF00]. One difficulty is that the channel and system capacity of ever more complex systems and technologies must be evaluated with ever more detailed and accurate propagation models, so as to not lead to erroneous conclusions.

4.5. Planning Methods and Tools

Thomas Kürner

An important task of COST 259 has been to contribute to the development of advanced systematic planning methods, which can be applied in radio network planning tools. A prerequisite for successful use of a radio planning tool is the availability of good geographical data. The classification criteria and requirements for these data are described in Section 4.5.1, together with the definition of proper data format and sophisticated processing techniques. Section 4.5.2 gives an overview on some state-of-the-art techniques for cellular planning, including some general aspects in terms of automatic planning algorithms. These techniques are applicable for both 2G and 3G networks. Some initial UMTS planning aspects are discussed in Section 4.5.3.

4.5.1. Geographic data

Thomas Kürner

For computer based radio network planning, digital terrain data of high quality is required. Depending on the propagation scenario, different data types and formats are used. The data can be organised roughly into three groups distinguishing low resolution raster data, street vector information and high resolution data including building information. Typical values describing the accuracy of such data are given for data sets available for Germany.

The input data for the traditional macro cellular propagation models consists of terrain height and land use information. Both data sets are typically applied with resolutions of 50×50 m^2 up to 200×200 m^2.

Digital Terrain Model (DTM):   the digital terrain model describes the ground height of the terrain without the land use. The accuracy of the DTM available for Germany is 3–20 m in height in 20–30 m in location accuracy [Krän95], [MEGR95].

Land use data: land use data describes the morphology of the surface. In [PaKr96] a classification based on Landsat thematic maps distinguishes 15 different classes and was realised with an original resolution of 25×25 m^2 for the whole of Germany.   Four urban classes (dense urban, urban, suburban and industry) are included. The definition of the 15 classes was made according to the requirements of radio network planning, requesting classification errors of less than 10 %. Extensive controls of the data used in this investigation revealed area-wide classification errors in the order of 5–10 %. If land use data is applied to resolutions other than the original one, appropriate aggregation and filtering of this data has to ensure consistency of land use data for all resolutions.

Typically, street vector information can be used as a background information layer when coverage maps are produced.   Street vector information is available for the whole of Germany. However, this data can be also used to improve the accuracy of predictions in urban areas [KüFa97]. The street vector data available for Germany distinguishes eight classes in economic areas (including urban areas) and five classes in country areas. The position accuracy of the street vector data is less than 10 m in economic areas and about 30 m in country areas. A maximum of six different vector classes are used for the predictions (motorway, two types of main roads, regional roads, side streets, other roads).

High resolution data consists of information about:

- (complex) building structures, including roof shapes
- relative and absolute building heights
- vegetation.

The corresponding propagation prediction models using this kind of data are able to process either raster and vector formats [Cich94], [Gsch95]. The preferred format depends on the processing algorithm. High resolution data is available for selected cities only. In [Kürn99b], data for the city of Cologne is used, which is available in both data formats. The accuracy for the vector data and the raster data (resolution 1×1 m^2) is 1–2 m both in height and position. The raster data with a resolution of 5×5 m^2 is derived from these data sets.

In [Basi99], a new database has been defined in order to improve the precision of urban predictions, while using statistical database with no specific information about building heights and positions. Therefore, a link between the clutter (or land use) codes and the ideal setting of the clutter parameters is created. The new clutter database has been produced starting from high resolution clutter and building height databases with 1 m resolution. The map data is re-sampled to areas of 25×25 m^2 square size by the calculation of average building heights and average building densities. The clutter code assigned to each pixel is defined as the combination of the two clutter classes, expressed as building height and density classes.

A couple of data formats for high resolution building data, as well as for the storage of measurement data from such environments, have been proposed. In [DHLZ98], the proposed data format is based on the data drawing exchange format (DXF). It is shown how this data format can be used for rural, urban and indoor environments. The TWP-standard [NiML98] of the National Radio Propagation Committee of the United Kingdom is based on the DXF format as well. The standard defines a detailed format for 3D-building data and provides guidelines on the collection. Furthermore, a number of data collection anomalies that are often monitored cause problems for ray-tracing or rasterisation.

In [PoSt98] a universally applicable file format, called CSM (Channel Sounder Measurement format), is proposed. It is the result of the CEC-SMT project METAMORP (Measurements, Testing and Calibration of Advanced Mobile Radio-Channel Test Equipment),. The key features of CSM are:

- Different data types: CSM is suitable for different types of data, like channel measurements, evaluation results, simulation data, and scenario descriptions (measurement devices, simulation set up, etc.).
- Easy data handling: a data catalogue provides an overview of several characteristic parameters of the stored data. Since it is a structured text file, there is no need for specific software to access this overview information. CSM is accompanied by a software library called METABASE, written in ANSI-C, for easy access to the CSM data files.
- Flexibility: CSM files are built from records, which structure is defined in a specific 'set-up file'. Additional data entities can be defined without changing the structure of CSM files by just adding their description to the set-up file. Thus, there is no need for new versions.
- Storage efficiency: CSM is a relational database with index files, data compression, and binary records. The relational database concepts ensures the re-usability of information, like antenna characteristics or environment parameters (geographic information, building materials). Index files allow fast access to any record within the data files. Binary records and (optional) data compression reduce the storage capacity needed by approximately 50 % compared to pure ASCII data files.

Deterministic 3D-modelling of wave propagation in urban areas has become a widely used technique for the planning of micro-cells and small macro-cells. However, the practical use of these deterministic models has been limited due to their computational demands. Therefore, a couple of intelligent processing techniques yielding a reduction of computing time have been discussed within COST 259.

- Pre-processing of raster data: the availability of detailed building information enables us to extract site-specific input parameters for some Walfisch-type models, i.e. an individual set of the parameters street width, street orientation, building height and building separation for each profile between BS and MS. Whereas the mean building height and mean building separation can be extracted directly from the terrain profile, the calculation of street width and street orientation requires the analysis of a two-dimensional area around the MS, which is a time consuming process. As these two parameters are independent from the BS position, it is appropriate to generate two additional raster layers during pre-processing containing these parameters [KüFa97].

- Pre-processing of vector data: in [HoWL99], a single intelligent pre-processing of the vector building database is performed. In this pre-processing the mutual visibility conditions between the walls and edges of the buildings are determined and stored. These visibility conditions are independent of the positions of BS and MS. The walls of the buildings are divided into tiles and the edges into vertical and horizontal segments. After this discretisation of the database, the visibility relations of the different elements are determined and stored in a file. Therefore, all elements are represented by their centres. In [Kürn99b], an object oriented vector data format is applied to store pre-processed information like bounding boxes, bounding spheres and normal vectors for each of the buildings. If this pre-processed data is used in conjunction with methods typically applied in computer graphics [Fole93], a reasonable acceleration of 3D predictions is achieved.

- Fast visibility algorithms in 3D vector data: an approach described in [MDDW00] is based on a successive reduction of the problem. In a first step, the whole 3D scenario is projected perspectively in several 2D planes. The resulting 2D polygons are tested for overlap using a sweep-line algorithm. With knowledge of the relative surface positions, a graph theoretical approach (polygon subtraction) is applied to determine the visible surfaces. The resulting visible surfaces are subject to the imaging method at highly reduced complexity.

4.5.2. Methods for optimised planning

Paolo Grazioso

Some state-of-the-art techniques for cellular planning in mobile radio systems and their consideration in planning tools are discussed in this section. An ample, though not exhaustive, overview of such techniques is given in [FRGF97]. The authors discuss several techniques that allow us to perform optimised resource planning in cellular systems. Even though such techniques may become common in third generation systems, they could be applied to current systems as well, without requiring substantial modifications to the current network entities and functionality, nor to the radio interface.

The first techniques that are analysed in [FRGF97] are those that could be applied in current systems; actually, some of them are already in use in congested areas.

- Frequency Hopping (FH) is a well known technique, already widely in use in areas with high traffic density. As each call uses a different hopping sequence, the interference pattern changes once per hop: therefore, interference is averaged and the system performance improves. This in turn leads to a remarkable capacity gain or, conversely, a decrease in call dropping rate and an improvement in estimated quality.
- Discontinuous Transmission (DTX) consists of transmitting a carrier only when there are actually some bits to be TX, i.e. the TX is silent when a user is not talking, thus leading to a reduction of the overall interference. Typical activity factors for human conversations are around 0.4, and even lower values are expected for data communications. Apart from the obvious advantages in terms of power consumption and battery life, DTX allows us to improve system quality and/or capacity. Its benefits are augmented when it is used jointly with FH: in this case, the reduced interference is also averaged among users.
- Adaptive antenna arrays: the use of adaptive antennas can further reduce interference levels, thus allowing tighter re-use, due to their capability of removing the strongest interferers. In this way, a further improvement in capacity and/or a tighter re-use can be obtained through.
- Fractional loading is a technique consisting of assigning more bandwidth than is strictly needed to cope with its traffic load to every cell, using a low cluster size. If the system were fully loaded, this would result in unacceptable quality, so channel occupancy must be kept at the desired level by means of an appropriate Call Admission Control (CAC). The authors show that fractional loading with CAC may allow us to improve frequency re-use, and in certain cases a unitary re-use can be achieved.

In [FRGF97], also frequency planning techniques in layered cellular architectures are discussed. In such architectures, micro-cells are deployed in hot spots, while universal coverage is provided by macro-cells. Planning in this kind of environment is not an easy task; one possible solution is to split the total available bandwidth into two sub-bands, for macro- and micro-cell, respectively, which would allow independent planning of the two layers. An optimum splitting can be found as a function of the expected traffic densities in hot spots and in other areas. As far as micro-cellular planning is concerned, the authors propose a self-learning procedure, based on the Channel Segregation algorithm, to perform frequency planning in a micro-cellular environment without manual intervention of the operator (apart from BS placement). The algorithm leads to satisfactory results in a realistic case, thus proving its viability as a planning procedure for micro-cells. Similar concepts were also discussed and used within the STORMS project, in the framework of ACTS activities [MePi99]. These results were also published within COST 259 [Fall98], [Meno98].

[Fall98] presents the automatic planning tool developed within STORMS, which determines the optimum subset of BS sites and configurations (antenna heights and radiation patterns), choosing them from among a larger set of candidate BSs. A brief description of the optimisation algorithm follows. The initial operating assumptions were the following:

- discretised geographical area: the area under study can be considered as a finite set of locations;
- uniform traffic: the offered traffic density is uniform throughout the service area;
- constant cost for the BSs: therefore, the total cost of a network is proportional to the number of required BSs.

The goal of the optimisation procedure is to determine a subset of the user-provided candidate BS locations and configurations, ensuring the required coverage and capacity necessary. The initial set can be either determined by an expert operator or by an automatic engine. More precisely, two modes of optimisation are supported [Meno98]:

- constrained: the operator states the total available bandwidth, and the tool looks for the optimum plan respecting all the constraints;
- unconstrained: the operator fixes only the bandwidth upper bound, and the tool looks for the plan respecting all the constraints while minimising the number of required carriers.

The optimisation procedure starts by building a graph to model the relationship between a BS and its cell. This graph, denoted by $G(V,E)$, is defined as follows: $V = A \cup B$, where A is the set of all possible BSs and B is the set of all potentially covered locations, while E is the set of edges: an edge $(u,v) \in E$ exist if $u \in A$, $v \in B$, and the BS u covers the location v. The problem to be solved corresponds to the well known problem of finding a minimum dominating set of a graph. This problem is NP-complete, i.e. it requires a non-polynomial computation time, and therefore the exact solution cannot be found in a reasonable time for the typical dimensions of a cellular planning. For this reason, the optimisation tool looks for satisfactory, though possibly sub-optimal, solutions by means of a genetic algorithm. In order to avoid too quick convergence, random *mutations* were introduced into the algorithm. Preliminary results were encouraging, allowing a reduction in the number of required BSs and a minimisation of locations served by more than one BS. However, the algorithm in its initial form was still too slow for practical use by an operator. For this reason, the authors introduced the *island concept*, which consists of splitting the total population into sub-populations called *islands*, so that the genetic algorithm can be run in parallel in every island. The authors also introduced the possibility of *migrations* of users between islands, in order to let islands benefit from information found by others.

In order to further speed up the computation, the authors introduced the concept of *inter-cell graphs*, which consists of grouping together all the locations which are covered by exactly the same BSs, thus reducing the graph cardinality. Each of these sub-sets is called an inter-cell. To further satisfy the operators' needs, the authors also introduced the possibility of considering technical constraints into the tool (e.g. site availability, accessibility, etc.), as well as capacity constraints, in order to guarantee that the necessary traffic be served. The results were encouraging, and the modularity of the tool also allowed us to introduce advanced features such as adaptive antennas and super-conducting technology, in order to assess their benefits in the framework of a realistic network planning [Meno98].

The assessment was conducted by covering the same geographical reference area with standard and adaptive antennas; the parameters that were monitored were the number of BTSs necessary to cover the same area, the overall interference figure after the frequency planning, and the achieved frequency re-use. The benefits achievable by adaptive antennas can be distinguished as follows:

- High Sensitivity RX (HSR): requires switched beam or beam-forming in the uplink only; it allows coverage extension, and can be used in rural or suburban low traffic areas mainly to extend coverage;
- Spatial Filtering for Interference Reduction (SFIR): requires switched beam or beam-forming in both the uplink and the downlink; it allows a tighter frequency re-use, hence a capacity increase is achieved; it can be used in urban or suburban high traffic areas, where an extensive BTS deployment is foreseen;
- Space Division multiple Access (SDMA): requires beam-forming only, in both the uplink and the downlink; it allows frequency re-use within the same cell, thus leading to a capacity increase; its use is mainly for urban high traffic zones, with isolated deployment covering hot spots.

The capabilities of all these techniques were assessed with the planning optimisation tool, which showed their advantages.

The same technical approach was adopted to assess the benefits of BTS super-conductivity technology. The main advantages of high temperature super-conducting are:

- sensitivity: the decreased surface resistance directly translates into reduced insertion loss of band pass filters and higher conductor quality factors;
- selectivity: for a given maximum band pass insertion loss much steeper filter skirts can be achieved for reducing interference;
- size reduction: a number of planar HTS filters can be integrated with one cooler leading to a size reduction with respect to conventional wave guide technology.

Simulation again showed benefits both in terms of coverage and in interference (which translates in an increased capacity), encouraging us to explore this technique further.

4.5.3. UMTS planning

Thomas Kürner and Thomas Neubauer

4.5.3.1. Problem definition

Radio planning in UMTS will be significantly different from radio planning in GSM networks. The key characteristics for UMTS planning are:

- CDMA-specific issues like cell-breathing, pole capacity and soft-HO mode;
- multiple services/multiple data rates each described by a different inhomogeneous traffic distribution;
- frequency blocks of 5 MHz.

As a consequence of these key characteristics, cell size does not depend solely on propagation conditions. Due to cell breathing, the coverage at the cell edge varies with the traffic load in the cell. Quality requirements differ significantly from the different services (e.g. BER of 10^{-3} for speech and BER = 10^{-6} for long constraint delay data [ITUR97]) causing a variation of maximum cell ranges for different services and/or data rates [ETSI98c]. The low granularity of frequency blocks (5MHz) leave only 2–5 blocks per operator giving low flexibility in adjusting the radio resources to the inhomogeneous traffic conditions. Consequently, an integrated coverage and capacity planning approach is required, where the selection of potential cell sites is not only determined by propagation conditions [MePi99].

An integrated coverage and capacity planning approach has to be based on the detailed estimation of the interference situation. Figure 4.5.1 shows an exemplary uplink interference situation. Each mobile that is not linked to a certain BS can be considered as an interferer for this BS. Due to the possibility of soft-HO, a single MS can be linked simultaneously to more than one BS.

In order to assign the BSs to each mobile, the basic CDMA link equation has to be fulfilled [Oksa99]:

$$\frac{W}{R}\left(\frac{P_{tx}/L}{I+N}\right) \geq \left(\frac{E_b}{N_0}\right)_{required},$$

(4.5.1)

where W is the modulation bandwidth, R is the bit rate, I is the interference power, N is the thermal noise power, L is the path loss, P_{tx} is the transmit power and $(E_b/N_0)_{required}$ is the minimum bit energy to noise density ratio.

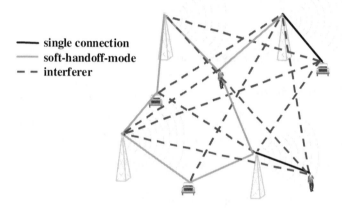

Figure 4.5.1: Uplink interference situation [Kürn99d].

From (4.5.1), it can be derived that interference depends on the transmitted power (adjusted by PC) of every link, and the transmitted power depends on the interference. Therefore, it is not known *a priori* whether there is enough power for a connection. The parameters affecting the total interference are:

- propagation conditions;
- traffic;
- radio resource parameters relevant for PC and (soft-) HO.

These influences have to be considered simultaneously within the radio planning process, which differs significantly when compared to GSM planning (see Figure 4.5.2). In GSM planning, the influence of propagation conditions, traffic forecast and radio resource parameters are planned in subsequent steps. Consequently, an iterative simulation approach is required [WLSJ99], [Pott99], [Oksa99], [HoTo00].

State-of-the-art techniques for CDMA planning [OjPr98], [Yang98] are using Monte-Carlo (MC) simulation for this complex iteration process, causing extremely long run-times combined with very high memory requirements. Therefore, approximations are required in order to reduce the complexity of the problem. Possible approximations investigated within COST 259 are the reduction of the network size by considering sub-networks during simulation [NeBa00], and the reduction of equations to solve the MS-BS assignment [JCAB00].

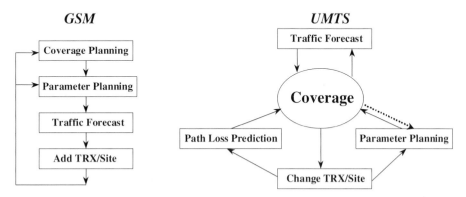

Figure 4.5.2: Comparison GSM planning vs. UMTS planning [Kürn99d].

4.5.3.2. Reduction of network size

In [NeBa00], a system simulator for the uplink in the UMTS FDD mode is described. The system simulations are based on a stationary MC approach and they refer to link level simulation results of standardisation bodies. The capabilities of this simulator are shown for a multi-service scenario that includes both speech and data users with variable data rates of up to 384 kbit/s. Soft HO (SHO), transmit PC, variable antenna pattern at the BS and variable user distributions are included. Due to the modular architecture, it can easily be adapted to any kind of system performance investigations in FDD uplink. The system simulation is split into three phases: initialisation, iteration and evaluation.

In the initialisation phase, all relevant input parameters are set. User equipment (UE) is randomly positioned within the disk. In order to satisfy a certain traffic situation, the amount of UE can be chosen individually for each service (speech, 64 kbit/s, 144 kbit/s and 384 kbit/s Low Constraint Data – LCD). For the initial call set-up, UE TX power levels are chosen according to a path loss criterion.

The schematic flowchart of the iteration process in the system simulator is shown in Figure 4.5.3. After system initialisation, the received power levels in each sector are calculated for each UE. The resulting interference at the best server, i.e. the BS with the best signal-to-interference ratio, is estimated for each communication link. The termination criterion of a single snapshot is based on the UE transmit powers, i.e. the variations in the average transmit power of all UEs become stable after a number of iterations. This is called a stationary situation. Figure 4.5.4 shows the necessary number of iteration steps for a single snapshot with slightly different, but fixed loading situations (a situation like the one during the first 15 steps in Figure 4.5.4 will never occur in a real system, since the initial call set up will never be done simultaneously by the whole system; for simulation purposes, this is necessary). Note that we assumed perfect PC in here, but still we need about 15–20 iteration steps to have UE TX power fluctuations of less than 2 %. In the evaluation phase the results from the iteration phase are determined by post-processing. Outputs are plots of the CDFs and the PDFs of the TX power of the UEs, the measured quality in terms of $E_b/(N_0+I_0)$ and the Total-, Inter-cell- and Intra-cell-Interferences in each cell, respectively.

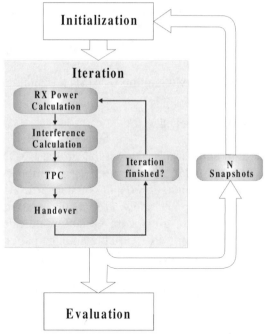

Figure 4.5.3: Schematic flowchart of the system simulator.

Since UE TX power levels and $E_b/(N_0+I_0)$ requirements are different for each service, the evaluation of TX power, etc. is done separately for each service. Furthermore, throughput and cell loading parameters are evaluated in order to derive coverage and capacity figures. Collecting a high number (e.g. $N = 1\,000$) of individual snapshots for a single business case (which describes the number of users for the various services) allows us to take a closer look on the statistics in the evaluation. The investigations presented in [NeBa00] focus on the minimal necessary network size that has to be implemented for single cell investigations. Since system simulations require very long run-times, the minimal network size for that has to be implemented for convincing conclusions is of great interest. In Figure 4.5.5, the total interference of both 7- and 19-site networks is shown.

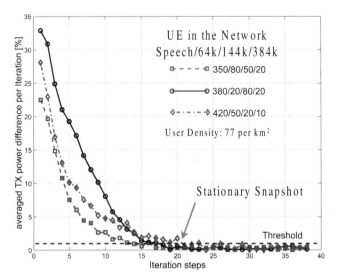

Figure 4.5.4: The number of iterations vs. the relative difference in the average UE TX power. The curves indicate various loading situations speech/64k/144k/384k.

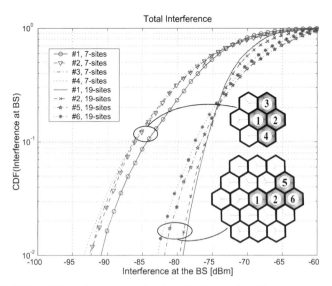

Figure 4.5.5: Total interference in both a 7-site and a 19-site network.

It can be clearly seen that the total interference in the 7-site network is much lower than the total interference in the 19-site network. Since we used the same UE density for both networks, it turns out that a 7-site scenario would lead to a clear underestimation of the total interference at the inner BS site. From the simulation results for different network implementations, it was concluded that a 7-site-network leads to a clear underestimation of the total interference, and thus an overestimation of system capacity, whereas a 19-site-network is sufficient for the estimation of reliable inter-cell interference levels for single site/cell investigations.

4.5.3.3. Reduction of equations

In [JCAB00], a simplified approach for simulating CDMA has been proposed. The proposed technique implies a reduced number of equations to solve the problem of assigning MS to BS in CDMA planning.

In the simplification an instantaneous PC algorithm is assumed. In this case, it is possible to directly evaluate the required power needed for every transceiver to achieve a P_{tx} value for the required E_b/N_0 in (4.5.1). The resulting linear equation system has the dimension $MS \times NS$ (number of MSs in the system multiplied by the number of services) since one MS may have more than one service. Those transceivers requiring a transmitted power over the maximum available (or negative) would be considered to be in outage. However, in order to determine the links in outage it is necessary to enter the MS one by one and try every possible MS-BS assignment. If the number of users in the system is small enough, it is possible to solve the linear equation system for all users in a single step, yielding several difficulties in that simplification:

- If any of the required transmitted powers obtained as a solution of the linear equation system is less than zero, it is not possible to discriminate which user or users are responsible for the outage situation.
- PC errors cannot be implemented directly.
- It is necessary to know beforehand what MS are assigned to a BS.
- Macro diversity cannot be handled directly.

A further – very significant – simplification is used for the uplink, which is usually the limiting link. If all users are assumed to be perfectly power controlled, it is not necessary to evaluate the transmitted power at every mobile. In that case, it is enough to calculate the received power at the BS. The transmitted power can be easily obtained by multiplying this value by the attenuation for each MS. This yields a much simpler linear system, since it has only $BS \times NS$ equations. Unfortunately, that approach has the limitations mentioned above and is only applicable to the uplink.

In order to use the simplified approaches, it is essential to determine before hand what MSs are assigned to a BS. The most direct solution is to assign a MS to a BS that has the least attenuation (the so-called 'best server'). That can be done before any evaluation, just using the MS position. If the attenuation between MS and BS includes log-normal fading, a best server assignment is equivalent to assuming the BS is always connected to the lowest attenuation BS. However, the results are optimistic, since it does not take into account the effect of fading time-variance and the limited number of BSs that can be included in the active set (number of stations in soft HO). To circumvent this, it is possible to simulate just the assignment procedure independently and then solve the simplified linear equation.

Furthermore, in [JCAB00] the use of the network load as a comparison parameter is suggested. An important problem of simulating a CDMA network is that the usual quality measure (outage probability) should be a small value (normally around 1–5 %). That means that a very large number of simulations [JeBS92] had to be made in order to obtain a reasonable accuracy. When studying a large network, it is desirable to define a quality indicator of the proposed solution requiring less simulations. Therefore a cell load distribution is used. The cell load is defined for every cell and service. It is shown in [JCAB00] that the cell load can be derived using the linear equation system mentioned above. The main advantage of the cell load indicator is that the expected value should be around 0.5–0.7 (for optimum network service) and requires a small number of realisations to have a good statistical indicator. Besides, the cell load has very low variance. As an example, the procedure has been used to analyse the advantages of two frequencies for hot spot coverage, Figure 4.5.6.

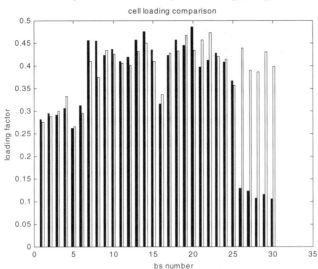

Figure 4.5.6. Relative loading of BSs when using one (white bars) or two (black bars) frequencies.

It can be concluded, that if the hot spot is small, using a different frequency is very advantageous since repeating the same frequency results in a very inefficient use of network resources.

4.5.3.4. Planning of TDD indoor systems

In [Gehr00], a system simulator for both TDD uplinks and downlinks is presented. The simulator was used to investigate the effect of BS location on the transmission power levels of an indoor TDD environment. In the simulation the environment described in [ETSI98b] has been implemented. It consists of an office building with three floors. Both BS and MS are located indoors. Each floor houses 40 office cubicles that are separated by two corridors. A couple of arrangements with one or two BSs, which are located in different floors, have been investigated. Furthermore, the transmission power levels have been used to evaluate the system capacity. It turned out that the BS location is a very important parameter in terms of system capacity of TDD indoor systems. For low transmission power levels the following two constraints should be fulfilled:

- The path loss between unsynchronised BSs should be equal. Otherwise, the interfering pilot power of two co-located BSs reduces the link quality of all connected uplink users.
- There should be one dominant path for the majority of MSs. It must be avoided making MS path loss to two or more BSs low. If that constraint is not fulfilled, the interfering signals from other users are always received nearly as strongly as the signals desired, and causes strong interference.

The simulations showed that the transmission power levels both in uplinks and downlinks are minimum for an environment in which these constraints are fulfilled. Since the transmission power is linked to system capacity, it can be concluded that the capacity of TDD systems can be optimised by proper positioning of BSs.

4.6. Efficient Protocols for High Data Rates

Silvia Ruiz

An important goal of COST 259 has been to contribute to the development of efficient protocols for high data rate networks. Evolving from 2G networks, future mobile networks (UMTS, GPRS) will operate in a multi-service environment, supporting both circuit-switched and packet-switched data, with different transmission rates. Working Groups 1 and 3 have been very active in studying and comparing different protocols for multimedia systems, offering voice and video integration, and interconnectivity with

ATM access. In Section 4.6.1 there is a short mention of some key points that have been exhaustively studied by COST259 authors. A deeply insight is given later in Section 4.6.2, which deals with Wireless ATM, and Section 4.6.3, devoted to Packet Reservation Multiple Access (PRMA) protocols.

4.6.1. High data rate protocols: general aspects

Silvia Ruiz

Recent developments in high data rate protocols studied by COST members have been grouped in five points: Access Schemes for WPN's, Wireless LAN's, Scheduling, Comparison between Protocols and Wireless ATM.

- Access schemes for Wireless Personal Networks (WPN): Some authors propose the use of new efficient algorithms, designed in terms of limiting packet dropping probability for ongoing calls to a selected threshold, while offering high throughput. Other authors develop analytical or software tools that allow a deep study of the performance of known access schemes.

  [KhZH97] presents two simple Distributed-Queuing (DQ) Call Admission Control (CAC) strategies. The first limits the number of active communications by using centralised control in the BS, while the second uses a time-out mechanism to control the maximum number of active connection in a given cell. With this capacity gains are roughly two to three times greater than in traditional circuit switched systems.

  [Mast97a] and [Mast97b] study the influence of the congestion states on the performance of a system based PRMA, developing an original analytical approach that has been used to confirm the result published by the European RACE project R2008.

  [SaAg98] presents the performance of an adaptive Slotted-Aloha DS-CDMA packet random access scheme, maximising the throughput by the use of a MS based algorithm and without the intervention of the BS.

  [RuCO98] simulates the impact of real channel characterisation on a S-ALOHA channel access scheme for different retransmission strategies, and uses spatial diversity and coding to improve system performance.

  [RCGO99] studies and simulates a multiple access scheme based on multi-code DS-CDMA, showing its capacity to support non-homogenous traffic achieving a good QoS.

- Wireless LANs: The influence of packet length, number of nodes and lifetime on expected user data rate, on the upper boundary of the HIPERLAN channel access mechanism is studied by [HaCi97], showing that it is very robust under heavy load conditions. It is also shown that data rate for bounded services, can be increased by adding a buffer to the HIPERLAN MAC user interface. A new protocol based on the standard RTS/CTS that is expected to increase throughput on a data-per-area basis, compared with HIPERLAN, has been developed by [PhOB99a].

- Scheduling: several BS controlled scheduling algorithms have been simulated in [RCGO99] and [CoRF99] in a multi-rate multi-code DS-CDMA system with Distributed Queuing Request Update Multiple Access (DQRUMA), which offers a high throughput while maintaining a low packet delay.

- Comparison between protocols: a comparison between PRMA++ and Adaptive S-ALOHA DS-CDMA is given in [SaAg98], showing that while PRMA++ is optimised for voice traffic, it is sensitive to different traffic statistics, and inefficient for integrating different types of traffic. In these circumstances S-ALOHA DS-CDMA outperforms PRMA++.

- Wireless ATM access: Data Link Control (DLC) is essential for wireless ATM systems to overcome the incompatibility between ATM design assumptions and the characteristics of the physical radio channel. The DLC scheme developed for the ACTS project Magic WAND is evaluated in [MeBe99], and the transmission scheme without error control is compared to a scheme with GBN ARQ mechanism that was exhaustively described in [MeBe97]. These works allow us to determine for which services ARQ schemes are beneficial, when considering network services with different maximum delay requirements. HO in wireless ATM is studied in [TeVV98], offering the following requirements: cell sequence preservation, minimal cell loss and HO delay and integration with standard wired large-scale ATM networks. [VeOl99] proposes a hybrid type II FEC/ARQ protocol for the Link Layer Control (LLC), which uses punctured Reed-Solomon codes for a wireless ATM system based on OFDM.

4.6.2. Wireless ATM

Michel Teughels and Emanuel Van Lil

ATM was never designed with wireless mobile applications in mind. Indeed, the procedure of setting up connections is still an inheritance of the circuit-switching technology, even if it has been optimised to the maximum, to reduce the switching overhead associated with the POTS. Studies have been performed on different algorithms that integrate mobile wireless applications, and more specially the HO. Indeed, one could be inspired by the actual mobile principles and keep a centralised track of the topology of the network, also requiring special switches (roots) that can deal with mobility requirements. The root would then reserve resources for every leaf

(Virtual Connection Tree). Another would use full meshing between the access points and enhanced switches called anchor points. It is obvious that this procedure cannot be scaled to large networks. The most promising protocol investigated at K.U.Leuven in detail is the Leaf Initiated Join (LIJ) HO concept. In this case, only the mobile terminals and the mobile BSs need a specific radio interface. This is based on the UNI 4.0 protocol, which is originally implemented to minimise bandwidth requirements for multicast (or point-to-multipoint) connections (late join source route trees). There, a leaf can request to join the tree, without any intervention of any centralised root or anchor point. The different possibilities of HO are investigated and compared (e.g. forward, backward, soft, hard), from the viewpoint of recovering the order of the cells and the delays involved. All options are only making use of the LIJ capabilities of the switches. Simulations on 60 GHz and 155 Mbit/s data rate show that QoS can be maintained using standard buffers (with delays less than 1 ms in a 71 % load situation) [TDVV99]. The delay was essentially due to the number of switches between the branches, and not the size of the packets. Further work, which is performed within the framework of the MEDIAN ACTS project and also supported by the network technologies programme of the Flemish government, is focusing on the release algorithms for the worn connection, as well as more complex traffic configurations than a mere HO.

4.6.3. Packet reservation multiple access

Paolo Grazioso

In this section we will briefly illustrate the effort devoted to packet-based multiple access protocols, giving an account of the main findings of this effort. For the sake of brevity and readability, we do not enter into mathematical demonstrations or discuss technical details in depth; the interested reader should refer to the original works cited. Packet protocols are widely considered for third generation systems due to their inherent flexibility and capability to support variable bit-rate and multimedia communications. The multiple access protocol is particularly important in the uplink, because users can transmit data independently from one another, and service must be fairly given to all users, meeting the QoS constraints.

4.6.3.1. S-ALOHA

One of the oldest and most widely used packet access protocols is the well known S-ALOHA, where users transmit packets over a common radio channel in pre-defined time slots without any mutual control or centralised regulation except for a positive acknowledgement of reception. This classical protocol is being currently considered for future mobile radio systems, and a detailed study of its capabilities was performed in [RuCO98] and [RCGO99], mainly with the aid of simulation due to the complexity of the equation describing the system and to the difficulty of solving them in practical cases. The former paper focuses on the performance of S-ALOHA in a wireless packet network when propagation and spatial distribution of users are taken into account; different retransmission policies after collision were compared and performance assessed in terms of throughput and delay. More precisely, the simulations addressed the following points.

Among the various algorithms proposed in the literature, *ideal retransmission* and *exponential back-off* was selected in [RuCO98].

Ideal retransmission keeps a constant retransmission probability p for all subsequent trials to transmit a packet after collision; this probability is generally equal to the probability of the generation of a new packet. This scheme is known to become unstable when the offered traffic grows. To stabilise the system, it is possible to dynamically adjust the retransmission probability according to the number of terminals in backlog mode (i.e. with packets to retransmit). One possibility is to resort to the Exponential Back-off (EB) scheme, where the retransmission probability is updated according to the feedback received about the status of the transmission in the previous slot: the BS issues a feedback message about the channel status (idle, success, collision). The authors implemented two versions of EB, namely:

- EB algorithm of type a: after a collision state, all the terminals in backlog mode decrease p, while in idle state all the mobiles increase p.
- EB algorithm of type b: after a collision state, only the terminals that actually transmitted decrease p, while in idle state all the mobiles increase p.

Three types of channels were simulated:

- Slow fluctuations: the received field strength is affected by log-normal shadowing and by Rayleigh fading, but its value can be considered constant throughout a packet (this corresponds to slowly moving or stationary terminals).
- Fast fluctuations: two adjacent bits of a packet are affected by independent Rayleigh fading.
- Fast protected: there are fast fluctuations, but the channel is protected with diversity and information is protected by FEC codes.

In case of a collision, it is possible for the BS to receive successfully one of the colliding packets, provided that its power is sufficiently higher than the sum of all the others. This *capture effect* allows a significant increase in the system throughput; it is heavily affected by the spatial distribution of users, and its benefits are reduced if adaptive PC is used in the uplink.

Three possibilities were investigated concerning the spatial users distribution: uniform, bell-shaped and uni-modal (i.e. users are concentrated around a certain distance from the BS). The uni-modal distribution is not realistic, but it can be used to model interferers in a multi-cell environment, or to model a cell where adaptive PC is used, so that the effects of different distances are filtered out.

The outputs of the simulations were: throughput, average delay, and mean number of mobile terminals in backlog mode. The main results are that the EB retransmission scheme is stable and outperforms the ideal retransmission; furthermore, the capture effect results in a remarkable increase of system throughput.

The authors also investigated the possibility of using S-ALOHA as the request access protocol in a mobile radio system using Multi Code DS-CDMA where multimedia applications are required. This work was expanded in the subsequent document [RCGO99], where they also addressed the issue of the slot allocation and scheduling algorithms enforced by the BS, in order to optimise throughput and performance while guaranteeing a fair service to all users.

Simulations for a mixture of terminals with different data rates confirmed the suitability of S-ALOHA as access request protocols. Furthermore, different options for scheduling were all found to meet the quality constraints (in terms of delay and throughput) for different mobile terminal classes, thus confirming that Multi Code DS-CDMA is a suitable protocol for multimedia wireless systems, such as third generation mobile radio systems.

4.6.3.2. ARQ in SFH radio systems

Following a similar concept, i.e. assessing the performance of packet access protocols under realistic assumptions for channel modelling, [Chia97] studies the error distribution for packet transmissions over mobile radio systems employing SFH. Furthermore, a new analytical methodology was presented to evaluate ARQ techniques, taking into account FEC and non-ideally interleaved SFH over Rayleigh fading channels.

Two ARQ techniques were considered in the work, namely:

- Pure Selective Repeat (SR): a block of information symbols is encoded using an error detection code to form a packet that is transmitted through the channel; at the RX, if any errors have been detected, the whole packet is discarded and a Negative Acknowledgement (NACK) is issued towards the TX, which will have to retransmit that packet.

- Selective Repeat Hybrid Type I ARQ/FEC (ARQ/FEC-I): with this scheme an error correction code is used; if the RX detects some errors in the received packet, it tries to correct them, and only if the correction fails a NACK is issued.

Both techniques were compared for a varying number of bits per hop, so gradually passing from the non-interleaved non-SFH case to the perfectly interleaved case. It was found that the use of SFH and/or interleaving deteriorates the efficiency of the transmission protocol: in other words, when ARQ is applied to slow fading channels, the best performances are obtained without SFH or interleaving. Furthermore, the hybrid ARQ/FEC technique gives marginal improvements with respect to pure SR only when perfect interleaving and SFH are applied.

4.6.3.3. PRMA++

A quite different approach, neglecting channel characterisation to concentrate on the inherent protocol peculiarities, is made in the two companion papers [Mast97a] and [Mast97b], which actually started the study of the Packet Reservation Multiple Access (PRMA) protocol, and of its evolved variant PRMA++, within COST 259.

[Mast97a] presented an original analytical approach to study the influence of the congestion states on the performance of a system based on the PRMA protocol. A lot of previous analytical studies were proposed in the literature, but none of them clearly showed the effect exerted by the congestion states on PRMA operation. The paper quoted, on the other hand, considers an ideal PRMA protocol, where the effect of collisions has been removed, so that the system performance is only limited by congestion states. The analytical model was then compared with simulation results for a realistic PRMA system, i.e. including the effect of collisions, taking packet dropping probability as the sole quality indicator. From the comparison, apart from the obvious result that the ideal PRMA without collisions performs better than the actual PRMA with collisions, one can observe that:

- when the number of users is small (of the order of the number of slots per frame) the packet dropping in the two cases differ significantly, hence the actual packet dropping is mainly due to collisions;
- increasing the number of active users the two packet dropping curves tend to coincide, i.e. the packet dropping can be ascribed substantially to the congestion states, whereas the packet collisions can be neglected;
- taking a 1 % average dropping as the required QoS, the capacity of both ideal and real PRMA are nearly equal.

To further validate the model, the author applied it also to the PRMA++ protocol adopted within the RACE project R2008, ATDMA. The excellent agreement between analysis and the independent simulations published by ATDMA confirmed the validity of the proposed analytical model.

[Mast97b] finally, deals with a demonstration of a theorem on the maximum efficiency of PRMA-based systems. Although this limit was already assumed in the literature to be equal to $1/\psi$, where ψ is the speech activity factor, no analytical demonstration had been previously provided of this fact. It had been always justified on the basis of simple intuitive considerations, lacking a rigorous and sound demonstration. This also prevented us from understanding the real meaning of the parameter $1/\psi$, sometimes leading to erroneous interpretations. The demonstrated theorem clarifies the meaning of the maximum efficient and the role that it plays for the access techniques belonging to the PRMA family. As a matter of fact, let η be the system efficiency defined as the ratio between the system capacity (in terms of maximum number of users) and the allocated resources (time slots per frame). In other words, it represents the statistical multiplexing gain of PRMA with respect to circuit switched TDMA. The theorem shows that $\eta_{max} = 1/\psi$ is the maximum efficiency that can be obtained while keeping an arbitrarily low packet dropping probability. Of course, it is possible to exceed this limit (e.g. when the system is congested), but then the packet dropping probability is no more controllable.

4.7. References

[AaKo89]   Aarts,E. and Korst,J., *Simulated Annealing and Boltzman Machines*, John Wiley, New York, New York, USA, 1989.

[ACTV92]   Andrisano,O., Chiani,M., Tralli,V. and Verdone,R., "Impact of Co-channel Interference on Vehicle to Vehicle Communications at Millimetre Waves", in *Proc of ICCS'92 – IEEE International Conference Communications Systems*, Singapore, Nov. 1992.

[ADFV94]   Andrisano,O., Daniele,P., Frullone,M. and Verdone,R., "Roadside to Vehicle Communications at 60 GHz", in *Proc. of EuMC'94 – 24th European Microwave Conference / Short Range Communications Workshop*, Cannes, France, Sep. 1994.

[AlQC99]   Almeida,S., Queijo,J. and Correia,L.M., "Spatial and Temporal Traffic Distribution Models for GSM", in *Proc. of VTC'99 Fall – 50th IEEE Vehicular Technology Conference*, Amsterdam, The Netherlands, Sep. 1999 [also available as TD(99)037].

[AnDV95]   Andrisano,O., Dardari,D. and Verdone,R., "Short Range Communication Systems for Vehicle-to-Roadside in Urban Environment", in *Proc. of ROVA International Conference*, Bolton, UK, Sep. 1995.

[AsFu93]   Asakura,H. and Fujii,T., "Combining Micro and Macro Cells in a Cellular System", in *Proc. of VTC'93 – 43rd IEEE Vehicular Technology Conference*, Secaucus, New Jersey, USA, May 1993.

[BaSa97]   Bachelier,T. and Santé, J.-F., "Capacity increase in cellular PCS by smart antennas", in *Proc. of IEE Symposium on Intelligent Antenna*, Guildford, UK, 1997 [also available as TD(97)048].

[Basi99]   Bassile,M., "Obtaining accurate predictions with an untuned semi-deterministic model used with an advanced statistical database", *COST 259*, TD(99)031, Vienna, Austria, Apr. 1999.

[BEGM98]   Borndörfer,R., Eisenblätter,A., Grötschel,M. and Martin,A., "Frequency assignment in cellular phone networks", *Annals of Operations Research,* Vol. 76, No. 1, 1998, pp. 73–93.

[BeKi99a]   Beckmannn,D. and Killat,U., "Frequency Planning with respect to Interference Minimization in Cellular Radio Networks", *COST 259*, TD(99)032, Vienna, Austria, Jan. 1999.

[BeKi99b]   Beckmann,D. and Killat,U., "A New Strategy for the Application of Genetic Algorithms to the Channel-Assignment Problem", *IEEE Trans. Veh. Technol.,* Vol. 48, No. 4, July 1999, pp. 1261–1269.

[Benv96]   Benveniste,M., "Probability Models of Microcell Coverage", *Wireless Networks*, Vol. 2, No. 2, 1996, pp. 289–296.

[BGMM98] Berruto,E., Gudmunson,M., Menolascino,R., Mohr,W. and Pizarosso,M., "Research Activities on UMTS Radio Interface, Network Architectures, and Planning", *IEEE Commun. Magazine*, Vol. 36, No. 2, Feb. 1998, pp.82–95.

[Brél79]   Brélaz,D., "New methods to color vertices of a graph", *Communications of the ACM,* Vol. 22, No. 4, 1979, pp. 169–174.

[BrPB97]   Bratanov,P., Paier,M. and Bonek,E., "Mobility Model of vehicle-borne subscriber units in urban cellular systems", *COST 259*, TD(97)043, Lisbon, Portugal, Sep. 1997.

[BrVe96]   Brázio,J. and Velez,F., "Design of Cell Size and Frequency Reuse for a Millimetre Wave Highway Coverage Cellular Communication System", in *Proc. of PIMRC'96 – 7th IEEE International Symposium on Personal Indoor, and Mobile Radio Communications*, Taipei, Taiwan, Oct. 1996.

[BuMZ99]   Burer,S., Monteiro,R.D.C. and Zhang,Y., *Interior-Point Algorithms for Semidefinite Programming Based on a Nonlinear Programming Formulation*, Technical Report 99-27, Dept. of Comput. and Appl. Mathematics, Rice University, Houston, Texas, USA, Dec. 1999.

[Burr00]   Burr,A., "Channel capacity evaluation of multi-element antenna systems using a spatial channel model", *COST 259*, TD(00)006, Valencia, Spain, Jan. 2000.

[Cami99]   Caminada,A., "Benchmarking on Frequency Planning", Personal Communication, *CNET – France Telecom*, Belfort, France, Oct. 1999.

[CCAA97]   Chiani,M., Conti,A., Agrati,E. and Andrisano,O., "An Analytical Approach to Evaluate Service Coverage in Slow Frequency Hopping Mobile Radio Systems", *COST 259*, TD(97)065, Lisbon, Portugal, Sep. 1997.

[CCVZ99]   Chiani,M., Conti,A., Verdone,R. and Zanella,A., "Signal-Level-Based Power Control over Slow Frequency Hopped Mobile Radio Systems", in *Proc. of VTC'99 Spring – 49th IEEE Vehicular Technology Conference*, Houston, Texas, USA, June 1999 [also available as TD(99)022].

[CEPT91]   CEPT, *Designation of a harmonized frequency band for Multipoint Video Distribution Systems in Europe*, CEPT, Recommendation T/R 52-01, 1991.

[Cerný85]   Cerný,V., "Thermodynamical Approach to the Travelling Salesman Problem: An Efficient Simulation Algorithm", *Journal of Optimization Theory and Applications*, Vol. 45, No. 1, 1985, pp. 41–51.

[ChCA99a]   Chiani,M., Conti,A. and Andrisano,O., "Up-link Analytical Outage Evaluation for Slow Frequency Hopping Mobile Radio Systems", in *Proc. of ICC'99 – IEEE Conference on Communications*, Vancouver, Canada, June 1999 [also available as TD(98)080].

[ChCA99b]   Chiani,M., Conti,A. and Andrisano,O., "Outage Evaluation for Slow Frequency Hopping Mobile Radio Systems", *IEEE Trans. Commun.*, Vol. 47, No. 12, Dec. 1999, pp. 1865–1874.

[ChCo99]   Chiani,M. and Conti,A., "Impact of Modulation and Coding on the Outage Probability for Mobile Radio Systems over Block Fading Channels", in *Proc. of GLOBECOM'99*, Rio de Janeiro, Brazil, Dec. 1999.

[Chia00]   Chiani,M., "Throughput Evaluation for ARQ Protocols in Finite-interleaved Slow Frequency Hopping Mobile Radio Systems", *IEEE Trans. Veh. Tech.*, Vol. 49, No. 2, Mar. 2000, pp. 576–581.

[Chia97]   Chiani,M., "Errors Distribution in SFH mobile radio systems with application to ARQ", in *Proc. of GLOBECOM'97*, Phoenix, Ariz., USA, Nov. 1997 [also available as TD(98)042].

[Chia98]   Chiani,M., "Error Probability for Block Codes over Channels with Block Interference", *IEEE Trans. Inform. Theory,* Vol. 44, No. 7, Nov. 1998, pp. 2998–3008.

[Cich94]   Cichon,D.J., *Ray-Optical Wave Propagation Modelling in Urban– and Picocells* (in German), Ph.D. Thesis, Univ. of Karlsruhe, Karlsruhe, Germany, 1994.

[CMVZ98] Chiani,M., Monguzzi,G., Verdone,R. and Zanella,A., "Analytical Modeling of Handover Algorithm Performance in a Multicell Urban Environment with Frequency Hopping", in *Proc. of VTC'98 – 48th IEEE Vehicular Technology Conference*, Ottawa, Canada, May 1998 [also available as TD(98)015].

[CoFr94]   Correia,L.M. and Francês,P.O., "A Propagation Model for the Average Received Power in an Outdoor Environment in the Millimetre Wave Band", in *Proc. of VTC'94 – 44th IEEE Vehicular Technology Conference*, Stockholm, Sweden, Jun. 1994.

[CoLR90]   Cormen,T.H., Leiserson,C.E. and Rivest,R.L., *Introduction to Algorithms*, McGraw-Hill, New York, New York, USA, 1990.

[CoRe73]   Cox,D.C. and Reudink,D.O., "A Comparison of Some Channel Assignment Strategies in Large-Scale Mobile Communications Systems", *IEEE Trans. Commun.*, Vol. COM-21, No. 11, Nov. 1973, pp. 1302–1306.

[CoRF99]   Covarrubias,D., Ruiz,S. and Fernandez,A., "A decentralised channel load sensing protocol for wide-band CDMA systems", *COST 259*, TD(99)111, Leidschendam, The Netherlands, Sep. 1999.

[COST99] COST 231, *Digital mobile radio towards future generation systems*, Final report, COST Telecom Secretariat, European Commission, Brussels, Belgium, 1999.

[Costa93] Costa,D., "On the use of some known methods for T-coloring graphs", *Annals of Operations Research*, Vol. 41, No. 1–4, 1993, pp. 343–358.

[deUr99]   de Urries,L.J., "A Simulated Annealing Implementation for Frequency Planning", Personal Communication, Telefonica, Madrid, Spain, Nov. 1999.

[DHLZ98] Didascalou,D., Horsmannheimo,S., Lähteenmaki,J. and Zwick,T., "Recommendations for a common 3D-format for wave-propagation modelling tools", *COST 259*, TD(98)044, Bern, Switzerland, Feb. 1998.

[DMSV97] De Coster,I., Madariaga,G.A., Souto,B.P., Van Lil,E. and Perez-Fontan,F., "An Extended Propagation Software Package for Indoor Communication Systems", in *Proc. of ICAP'97 – IEE International Conference on Antennas and Propagation*, Edinburgh, UK, Apr. 1997.

[DoVe87]   Dornstetter,J. and Verhulst,D., "Cellular Efficiency with Slow Frequency Hopping: Analysis of the Digital SFH 900 Mobile

System", *IEEE J. Select. Areas Commun.*, Vol SAC-5, No. 5, June 1987, pp. 835–848.

[dRFG95]   del Re,E., Fantacei,R. and Giambene,G., "Handover and Dynamic Channel Allocation Techniques in Mobile Cellular Networks", *IEEE Trans. Veh. Tech.*, Vol. 44, No. 2, May 1995, pp. 229–237.

[dUDB00]   de Urries,L.J., Diaz,M.A. and Berberana,I., "Frequency planning using simulated annealing", *COST 259*, TD(00)054, Bergen, Norway, Apr. 2000.

[DuSc90]   Dueck,G. and Scheuer,T., "Threshold Accepting: a general purpose optimization algorithm appearing superior to simulated annealing", *Journal of Computational Physics*, Vol. 90, No. 1, 1990, pp. 161–175.

[EiKF98]   Eisenblätter,A., Kürner,T. and Fauß,R., "Radio Planning Algorithms for Interference Reduction in Cellular Network", in *Proc. of COST 252/259 Joint Workshop*, Bradford, UK, Apr. 1998.

[EiKF99]   Eisenblätter,A., Kürner,T. and Fauß,R., "Analysis of C/I Ratio Thresholds for Frequency Planning", *COST 259*, TD(99)012, Thessaloniki, Greece, Jan. 1999.

[EiKü00]   Eisenblätter,A. and Kürner,T., "Benchmarking for Frequency Allocation", *COST59*, TD(00)044, Bergen, Norway, Apr. 2000.

[Eise00]   Eisenblätter,A., "Obtaining lower bounds on co-channel interference in GSM-networks by Semidefinite Programming", *COST 259*, TD(00)043, Bergen, Norway, Apr. 2000.

[Eise98a] Eisenblätter,A., "Proposal: File Formats for the Standard Scenarios for Frequency Planning", *COST 259*, TD(98)048, Bern, Switzerland, Feb. 1998.

[Eise98b] Eisenblätter,A., "Combinatorial Lower Bounds for Co-Channel Interference in GSM networks", *COST 259*, TD(98)104, Duisburg, Germany, Sep. 1998.

[EpPe53]   Epstein,J. and Peterson,D.W., "An experimental study of wave propagation at 850 Mc/s", *Proc. Inst. Radio Eng.*, Vol. 41, No. 5, 1953, pp. 595–611.

[ERC97a]   ERC, *Sharing and Compatibility of UMTS with Adjacent Services: DECT (1880–1900 MHz)*, ERC, ERC/TG1(97)18, London, UK, Aug. 1997.

[ERC97b]   ERC, *ERC Decision of 30 June 1997 on the frequency bands for the introduction of the Universal Mobile Telecommunications System (UMTS)*, ERC, Decision ERC/DEC(97)07, The Hague, The Netherlands, June 1997.

[ESRF00]   Elouardi,N., Steuer,J., Radimirsch,M., Fette,H. and Jobmann,K., "Proactive Dynamic Resource Distribution Based on Borrowing and Control Engineering Methods for GSM-like Systems", *COST 259*, TD(00)029, Valencia, Spain, Jan. 2000.

[ETSI92]   ETSI, *Digital European Cordless Telecommunications (DECT) Common Interface, Part2: Physical Layer,* ETSI – Radio Equipment and Systems (RES), EN 300 175-2, Ed. 1, Sophia-Antipolis, France, Oct. 1992.

[ETSI97]   ETSI, *Concept Group Alpha – Wideband Direct-Sequence CDMA: Guard Bands for WCDMA*, ETSI, TDoc. SMG2 382/97, Cork, Ireland, Dec. 1997.

[ETSI98a]   ETSI, *Digital Video Broadcasting (DVB); Interaction channel for Local Multi-point Distribution Systems (LMDS)*, ETSI, DVB-RCL, Draft EN 301 199 V1.1.1, Sophia-Antipolis, France, Jan. 1998.

[ETSI98b]   ETSI, *Selection Procedures for the Choice of Radio Transmission Technologies of the Radio Transmission Technologies of the UMTS*, ETSI-SMG, Technical Report TR 101, 112, v. 3.2.0, Sophia-Antipolis, France, Apr. 1998.

[ETSI98c]   ETSI, *The ETSI UMTS Terrestrial Radio Access (UTRA) ITU-R RTT Candidate Submission*, Technical Report, Attachment 5 ("Updated performance results"), ETSI-SMG, Sophia-Antipolis, France, June 1998.

[Fall98]   Fallot-Josselin,S., "Automatic Radio Network planning in the context of 3rd generation mobile systems", *COST 259*, TD(98)102, Duisburg, Germany, Sep. 1998.

[FAPL00]   COST 259, Sub-Working Group 3.1: Standard Scenarios for Frequency Planning, URL: http://fap.zib.de/, 2000.

[Faru96]   Faruque,S., *Cellular Mobile Systems Engineering*, Artech House, London, UK, 1996.

[Fern95]   Fernandes,L., "Developing a System Concept and Technologies for Mobile Broadband Communications", *IEEE Personal Commun.*, Vol. 2, No. 1, Feb. 1995, pp. 54–59.

[FeVe99]   Ferracioli,M. and Verdone,R., "Seamless Multimedia Service for Third Generation Mobile Radio Networks", in *Proc. of MMT'99 – Multiaccess, Mobility and Teletraffic for Wireless Communications,* Venice, Italy, Oct. 1999 [also available as TD(99)090].

[FGPR99]   Frullone,M., Grazioso,P., Passerini,C. and Riva,G., "Adaptive antennas in adaptive systems: a plus or a mess?", in *Proc. of 4th ACTS Mobile Communications Summit*, Sorrento, Italy, June 1999.

[Fole93]   Foley,J.D. *et al.*, *Computer Graphics – Principles and Practice*, Addison-Wesley, Reading, Mass., USA, 1993

[FRGF96]   Frullone,M.   Riva,G.,   Grazioso,P.   and   Falciasecca,G.,
            "Advanced Planning Criteria for Cellular Systems", *IEEE
            Personal Commun.*, Vol. 3, No. 6, Dec. 1996, pp. 10–15.

[FRGF97]   Frullone,M.,   Riva,G.,   Grazioso,P.,   Falciasecca,G.   and
            Barbiroli,M., "Advanced Cell Planning Criteria for Personal
            Communications Systems", *COST 259*, TD(97)092, Lisbon,
            Portugal, Sep. 1997.

[FuKB97]   Fuhl,J., Kuchar,A. and Bonek,E., "Capacity increase in cellular
            PCS by smart antennas", in *Proc. of VTC'97 – 47th IEEE
            Vehicular Technology Conference*, Ottawa, Canada, May 1997
            [also available as TD(97)015].

[Gams88]   Gamst,A., "A Resource Allocation Technique for FDMA
            Systems", *Alta Frequenza*, Vol. LVII, Feb./Mar. 1988, pp. 89–
            96.

[GBSZ97]   Gamst,A., Beck,R., Simon,R. and Zinn,E.-G., "The effect of
            handover algorithms with distance measurement on the
            performance of cellular radio networks", in *Proc. of
            International Conference on Digital Land Mobile Radio
            Communication*, Venice, Italy, 1997.

[GDVP99]   Gavilanes,B., De Coster,I., Van Lil,E. and Pérez-Fontán,F.,
            "Indoor Penetration and Interference Studies", *COST 259*,
            TD(99)019, Thessaloniki, Greece, Jan. 1999.

[Gehr00]   Gehring,A., "Effect of base station location on the transmission
            power levels of an indoor TDD environment", *COST 259*,
            TD(00)46, Bergen, Norway, Apr. 2000.

[Goda97]   Godaral,L.C., "Applications of antenna array to mobile
            communications. Parts 1 & 2", *Proc. IEEE*, Vol. 85, No. 5 & 6,
            Jul. & Aug. 1997, pp. 1031–1060 & pp. 1195–1245.

[GoGR97]   Gotzner,R., Gamst,A. and Rathgeber,R., "Spatial Traffic
            Distribution in Cellular Networks", in *Proc. of VTC'97 – 47th
            IEEE Vehicular Technology Conference*, Ottawa, Canada, May
            1997.

[Grace99]   Grace,D., *Distributed Dynamic Channel Assignment for the
            Wireless Environment*, Ph.D. Thesis, University of York, York,
            UK, 1999, http://www.amp.york.ac.uk/external/comms/theses/
            grace.shtml.

[GRBT97]   Grace,D., Burr,A.G. and Tozer,T.C., "The Effects of
            Interference Threshold and SNR Hysteresis on Distributed
            Channel Assignment Algorithms for UFDMA", *COST 259*,
            TD(97)067, Lisbon, Portugal, Sep. 1997.

[GrBT98]   Grace,D., Burr,A.G. and Tozer,T.C., "Performance of a
            Distributed Dynamic Channel Assignment Algorithm
            Incorporation Power Control in a Wireless Environment", in
            Proc. of GLOBECOM'98, Sydney, Australia, Nov. 1998.

[Gsch95]   Gschwendtner,B.E., *Adaptive wave propagation models for radio network planning* (in German), Ph.D. Thesis, University of Stuttgart, Stuttgart, Germany, Nov. 1995

[HaCi97]   Hagenauer,M. and Cis,R., "HIPERLAN type I Quality of Service Medium Access Control User Data Rate", *COST 259*, TD(97)080, Lisbon, Portugal, Sep. 1997.

[HaJa99]   Hayn,A. and Jakoby,R., "Propagation and Standardization Issues for 42 GHz Digital Microwave Video Distribution System (MVDS)", *COST 259*, TD(99)021, Thessaloniki, Greece, Jan. 1999.

[HeHe00]   Hellebrandt,M. and Heller,H., "A New Heuristic Method for Frequency Assignment", *COST 259*, TD(00)003, Valencia, Spain, Jan. 2000.

[HeKi99]   Helmberg,C. and Kiwiel,K.C., *A Spectral Bundle Method with Bounds*, Technical Report 99-37, Konrad-Zuse-Zentrum für Informationstechnik Berlin, Berlin, Germany, 1999.

[HeMS97]   Hellebrandt,M., Mathar,R. and Scheibenbogen,M., "Estimating Position and velocity of Mobiles in a Cellular Radio Network", *IEEE Trans. Veh. Tech.*, Vol. 46, No. 1, Feb. 1997, pp. 65–71.

[HeRe97]   Helmberg,C. and Rendl,F., *A Spectral Bundle Method for Semidefinite Programming*, Technical Report 97-37, Konrad-Zuse-Zentrum für Informationstechnik Berlin, Berlin, Germany, 1997

[HLMN99] Hellebrandt,M., Lambrecht,F., Mathar,R., Niessen,T. and Starke,R., "Frequency allocation and linear programming", in *Proc. of VTC'99 Spring – 49th IEEE Vehicular Technology Conference*, Houston, Texas, USA, May 1999.

[HoTo00]   Holma,H. and Toskala,A. (eds.), *WCDMA for UMTS*, John Wiley, New York, New York, USA, 2000.

[HoWL99]   Hoppe,R., Wöfle,G. and Landstorfer,F.M., "Fast 3D Ray Tracing for the Planning of Microcells by Intelligent Preprocessing of the Data Base", in *Proc. of EPMCC'99 – 3rd European Personal Mobile Communications Conference*, Paris, France, Mar. 1999.

[ITUR97]   ITU-R, *Request for Submission of Candidate Radio Transmission (RTTs) for the IMT–2000/FPLMTS Radio Interface*, Circular Letter 8/LCCE/47, Apr. 1997.

[Jabb96]   Jabbari,B., "Tele-traffic Aspects of Evolving and Next-Generation Wireless Communication Networks", *IEEE Personal Commun.*, Vol. 3, No. 6, Dec. 1996, pp. 4–9.

[JAGB99]   Jiménez,J., Alonso,E., Gonzalez,H. and Berberana,I., "Call segregation techniques for GSM", *COST 259*, TD(99)086, Leidschendam, The Netherlands, Sep. 1999.

[JaGr96]   Jakoby,R. and Grigat,M., *MMDS to extend Broadband Cable Networks* (in German), Deutsche Telekom TZ, FZ232b and FZ231b, Darmstadt, Germany, May 1996.

[JAGR99]   Jiménez,J., Alonso,E., Gonzalez,H., Romero,M. and Berberana,I., "Mobile Location using coverage information: Theoretical analysis and results", *COST 259*, TD(99)043, Thessaloniki, Greece, Jan. 1999.

[JAMS91]   Johnson,D.S., Aragon,C.R., McGeoch,L.A. and Schevon,C., "Optimization by Simulated Annealing: An experimental evaluation; Part II, Graph Coloring and number partitioning", *Annals of Operations Research,* Vol. 39, No. 3, 1991, pp. 378–406.

[JBGA98]   Jiménez,J., Berberana,I., Gonzalez,H. and Alonso,E., "Mobile location using coverage information", *COST 259*, TD(98)084, Duisburg, Germany, Sep. 1998.

[JCAB00]   Jiménez,J., Canales,C., Alonso,E. and Berberana,I., "CDMA network planning: simplified procedures", *COST 259*, TD(00)053, Bergen, Norway, Apr. 2000.

[JeBS92]   Jeruchim,M., Balaban,P. and Shanmugan,K., *Simulation of Communication System*s, Plenum Press, New York, New York, USA, 1992

[JeLe96]   Jedrzycki,C. and Leung,V.C.M., "Probability Distribution of Channel Holding Time in Cellular Telephony Systems*", in Proc. of VTC'96 – 46th IEEE Vehicular Technology Conference*, Atlanta, Georgia, USA, May 1996.

[JiGA98]   Jiménez,J., González,H. and Alonso,E., "Mobile Location for Hot Spot Identification", *COST 259*, TD(98)012, Bern, Switzerland, Feb. 1998.

[JiGB98]   Jiménez,J., Gonzalez,H. and Berberana,I., "Optimising A SFH Frequency Assignment", *COST 259*, TD(98)014, Bern, Switzerland, Feb. 1998.

[JiGo98]   Jiménez,J. and González,H., "Use Of Compatibility Matrix For Optimising A SFH Frequency Assignment", *COST 259*, TD(98)068, Bradford, UK, Apr. 1998.

[Jimé97]   Jiménez,J., "Notes on the application of SFH to GSM networks", *COST 259*, TD(97)044, Lisbon, Portugal, Sep. 1997.

[JoSk95]   Jones,B.C. and Skellern,D.J., "Outage Contours and Cell Size Distributions in Cellular and Microcellular Networks", in *Proc. of VTC'95 – 45th IEEE Vehicular Technology Conference*, Chicago, Illinois, USA, May 1995.

[KaMo98]   Karray,M.K. and Molina,A.O., "Optimum Solution Accessibility of Some Graph Coloring Techniques", *COST 259*, TD(98)081, Duisburg, Germany, Sep. 1998.

[KeDK98]   Kennedy,K., De Vries,E. and Koorevaar,P., "Performance of a Distributed DCA Algorithm Under Inhomogeneous Traffic Modelled from an Operational GSM Network", in *Proc. of VTC'98 – 48th IEEE Vehicular Technology Conference*, Phoenix, Ariz., USA, May 1998 [also available as TD(98)061].

[KeOB99]   Kemp,A.H., Orriss,J. and Barton,S.K., "The Impact of Base Station Diversity on CDMA system capacity", *COST 259*, TD(99)089, Leidschendam, The Netherlands, Sep. 1999.

[KHGK96] Kuhn,A., Haggerty,W., Grage,U., Keller,M., Reyering,C., Arndt,F. and Scholtholt,W., "Validation of the Future Frequency Hopping in a Live GSM Network", in *Proc. of VTC'96 – 46th IEEE Vehicular Technology Conference*, Atlanta, Georgia, USA, May 1996.

[KhZe97a] Khan,F. and Zeghlache,D., "Effect of Cell Residence Time Distribution on the Performance of Cellular Mobile Networks", *COST 259*, TD(97)087, Lisbon, Portugal, Sep. 1997.

[KhZe97b] Khan,F. and Zeghlache,D., "Performance Analysis of Link Adaptation in Wireless Personal Communications Systems", *COST 259*, TD(97)085 Lisbon, Portugal, Sep. 1997.

[KhZH97]   Khan,F., Zeghlache,D. and Hébuterne,G., "Distributed-Queuing Call Admission Control in Wireless Packet Communications", *COST 259*, TD(97)086, Lisbon, Portugal, Sep. 1997.

[KiGV83]   Kirkpatrick,S., Gelatt,C.D. and Vecchi,M.P., "Optimization by Simulated Annealing", *Science,* Vol. 220, 1983, pp. 671–680.

[KPOB99] Kemp,A.H., Phillips,A.R., Orriss,J. and Barton,S.K., "High Integrity wireless data communications over a channel exhibiting severe multi-path propagation", in *Proc. of VTC'99 Fall – 50th IEEE Vehicular Technology Conference*, Amsterdam, The Netherlands, Sep. 1999.

[Krän95]   Kränzle,H., "Remote sensing data and GIS in the design of a cellular communication network" (in German), *GEO-Information-Systems*, Vol. 8, No. 3, 1995, pp. 13–18.

[KrCo98]   Kristic,D. and Correia,L.M., "Optimisation of Micro Cellular Regions by Minimising the Number of Base Stations", in *Proc. of PIMRC'98 – 9th IEEE International Symposium on Personal, Indoor and Mobile Radio Communications,* Boston, Mass., USA, Sep. 1998 [also available as TD(98)080].

[KrCo99]   Kristic,D. and Correia,L.M., "Influence of Mobility from Handover in Cellular Planning", in *Proc. of PIMRC'99 – 10th IEEE International Symposium on Personal, Indoor and Mobile Radio Communications,* Osaka, Japan, Sep. 1999 [also available as TD(99)082].

[KüFa94] Kürner,T. and Fauß,R, "Investigation of Path Loss Algorithms at 1900 MHz", *COST 231*, TD(94)109, Darmstadt, Germany, Sep. 1994.

[KüFa97] Kürner,T. and Fauß,R, "Impact of digital terrain databases on the prediction accuracy in urban areas – comparative study", in *Proc. of EPMCC'97 – 2nd European Personal Mobile Communications Conference*, Bonn, Germany, Oct. 1997.

[KüFW96] Kürner,T., Fauß,R. and Wäsch,A., "A hybrid propagation modelling approach for DCS 1800 macro cells", in *Proc. of VTC'96 – 46th IEEE Vehicular Technology Conference*, Atlanta, Georgia, USA, May 1996.

[Kürn98] Kürner,T., "Analysis of Cell Assignment Probability Predictions", *COST 259*, TD(98)005, Bern, Switzerland, Feb. 1998.

[Kürn99a] Kürner,T., "Randomised Generation of Realistic Spatial Teletraffic Distributions for Frequency Planning Simulations", *COST 259*, TD(99)011, Thessaloniki, Greece, Jan. 1999.

[Kürn99b] Kürner,T., "A run-time efficient 3D propagation model for urban areas including vegetation and terrain effects", in *Proc. of VTC'99 Spring – 49th IEEE Vehicular Technology Conference*, Houston, Texas, USA, May 1999 [also available as TD(99)047].

[Kürn99c] Kürner,T., "Optimising Spectrum Efficiency", *in Proc. of UMTS'99*, Monte Carlo, Monaco, June 1999 [also available as TD(99)095].

[Kürn99d] Kürner,T., "UMTS Planning and Spectrum Allocation", in *Proc. of Optimising Base Station Subsystems*, London, UK, Dec. 1999.

[LaCW97] Lam,D., Cox,D.C. and Windom,J., "Teletraffic Modelling for Personal Communications Systems", *IEEE Commun. Magazine*, Vol. 35, No. 2, Feb. 1997, pp. 79–87.

[Lope96] Lopez,A.R., "Performance prediction for cellular switched-beam intelligent antenna systems", *IEEE Commun. Magazine*, Vol. 34, No. 10, Oct. 96, pp. 152–154.

[MaHe98] Mathar,R. and Hellebrandt,M., "Fast Algorithms for Optimal Channel Assignment in Cellular Radio Networks", in *Proc. of GLOBECOM'98*, Sydney, Australia, 1998 [also available as TD(97)081].

[Mark98] Markoulidakis,Y., "Location and Paging Area Planning in UMTS", *COST 259*, TD(98)101, Duisburg, Germany, Sep. 1998.

[Mast97a] Mastroforti,M., "Congestion states analysis of a PRMA/PRMA++ based system", *COST 259*, TD(97)074, Lisbon, Portugal, Sep. 1997.

[Mast97b]   Mastroforti,M., "Fundamental theorem of Packet Reservation Multiple Access Techniques", *COST 259*, TD(97)075, Lisbon, Portugal, Sep. 1997.

[MDDW00] Maurer,J., Drumm,O., Didascalou,D. and Wiesbeck,W., "A novel approach in the determination of visible surfaces in 3D vector geometries for ray-optical wave propagation modelling", in *Proc. of VTC'2000 Spring – 51st IEEE Vehicular Technology Conference*, Tokyo, Japan, May 2000 [also available as TD(00)014].

[MeBe97]   Meierhofer,J. and Bernhard,U.P., "Go-Back-N Retransmission Scheme for wireless Access to ATM Networks", *COST 259*, TD(97)050, Lisbon, Portugal, Sep. 1997.

[MeBe99]   Meierhofer,J. and Bernhard,U.P., "System Level Channel Model for Wireless ATM Performance Evaluation", *COST 259*, TD(99)051, Vienna, Austria, Apr. 1999.

[MEGR95] MEGRIN – Information Document, *Description for digital elevation model M745 Federal Republic of Germany (grid) – preliminary issue* (in German), Institut für Angewandte Geodäsie, Frankfurt, Germany, 23.11.95.

[Meno98]   Menolascino,R., "Cellular Coverage Engineering in Future Mobile Networks", *COST 259*, TD(98)095, Duisburg, Germany, Sep. 1998.

[MePi99]   Menolascino,R. and Pizarroso,M., *Software Tools for the Optimisation of Resources in Mobile Systems*, ACTS–AC016 STORMS Final Report, ACTS Office, Brussels, Belgium, Apr. 1999

[MIPL96]   Library of (Mixed) Integer Programming Problems, 1996. http://www.caam.rice.edu/~bixby/miplib/miplib.html.

[MoWF95] Mogensen,P., Wigard,J. and Frederiksen,F., "Performance of Slow Frequency Hopping in GSM Link-level", *COST 231*, TD(95)121, Poznan, Poland, Sep. 1995.

[MRRT53] Metropolis,N., Rosenbluth,W., Rosenbluth,A., Teller,A. and Teller,E., "Equations of state calculations by fast computing machines", *J. Chem. Phys.*, Vol. 21, 1953, pp. 1087–1091.

[Nand93]   Nanda,S., "Teletraffic Models for Urban and Suburban Microcells: Cell Sizes and Handoff Rates", *IEEE Trans. Veh. Tech.*, Vol. 42, No. 4, Nov. 1993, pp. 673–682.

[NeBa00]   Neubauer,T. and Baumgartner,T., "Required Network Size for System Simulations in UMTS FDD Uplink", *COST 259*, TD(00)026, Valencia, Spain, Jan. 2000.

[NeWo88]   Nemhauser,G.L. and Wolsey,L.A., *Integer and Combinatorial Optimization*, John Wiley, New York, New York, USA, 1988.

[NiML98]   Nix,A., Murphy,S. and Lister,D., "The TWP Format for a 3D-Geographical Dataset", *COST 259*, TD(98)058, Bradford, UK, Apr. 1998.

[NiWi00]   Nielsen,T.T. and Wigard,J., *Performance Enhancements in a Frequency Hopping GSM Network*, Kluwer, Amsterdam, The Netherlands, 2000

[NWMM99] Nielsen,T.T., Wigard,J., Michaelsen,P.H. and Mogensen,P., "Resource Allocation in a Frequency Hopping PCS1900/GSM/-DCS1800 Type of Network", in *Proc. of VTC'99 Spring – 49th IEEE Vehicular Technology Conference*, Houston, Texas, USA, May 1999.

[OjPr98]   Ojanperä,T. and Prasad,R., *Wideband CDMA for Third Generation Mobile Communications*, Artech House, London, UK, 1998.

[Oksa99]   Oksanen,L. "Radio Network Planning for WCDMA", in *Proc. of UMTS'99*, Monte Carlo, Monaco, June 1999.

[OOKK68] Okumura,Y., Ohmori,E., Kawano,T. and Fukuda,K., "Field strength and its variability in VHF and UHF land mobile radio service", *Review of Electrical Communication Laboratory*, Vol. 16, No. 9/10, Sep./Oct. 1968, pp. 825–873.

[OrBa00a]   Orriss,J. and Barton,S.K., "A Statistical Model for Connectivity Between Mobiles and Base Stations Part 2", *COST259*, TD(00)012, Valencia, Spain, Jan. 2000.

[OrBa00b]   Orriss,J. and Barton,S.K., "A Statistical Model for Connectivity Between Mobiles and Base Stations Part 3", *COST259*, TD(00)049, Bergen, Norway, Apr. 2000.

[OrBa99]   Orriss,J. and Barton,S.K., "A Statistical Model for Connectivity Between Mobiles and Base Stations", *COST259*, TD(99)094, Leidschendam, The Netherlands, Sep. 1999.

[OrPB99]   Orriss,J., Phillips,A.R. and Barton,S.K., "A Statistical Model for the Spatial Distribution of Mobiles and Base Stations", in *Proc. of VTC'99 Fall – 50th IEEE Vehicular Technology Conference*, Amsterdam, The Netherlands, Sep. 1999.

[Ostr96]   Ostrowski,M., "Efficient Transmission of Integrated Voice and Data in Wireless Networks", in *Proc. of ICC'96 – IEEE Conference on Communications*, Dallas, Texas, USA, June 1996.

[PaHW99]   Papathanassiou,A., Hartmann,C. and Weber,T., "Uplink Spectrum Efficiency and Capacity of TD-CDMA with Adaptive Antennas", *COST 259*, TD(99)113, Leidschendam, The Netherlands, Sep. 1999.

[PaKr96]   Patzig,S. and Kränzle,H., "Land use data for radio planning at E-Plus" (in German), in *Proc. International Society for*

            Photogrammetry and Remote Sensing, ISPRS Commission IV,
            Working Group 1, Vienna, Austria, July 1996.

[Pars92]   Parsons,J.D., *The Mobile Radio Propagation Channel*, Pentech
            Press, London, UK, 1992.

[Pavl94]   Pavlidou,F.N., "Two-Dimensional Traffic Models for Cellular
            Mobile Systems", *IEEE Trans. Commun.*, Vol. 42, No. 2/3/4,
            Feb./Mar./Apr. 1994, pp. 1505–1511.

[PeBT98]   Pearce,D.A.J., Burr,A.G. and Tozer,T.C., "Optimum Power
            Control Schemes for Cellular TDMA Systems Employing
            Adaptive Modulation and Coding", *COST 259*, TD(98)062,
            Bradford, UK, Apr. 1998.

[PhOB99a]   Phillips,A.R., Orriss,J. and Barton,S.K., "An update to a new
            protocol and a new approach towards analysing the
            performance of medium access control protocols for Wireless
            LANs", *COST 259*, TD(99)085, Leidschendam, The
            Netherlands, Sep. 1999.

[PhOB99b]   Phillips,A.R., Orriss,J. and Barton,S.K., "A New Protocol and a
            New Approach Towards Analysing the Performance of
            Medium Access Control Protocols for Wireless LANs", in
            *Proc. of VTC'99 Fall – 50th IEEE Vehicular Technology
            Conference*, Amsterdam, The Netherlands, Sep. 1999.

[PoGo96]   Pollini,G.P. and Goodman,D.J., "Signalling System
            Performance Evaluation for Personal Communications", *IEEE
            Trans. Veh. Tech.*, Vol. 45, No. 1, Feb. 1996, pp. 131–138.

[PoSt98]   Pospischil,G. and Steinbauer,M., "The CSM-Dataformat and
            the METABASE-Library", *COST 259*, TD(98)26, Bern,
            Switzerland, Feb. 1998.

[Pott99]   Potter,R., "Planning the UMTS radio network", in *Proc. of
            UMTS'99*, Monte Carlo, Monaco, June 1999.

[RACE91]   RACE, *Radiowave Propagation Model Document*, RACE,
            Mobile Telecommunication Project R1043 Report, Issue 1.1,
            RACE Office, Brussels, Belgium, Dec. 1991.

[Rapp93]   Rappaport,S.S., "Blocking, Hand-off and Traffic Performance
            for cellular Communication Systems with Mixed Platforms",
            IEE Proceedings-I, Vol. 140, No. 5, Oct. 1993, pp. 389–400.

[RCGO99]   Ruiz,S., Covarrubias,D., García,I. and Olmos,J., "Impact of real
            channel characterisation on wireless packet networks with S-
            ALOHA channel access", *COST 259*, TD(99)002, Thessaloniki,
            Greece, Jan. 1999.

[RuCO98]   Ruiz,S., Covarrubias,D. and Olmos,J., "Impact of real channel
            characterisation on wireless packet networks with S-ALOHA
            channel access", *COST 259*, TD(98)096, Duisburg, Germany,
            Sep. 1998.

[SaAg98]   Sallent,O. and Agustí,R., "Comparison between PRMA++ and adaptive S-ALOHA DS-CDMA", *COST 259*, TD(98)008, Bern, Switzerland, Feb. 1998.

[SASH99]   Seidenberg,P., Althoff,M., Schulz,M., Herbster,G. and Kottkamp,M., "Analysis of Mutual Interference in UMTS/IMT–2000 Mixed-Mode Scenarios", in *Proc. of GLOBECOM'99*, Rio de Janeiro, Brazil, Dec. 1999.

[SBEM90]   Swales,S.C., Beach,M.A., Edwards,D.J. and Mcgeehan,J.P., "The performance enhancement of multibeam adaptive base-station antennas for cellular land mobile radio systems", *IEEE Trans. Veh. Tech.*, Vol. 39, No. 1, Feb. 1990, pp. 56–67.

[SBGZ87]   Simon,R., Beck,R., Gamst,A. and Zinn,E.-G., "Influence of hand-off algorithms on the performance of cellular radio networks", in *Proc. of VTC'87 – 37th IEEE Vehicular Technology Conference*, Tampa, Flor., USA, May 1987.

[SFGC98]   Santos,L., Ferreira,L.S., Garcia,V. and Correia,L.M., "Cellular Planning Optimisation for Non-Uniform Traffic Distributions", in *Proc. of PIMRC'98 – 9th IEEE International Symposium on Personal, Indoor and Mobile Radio Communications,* Boston, Mass., USA, Sep. 1998 [also available as TD(98)009].

[SMHW92] Seskar,I., Maric,S.V., Holtzman,J. and Wasserman,J., "Rate of Location Area Updates in Cellular Systems", in *Proc. of VTC'96 Fall – 42nd IEEE Vehicular Technology Conference*, Denver, Color., USA, May 1992.

[Stee72]   Steele,R., *Mobile Radio Communications*, Pentch Press, London, UK, 1972.

[StGJ99]   Steuer,J., Giese,T. and Jobman,K., "Simple Smart Antenna Models for System Level Simulation of advanced Handover and Channel allocation Protocols", *COST 259*, TD(99)023, Thessaloniki, Greece, Jan. 1999.

[StLJ98]   Steuer,J., Lampe,M. and Jobman,K., "SDL-Simulation tool for Advanced Handover Protocols", *COST 259*, TD(98)010, Bern, Switzerland, Feb. 1998.

[Tang94]   Tangemann,M. *et al.*, "Introducing adaptive antenna array concepts in mobile communications systems", in *Proc. RACE Mobile Telecommunications Workshop*, Amsterdam, The Netherlands, May 1994.

[TDVV99]   Teughels,M., De Coster,I., Van Lil,E. and Van de Capelle,A., "Leaf Initiated Join Hand-Over Evaluation", accepted for publication in *Wireless Networks*, 2000.

[TeVV98]   Teughels,M., Van Lil,E. and Van de Capelle,A., "The Leaf Initiated Join Hand-over Concept", *COST 259*, TD(98)093, Duisburg, Germany, Sep. 1998.

[ThMG88] Thomas,R., Mouly,M. and Gilbert,H., "Performance Evaluation of the Channel Organisation of the European Digital Mobile Communication System", in *Proc. of VTC'88 - 38th IEEE Vehicular Technology Conference*, Philadelphia, Penns., USA, May 1988.

[TSPL95] Collection of "Travelling Salesman Problems", 1995. http://www.iwr.uni-heidelberg.de/iwr/comopt/soft/TSPLIB95/ TSPLIB

[UMTS98a] UMTS, *Minimum Spectrum Demand per Public Terrestrial UMTS Operator in the Initial Phase*, UMTS Forum, Report No. 5, London, UK, 1998. http://www.umts-forum.org.

[UMTS98b] UMTS, *UMTS/IMT–2000 Spectrum*, UMTS Forum, Report No. 6, London, UK, 1998. http://www.umts-forum.org.

[VaNi94] Van der Zee,T.M. and Niemegeers,I.G.M.M., "Case Study of Signalling Traffic Due to Pedestrians in Mobile Networks", in *Proc. of RACE Mobile Telecommunications Workshop*, Amsterdam, The Netherlands, May 1994.

[VeCo97] Velez,F. and Correia,L.M., "Optimisation Criteria for Cellular Planning of Mobile Broadband Systems in Linear and Urban Coverages", in *Proc. of ACTS Mobile Communications Summit*, Aalborg, Denmark, Oct. 1997 [also available as TD(97)053].

[VeCo98] Velez,F. and Correia,L.M., "Preliminary Report on Traffic from Mobility in Mobile Broadband Systems", *COST 259*, TD(98)053, Bradford, UK, Apr. 1998.

[VeMS84] Verhulst,D., Mouly,J. and Szpirglas,J., "Slow Frequency Hopping Multiple Access for Digital Cellular Radiotelephone", *IEEE J. Select. Areas Commun.*, Vol SAC 2, No.4, July 1984, pp. 563–574.

[VeOl99] Verikoukis,C. and Olmos,J.J., "Type II hybrid ARQ protocol for wireless ATM", *COST 259*, TD(99)105, Leidschendam, The Netherlands, Sep. 1999.

[Verd97] Verdone,R., "Performance Evaluation of Roadside-to-Mobile Communications at Millimetre Waves", *COST 259*, TD(97)030, Turin, Italy, May 1997.

[VeZa98] Verdone,R. and Zanella,A., "A Handover Algorithm to Counteract Corner Effects in Microcellular Mobile Networks", *IEE Electron. Lett.*, Vol. 34, No. 10, 14 May 1998, pp. 950–951.

[VeZa99] Verdone,R. and Zanella,A., "On the Optimization of Fully Distributed Power Control Techniques in Cellular Radio Systems", accepted for publication in *IEEE Trans. Veh. Tech.*, 2000.

[VJMW95] Vejlgaard,B., Johansen,J., Mogensen,P. and Wigard,J., "Capacity Analysis of a Frequency Hopping GSM System", *COST 231*, (TD)095.121, Poznan, Poland, Sep. 1995.

[Whit93] Whitehead,J.F., "Signal-Level-Based Dynamic Power Control for Co-channel Interference Management", in *Proc. of VTC'93 – 43rd IEEE Vehicular Technology Conference*, Secaucus, New Jersey, USA, May 1993.

[WLSJ99] Wacker A., Laiho-Steffens,J., Sipilä,K. and Jäsberg,M., "Static simulator for studying WCDMA radio network planning issues", in *Proc. of VTC'99 Spring – 49th IEEE Vehicular Technology Conference*, Houston, Texas, USA, May 1999.

[WNMM98] Wigard,J., Nielsen,T.T., Mogensen,P. and Michaelsen,P.H., "Frequency Planning for Frequency Hopping GSM Networks", *COST 259*, TD(98)083, Duisburg, Germany, Sep. 1998.

[XiGo93] Xie,H., and Goodman,D.J., "Mobility Models and Biased Sampling Problem", in *Proc. of ICUPC'93 – 2nd IEEE International Conference on Universal Personal Communications*, Ottawa, Canada, Oct. 1993

[Yaco93] Yacoub,M.D., *Foundations of Mobile Radio Engineering*, CRC Press, Boca Raton, Florida, USA, 1993.

[YaFo93] Yates,J.A. and Foose,W.A., "A Simplified Approach to Growth Planning for Cellular Systems", in *Proc. of VTC'93 – 43rd IEEE Vehicular Technology Conference*, Secaucus, New Jersey, USA, May 1993.

[Yang98] Yang,S.C., *CDMA RF System Engineering*, Artech House, London, UK, 1998.

[YeNa96] Yeung,K.L. and Nanda,S., "Channel Management in Microcell/Macrocell Cellular Radio Systems", *IEEE Trans. Veh. Tech.*, Vol. 45, No. 4, Nov. 1996, pp. 601–612.

[ZOBV00] Zanella,A., Orriss,J., Barton,S.K. and Verdone,R., "Performance of Directed Retry in Cellular Environments: Extension to the Non-Uniform Case", *COST259*, TD(00)049, Bergen, Norway, Apr. 2000.

[ZoDF95] Zonoozi,M.M., Dassanayake,P. and Faulkner,M., "Mobility Modelling and Channel Holding Time Distribution in Cellular Mobile communication Systems", in *Proc. of IEEE GLOBECOM'95*, Singapore, Nov. 1995.

[ZoDF96] Zonoozi,M.M., Dassanayake,P. and Faulkner,M., "Teletraffic Modelling of Cellular Mobile Networks", in *Proc. of VTC'96 – 46th IEEE Vehicular Technology Conference*, Atlanta, Georgia, USA, May 1996.

Annex I

List of Editors/Contributors

- Henrik Asplund, Ericsson Radio Systems, Sweden
- Dirk Beckmann, Technical University of Hamburg-Harburg (Department of Communication Networks), Germany
- Christian Bergljung, Telia Research, Sweden
- Ernst Bonek, Technical University of Vienna (Institut für Nachrichtentechnik und Hochfrequenztechnik) and Telecommunications Research Center Vienna (Forschungszentrum Telekommunikation Wien), Austria
- Conor Brennan, University of Dublin (Trinity College), Ireland
- Alister G. Burr, University of York, United Kingdom
- Narcís Cardona, Polytechnical University of Valencia, Spain
- Marco Chiani, University of Bologna, Italy
- Luis M. Correia, Technical University of Lisbon (Instituto Superior Técnico), Portugal
- Peter Cullen, University of Dublin (Trinity College), Ireland
- Andreas Czylwik, Deutsche Telekom (T-Nova Innovationsgesellschaft), Germany
- Dirk Didascalou, University of Karlsruhe (Institut für Höchstfrequenztechnik und Elektronik), Germany
- Patrick C.F. Eggers, Aalborg University (Center for PersonKommunikation), Denmark
- Andreas Eisenblätter, Konrad-Zuse-Zentrum für Informationstechnik Berlin, Germany
- Nabil Elouardi, University of Hannover (Institut für Allgemeine Nachrichtentechnik), Germany
- Carlos Fernandes, Technical University of Lisbon (Instituto Superior Técnico), Portugal
- José Fernandes, University of Aveiro, Portugal
- Cristophe Grangeat, Alcatel, France
- Paolo Grazioso, Fondazione Ugo Bordoni, Italy
- Ralf Heddergott, ETH Zurich (Institut für Kommunikationstechnik), Switzerland

- Hans Heller, Siemens, Germany
- José Jiménez, Telefonica I & D, Spain
- Gorazd Kandus, Institut Jozef Stefan, Slovenia
- Kimmo Kansanen, University of Oulu, Finland
- Peter Karlsson, Telia Research, Sweden
- Ralf Kattenbach, University of Kassel (Fachgebiet Hochfrequenztechnik / Kommunikationssysteme), Germany
- Mikael B. Knudsen, Bosch Telecom, Denmark
- Thomas Kürner, E-Plus Mobilfunk, Germany
- Emanuel Van Lil, Katholieke Universiteit Leuven, Belgium
- Preben E. Mogensen, Aalborg University (Center for PersonKommunikation), Denmark
- Andreas F. Molisch, Technical University of Vienna (Institut für Nachrichtentechnik und Hochfrequenztechnik) and Telecommunications Research Center Vienna (Forschungszentrum Telekommunikation Wien), Austria
- Thomas Neubauer, Technical University of Vienna (Institut für Nachrichtentechnik und Hochfrequenztechnik), Austria
- Bo Olsson, Telia Research, Sweden
- Jörg Pamp, IMST, Germany
- Gert F. Pedersen, Aalborg University (Center for PersonKommunikation), Denmark
- Klaus I. Pedersen, Aalborg University (Center for PersonKommunikation), Denmark
- António Rodrigues, Technical University of Lisbon (Instituto Superior Técnico), Portugal
- Silvia Ruiz Boque, Polytechnical University of Catalunya, Spain
- Laurent Schumacher, Université Catholique de Louvain, Belgium
- Peter Seidenberg, Aachen University of Technology (Kommunikationsnetze), Germany
- Martin Steinbauer, Technical University of Vienna (Institut für Nachrichtentechnik und Hochfrequenztechnik), Austria
- Jan Steuer, University of Hannover (Institut für Allgemeine Nachrichtentechnik), Germany
- Michel Teughels, Katholieke Universiteit Leuven, Belgium
- Roberto Verdone, CNR-University of Bologna (Centro di Studio per l'Informatica i Sistemi di Telecomunicazioni), Italy
- Jean-Frédéric Wagen, Fribourg School of Engineering and Architecture, Switzerland
- Martin Weckerle, University of Kaiserslautern (Lehrstuhl für hochfrequente Signalübertragung und –verarbeitung), Germany
- Thomas Zwick, University of Karlsruhe (Institut für Höchstfrequenztechnik und Elektronik), Germany

Annex II

List of Participating Institutions

Austria
- Mobilkom Austria
- Technical University of Vienna (Institut für Nachrichtentechnik und Hochfrequenztechnik)
- Telecommunications Research Center Vienna (Forschungszentrum Telekommunikation Wien)

Belgium
- Katholieke Universiteit Leuven
- Université Catholique de Louvain
- University of Ghent

Canada
- Communications Research Centre (Ottawa)

Denmark
- Aalborg University (Center for PersonKommunikation)
- Bosch Telecom
- L.M. Ericsson
- Telecom Denmark Mobile
- Telital

Finland
- Elektrobit
- Helsinki University of Technology
- Nokia
- University of Oulu
- VTT Information Technology

France
- Alcatel
- France Telecom R&D
- CEGETEL / SFR
- INSA de RENNES
- Institut National des Telécommunications

- TDF-C2R
- University of Science and Technology of Lille

Germany

- AWE Communications
- Aachen University of Technology (Institut für Hochfrequenztechnik; Kommunikationsnetze)
- Darmstadt University of Technology (Institut für Hochfrequenztechnik)
- Deutsche Telekom (T-Nova Innovationsgesellschaft)
- E-Plus Mobilfunk
- Ilmenau University of Technology
- IMST
- Konrad-Zuse-Zentrum für Informationstechnik Berlin
- Mannesmann Mobilfunk
- Siemens
- Technical University of Hamburg-Harburg (Department of Communication Networks; Department of Telecommunications)
- Munich University of Technology (Institute for Communications Engineering, LNT)
- University of Hannover (Institut für Allgemeine Nachrichtentechnik)
- University of Kaiserslautern (Lehrstuhl für hochfrequente Signalübertragung und -verarbeitung)
- University of Karlsruhe (Institut für Höchstfrequenztechnik und Elektronik)
- University of Kassel (Fachgebiet Hochfrequenztechnik / Kommunikationssysteme)
- University of Stuttgart (Institut für Hochfrequenztechnik)
- University of Ulm

Greece

- Aristotle University of Thessaloniki
- National Observatory of Athens (Institute for Space Applications and Remote Sensing)
- National Technical University of Athens (Mobile Radio Communications Laboratory)

Ireland

- Com Search
- University of Dublin (Trinity College)
- University of Limerick

Italy

- CNR (Centro di Studio per l'Informatica i Sistemi di Telecomunicazioni)
- CORITEL
- CSELT

- Ericsson Telecomunicazioni
- Fondazione Ugo Bordoni
- Italtel
- OMNITEL
- Siemens Information and Communication Networks
- Technical University of Milan (Dipartimento di Elettronica e Informazione)
- Telital
- University of Bologna (Dipartimento di Elettronica, Informatica e Sistemistica)
- University of Rome "La Sapienza" (Dipartimento di Informatica e Sistemistica)

The Netherlands
- Delft University of Technology
- KPN Research

Norway
- Telenor R&D

Poland
- Poznan University of Technology
- PTK Centertel

Portugal
- Optimus
- Technical University of Lisbon (Instituto Superior Técnico)
- Telecel
- TMN
- University of Aveiro

Slovenia
- Institut Jozef Stefan

Spain
- Polytechnical University of Catalunya
- Polytechnical University of Valencia
- Telefonica I & D
- University of Les Illes Baleares
- University of Vigo
- University School of Gandia

Sweden
- Allgon
- ComOpt
- Ericsson Radio Systems
- Lulea University of Technology
- Lund University
- Telia Research

- Uppsala University

Switzerland

- EPFL
- ETH Zurich (Institute for Quantum Electronics; Institut für Kommunikationstechnik)
- Fribourg School of Engineering and Architecture
- Swisscom
- Wavecall Wireless Consulting

United Kingdom

- Aethos Communications
- Department of Trade & Industry (Radiocommunications Agency)
- Metapath Software International
- Motorola
- Philips Research
- Roke Manor Research
- University of Bradford
- University of Bristol
- University of Leeds (Institute of Integrated Information Systems)
- University of Manchester
- University of York
- Vodafone

Index

DATE DUE
